T0122465

Studies in Fuzziness and Soft Computing

Volume 342

Series editor

Janusz Kacprzyk, Polish Academy of Sciences, Warsaw, Poland
e-mail: kacprzyk@ibspan.waw.pl

About this Series

The series "Studies in Fuzziness and Soft Computing" contains publications on various topics in the area of soft computing, which include fuzzy sets, rough sets, neural networks, evolutionary computation, probabilistic and evidential reasoning, multi-valued logic, and related fields. The publications within "Studies in Fuzziness and Soft Computing" are primarily monographs and edited volumes. They cover significant recent developments in the field, both of a foundational and applicable character. An important feature of the series is its short publication time and world-wide distribution. This permits a rapid and broad dissemination of research results.

More information about this series at http://www.springer.com/series/2941

Lotfi A. Zadeh · Ali M. Abbasov
Ronald R. Yager · Shahnaz N. Shahbazova
Marek Z. Reformat
Editors

Recent Developments and New Direction in Soft-Computing Foundations and Applications

Selected Papers from the 4th World Conference on Soft Computing, May 25–27, 2014, Berkeley

 Springer

Editors
Lotfi A. Zadeh
University of California
Berkeley
USA

Ali M. Abbasov
Ministry of Communications and
   Information Technologies of the Republic
   of Azerbaijan
Baku
Azerbaijan

Ronald R. Yager
Machine Intelligence Institute
New Rochelle, NY
USA

Shahnaz N. Shahbazova
Azerbaijan Technical University
Baku
Azerbaijan

Marek Z. Reformat
University of Alberta
Edmonton, AB
Canada

ISSN 1434-9922   ISSN 1860-0808   (electronic)
Studies in Fuzziness and Soft Computing
ISBN 978-3-319-81228-1   ISBN 978-3-319-32229-2   (eBook)
DOI 10.1007/978-3-319-32229-2

This Springer imprint is published by Springer Nature
The registered company is Springer International Publishing AG Switzerland

Contents

Part I
Educational and Human-Related Issues

Software Implementation Methodology of Intelligent Information Systems of Learning and Knowledge Control (IISLKC)

Ali M. Abbasov and Shahnaz N. Shahbazova

Abstract In this paper, research work allows making a statement which can be formulated as complex of models and methods. These models and methods are able to take on the function of whole intellectual complex which carries out the function of teaching and process of control knowledge with the minimal participation of teacher and education institutions.

Keywords Client–server software · Consciousness · Fuzzy logic · Fuzzy neural networks · Briefness · Complex systems · Decision making · Flexibility

1 Introduction

Intelligent information system of learning and knowledge control (IISLKC) was designed for the work in distributed environment of intranet work of high education institutions as the environment for the self-preparation of students, and it received development as a full-fledged education environment in future.

This system has been adjusted as client–server software with application of Internet browser as a thin client. Like in any other systems, the main principles of interface building remain the same—utilization of graphic user form of entry and work with information. This allows user to have visual presentation of form which contains components for entry of user's information as well as components for different actions with already entered information. This simplifies the user's work, because during the work process the user may clearly see the results coming from own actions and information changes. Essentially reduces likelihood of uncorrected

A.M. Abbasov (✉)
Minister of Communications and High Technologies of the Republic of Azerbaijan,
33 Z. Aliyeva str., Baku AZ1000, Azerbaijan
e-mail: abbasov@mincom.gov.az

S.N. Shahbazova
Department of Information Technology and Programming, Azerbaijan Technical University,
25 H. Cavid Ave, Baku AZ1073, Azerbaijan
e-mail: shahbazova@gmail.com

© Springer International Publishing Switzerland 2016
L.A. Zadeh et al. (eds.), *Recent Developments and New Direction in Soft-Computing Foundations and Applications*, Studies in Fuzziness and Soft Computing 342, DOI 10.1007/978-3-319-32229-2_1

3

entry data. Herewith, the student may completely not to see the presentation of data and their forms passing to the server during the education and knowledge control. These data are the only final results of student with the form and did not require the full in date of its all contents from server [1].

2 Software System Methodology

Basic functions of central system of educational process control of information interaction securities are as follows:

- Administration of users and user groups;
- Processing users sections;
- Provision of connection with database servers and realization of intellectual distribution of commitments between them;
- Provision of file-server synchronization;
- Support of channel connections;
- Provision of information allocations to different languages;
- Transaction management.

Human–machine interface provides connection between student and IISLKC complex and main principles of its building were as follows:

1. Consciousness. Interface complex of IISLKC was developed with the purpose of maximum access for students with any level of preparation and computer skills. In most cases, the system does not cause difficulties to student in searching of necessary directions (interface elements) for the management of process of accomplishing the task or searching of necessary information [2].
2. Consistency. If the student were used some work acceptance with certain functions in the work process with system, then other system parts of work access must be identical. The work in interface system is concurred with the principles and work acceptance of all distributed operating systems (OSs). Since the student may work in any OS environment, functional meaning of controlling elements of interface is not distinguished from the interface in Internet browser through which it is working with IISLKC system [3].
3. Briefness. It means that if student in IISLKC system enters only minimal necessary information for the work or in system management. For instance, student would not be asked to enter insignificant numbers and would not require entering the information which was entered earlier or automatically received from student's registration card or server. Unsung meaning "by default" is widely speeded and applied where it is possible to reduce the process of entry of information.
4. Direct access to the help system. Taking into account the fact that student can be anyone with poor computer skills and using the learning complex for the first

time, the system provides student with necessary instructions and explanations about the function of control element. Help system meets the following three aspects:

- Built-in help reference system;
- Full provision of managing orders with auxiliary information;
- Description of types and character as well as possible reasons of messages about errors and confirmation of system functioning.

Error messages and confirmation from the system server are created as auxiliary help element, and its result must be minimum quantity of repeating error functions.

5. Flexibility. For unexperienced and poor computer-skilled students, the interface with minimum quantity of controlling elements that simultaneously are located on the screen was created. In this case, the interface was organized as hierarchical menu structure. At the same time, for the advanced users (teachers, administrators, and student-testers) there is an interface with additions commands and possibility to control the interface with combination of keys which speed the work in the system. However, it will be factor of complicating the work and risk of accomplishing without requesting the commands for the rest of the users [4].

Intelligence of learning system depends directly on certainty of definition of current results of student by several key factors:

- Capacity of understanding;
- Understanding of previous materials;
- Current physiological state of student.

The research of experiments shows that by speed of riffling the information materials, speedy response, error frequency—current preparation of student to the learning and partial psychological state of student may be determined. Problem of closing future action strategies is decided by program on the basis of these factors. This might be as continuation of teaching of a new material and also the question from already passed learning material or the end of the education.

By the end of the process, relevant form might be given where current achievement and analysis result of previous achievements of student will be shown.

By the end of the process of knowledge control and after conducting analysis of his results, the program enters changes to the records of database about relevant student, updates the list of used questions, and saves coefficient per each of parameter of learning or testing.

By the end of the learning of students by the certain learning theme, the system analyzes the results with the purpose of decision-making decision about the knowledge of the student and its possibility to move to the next level of education. In the analysis, the information about the past material and results of evaluation of student are also mentioned [5].

3 Interaction of Teacher of Profiling Organization and Student with the System

As it was mentioned above, the strategy of interaction of IISLKC complex with the existed learning system is carried out on the basis of profiling organization object (school or university). Despite the fact that the system would recommend itself in the future, official certificate of education (school certificate or diploma) or license of certified organization is issued only after the confirmation of the student knowledge. It is possible where the student passing the full course of education himself applies to the relevant organization for the confirmation. However, the more convenient possibility has been provided in the complex—education in the IISLKC complex existed within the same profiling organization under the supervision of assigned teacher [6].

For the forming of the general structure of intelligent system works in the profiling organization environment, it is necessary to have a detailed look at the role of the teacher and student, as well as their interaction. In the system existed numbers of methods of interaction of teacher and student. It might be individual learning of teacher and student or work with all student simultaneously with taking into consideration that the main educational burden on the IISLKC complex. The most productive in terms of time saving and student learning quality is achieved using them together. This simultaneous method of education of group of students with individual control of each student with recommendations, advises, and notifications [7].

Let us look at the typical scheme of teacher's behavior in intelligent system. While accessing to the system, it is necessary to mention identical data—login and password. This allows teacher to add and amend own materials as well as access to the work files of students.

If teacher needs to conduct additional learning and testing of simultaneous students group (for instance, laboratory or practical assignments), it is necessary to enter the conception of general files. The general files are created by teachers by the concrete syllabus of subjects. By using these general files, the teacher may assign works and exercises to the group of students at the same time, as well as to check their decisions and results. The structure of general files allows simplifying the work with group of students; however, the access of teacher to the work files of students allows the teacher to work individually with each student. It can also be expressed in the recommendation to repeat taking the course, and in more detailed analysis of errors than it was made in the general files, and etc. The basis of wideness of the system is that the right to access by teacher allows to add and amend own materials, enter corrections according to the last achievements of science and culture.

The behavior of student in the system has its own principal differences from the behavior of teacher. Provision of security of information in the system is decided by offering to student the minimal advantages. In the access to the system, student is entitled with the right only to review and copy the system materials, and in work file— the right to add and amend. This given system of limitation allows securing the informational material of system from the unauthorized amendment and removal, and, at the same time allows to student to pass the course of education in the system efficiently.

For the purposeful passing of learning, the student may use the structure of general subjects. The student must chose the general subjects per interested by him in subjects and courses. After the registration, he is obliged to be on the track in chosen subjects by doing exercises and accomplishing the assignments leaving by the teacher of the system. Otherwise, he can be punished by rollback to the earliest learning materials, because already achieved knowledge would lose their efficacy [8].

The student must also visit his work file and check the availability of specific comments from teachers. In addition to this, any system materials are available to the student where he can find the information interested to him. This allows learning the student to the regular mental workload and successfully influencing to the increasing of knowledge. Like any other work, gaining the knowledge is possible only with interior and exterior motivation and wishing to receive the education is the one of the main guarantees for the success of this whole process of education.

4 General Model of Realization System of Control Knowledge

A focus on simplicity and ease of deployment of the software complex in any practical educational institution was emphasized during the choosing of a model for practical realization of the system of knowledge control. The only special equipment necessary is a hardware server database (DB) which may function as one server or as a cluster, consisting of multiple linked DB servers in the local network and operating under single management (Fig. 1).

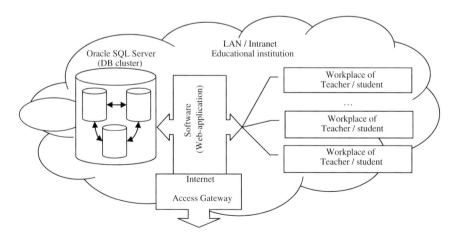

Fig. 1 Architecture of system

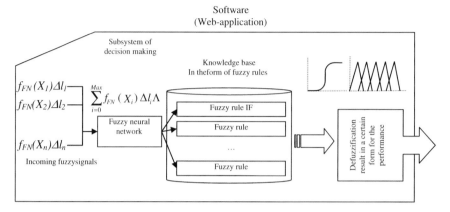

Fig. 2 Subsystem of decision making

Software is built on Web application technology [9, 10], one advantage of which is no client software. Any computer connected to the local institution network or through a gateway Internet access can immediately begin work in the system, after receiving the registration information and the level of access.

One of the major subsystems of Web application is a subsystem of knowledge control [11]. The most prominent algorithms of the intellectualization of information processes have been used in the development of the system. As mentioned above, the basis of its power and flexibility is the implementation of a combination of fuzzy logic and fuzzy neural networks (Fig. 2).

The knowledge base is a library of fuzzy rules, which are logically designed (1) and which interpret the signal received by a block of fuzzy neural networks [12–14].

$$
\begin{aligned}
&\text{IF } f_{\text{input}}(X_1)\text{THEN} f_{\text{output}}(Y_1)\\
&\text{IF } f_{\text{input}}(X_2)\text{THEN} f_{\text{output}}(Y_2)\\
&\qquad\qquad \ldots\\
&\text{IF } f_{\text{input}}(X_n)\text{THEN} f_{\text{output}}(Y_n)
\end{aligned}
\tag{1}
$$

Because the learning process is constantly evolving and being improved, the ability to modify the software was included.

A basic core of the system allows you to create on its basis the learning process that satisfy the needs of a wide range of educational institutions and learning materials, which can be formalized and the decision making expressed as logical expressions. The structure of access to the individual levels is built simply, but at the same time ensures the necessary security and flexibility [15, 16].

The mechanism of access control consists of two groups of three-level system privileges. This model is quite powerful and at the same time simple and functional, which fully satisfies the practical conditions and requirements (Fig. 3).

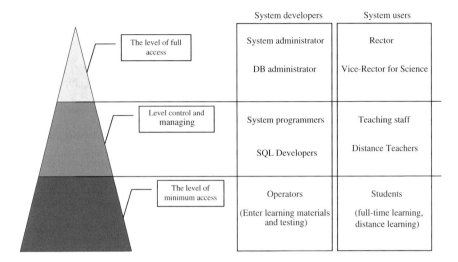

Fig. 3 The hierarchical system of control of level of access

5 Conclusion

Intelligent information system of learning and knowledge control allows to accomplish constant integrated educational process, whereas the student will pass the course under the control of intelligent computerized models and methods on the basis of application of modern mathematical technologies and by the analytical methods and informational process.

The methodology of applied research of programming IISLKC complex was suggested, where all aspects and system functioning were analyzed.

References

1. Shahbazova, Sh., Freisleben, B.: Network-based intellectual information system for learning and testing. In: Fourth International Conference on Application of Fuzzy Systems and Soft Computing, pp. 308–313. Siegen, Germany (2000)
2. Shahbazova, Sh.: Development of the knowledge base learning system for distance education. Int. J. Intell. Syst. **27**(4), 343–354 (2012)
3. Shahbazova, Sh.: Applied research in the field of automation of learning and knowledge control. In: Studies in Fuzziness and Soft Computing, Soft Computing: State of the Art Theory and Novel Application, pp. 223–240 (2012)
4. Shahbazova, Sh.: Application of fuzzy sets for control of student knowledge, ISSN 1683-3511. Appl. Comput. Math. Int. J. **10**(1), Special Issue on Fuzzy Set Theory and Applications. 195–208 (2011)
5. Bouchon-Meunier, B., Yager, R.R.: Fuzzy Logic and Soft Computing (Advances in Fuzzy Systems: Application and Theory), pp. 84–93, 103–119. World Scientific, Singapore (1995)
6. Jang, J.S.R., Sun, C.-T.: Neuro-Fuzzy modeling and control. Proc. IEEE **83**(3), 378–406 (1997)

7. Badami, Chiang, Khedkar, Marcelle, Schutten: Industrial Applications of Fuzzy Logic at General Electric. Bonissone Proc. IEEE **83**(3), 450–465 (1995)
8. Barsky, A.B.: Neural networks: recognition, management, decision-making. In Finance and Statistics, pp. 30–63. Moscow (2004)
9. Gorbunova, L.G.: On the realization of the rating system in pedagogical high schools. In: Proceedings of 2 International Technical Conference, University Education, Penza, Part 1. pp. 105–106 (1998)
10. Hanss, M.: Applied Fuzzy Arithmetic: An Introduction with Engineering Applications, 1 edn, pp. 100–116, 139–147. Springer, Berlin (2004)
11. Bernshteyn, L.S., Bojenyuk, A.V.: Fuzzy Models of Decision Making: Deduction, Induction, Analogy, pp. 78–99. University Tsure, Taganrog (2001)
12. Bellman, R., Zadeh, L.A.: Decision-Making in Ambiguous Circumstances, Issues Analysis and Decision-Making, pp. 180–199. Springer, Berlin (1976)
13. Yager, R., Filev, D. Essentials of Fuzzy Modeling and Control. Wiley, New York (1994)
14. Zadeh, L.A., Kacprzyk, J.: Fuzzy Logic for the Management of Uncertainty, First Printing edition, pp. 75–84. Wiley-Interscience, Hoboken (1992)
15. Zadeh, L.A.: A new approach to the analysis of difficulty systems and decision processes. Math. Today Knowl. 23–37 (1974)
16. Zadeh, L.A., Klir, G.J., Yuan, B.: Fuzzy Sets, Fuzzy Logic, and Fuzzy Systems: Selected Papers by Lotfi A. Zadeh, pp. 60–69. World Scientific, Singapore (1996)

Functioning of Control Module of Learning Materials

Shahnaz N. Shahbazova

Abstract The work presents the results of practical realization of database and methods. These results are the basic main element on which the work was done to determine the practical efficiency of application intelligent systems in learning process.

Keywords Artificial neural networks · Technical solutions · Fuzzy logic · Imitation of knowledge control procedures · Expert systems · Complex systems · Imitation of learning function · Decision making · Cybernetic simulation

1 Introduction

The learning system is in the constant development and advancing the methods of teaching, and knowledge control is the strategic task of almost all countries in the world. The existing reality demands constant increasing of qualification from the modern man and therefore relatively increases demands to available and effective learning system of education. One of the most perspective and reliable method is distant education [1].

The numerous system of distant education was designed, which is working perfectly on the base of traditional institutions of high and special education. However, they present itself as a development of informatization of traditional method of teaching, whereas connection of "teacher–student group" is translated to the virtual plane.

The big attention should be paid to the system of knowledge control where in majority it has existing system that was realized in the form of usual subprogram of testing.

S.N. Shahbazova (✉)
Department of Information Technology and Programming, Azerbaijan Technical University, 25 H. Cavid Ave, Baku, AZ 1073, Azerbaijan
e-mail: shahbazova@gmail.com

© Springer International Publishing Switzerland 2016
L.A. Zadeh et al. (eds.), *Recent Developments and New Direction in Soft-Computing Foundations and Applications*, Studies in Fuzziness and Soft Computing 342, DOI 10.1007/978-3-319-32229-2_2

11

In this work, special meaning was used on the mechanism of knowledge control and methods of its advancing, and it is one of the significant modules of teaching system. For example, models of conducting full cycle of teaching system with the minimum or completely without teacher's participation were developed on its basis.

2 The Principles of Presenting in the System of Information Resources and Functioning of Block Management of Teaching Materials

The presentation of information resources in automatization environment is the important implied task from the selected methods of decision where it depends on the flexibility and efficiency of functioning of the last product. Developed model of automatization educational system is aimed for the use in distributed environments of educational institutions, and the main users will be teachers and students with different levels of computer skills [2]. As the main distributors of informational materials, the teachers have a burden of intellectual analysis of learning materials with the purpose of dividing them on the maximum possible quantity of connecting teaching fragments [3, 4].

Resulted in the end numerous teaching pieces present itself as an initial step of translation of teaching materials in format necessary for the automatization system. As the automatization methods of intellectual analysis of educational materials are quite limited, the experts have a main burden [5]. The process of filing with teaching materials is illustrated in Fig. 1.

Learning materials management is conducted in 5 steps. The first step—layout of learning materials into minimum meaning fragments (Fig. 2) [6–8].

The second step—composing numerous questions determining the definition of learning fragments (Fig. 3) [9].

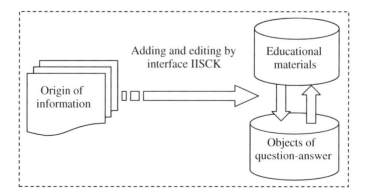

Fig. 1 The process of filing with learning materials

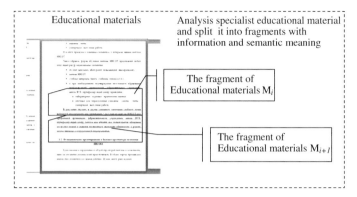

Fig. 2 Layout of teaching materials into minimum meaning fragments

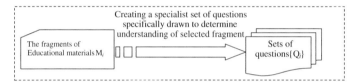

Fig. 3 Composing numerous questions determining the definition of learning fragments

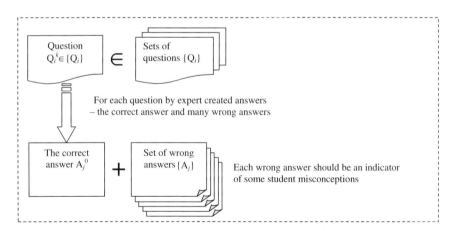

Fig. 4 Composing numerous wrong responses determining lack of understanding of learning material

The third step—composing numerous wrong responses determining lack of understanding of educational material (Fig. 4) [10].

The fourth step—creation of signs of wrong responses to the concrete pieces of learning material [11] (Fig. 5).

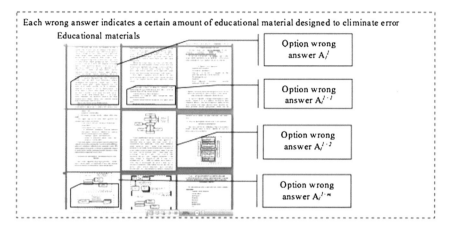

Fig. 5 Creation of signs of wrong responses to the concrete fragments of learning material

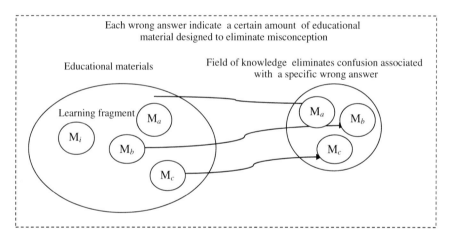

Fig. 6 Merging of numerous learning pieces in the knowledge volume eliminating concrete delusions

The fifth step—merging of numerous learning fragments in the knowledge volume eliminating concrete delusions [12] (Fig. 6).

The basis of knowledge system of student is the method of formalization of teaching process. Learning material, which is devoted to the some field of science and technology, is divided into the certain logically connected groups of learning courses [13, 14]:

- Multitude learning material
  - Learning course ∈ learning material
    - Multitude question–answer ∈ learning course

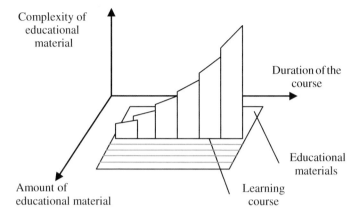

Fig. 7 Tripled presentation of learning system and testing

Fig. 8 General model of functioning of learning databases

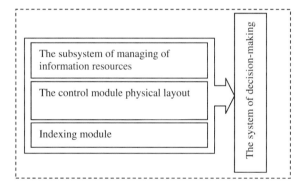

Multitude of learning materials (Ω) consist of submultitudes of learning courses (ω) and present complex field of crossed submultitudes.

Multitude of learning course (ω) consists of learning materials which might be described as chapters of relevant courses (Figs. 7, 8) [15, 16].

Knowledge control system is one of the main modules where the general achievement of research work is depending on as all other modules itself are relied on the results given by this current module.

With the practical viewpoint, the process of knowledge control leads to the questioning of students with the purpose of establishing their knowledge and skills in the field of tested subject [17, 18].

One of the more effective and qualified knowledge controls is willing to perform only teacher of relevant subjects. Therefore, selection as etalon system of teacher's behavior is more reasonable.

The analyses of teacher's behavior while performing the process of knowledge control determine the limited application of automatization system in the field of knowledge control of learning discipline, whereas the clear definition of right answers is possible [19, 20].

3 Conclusion

These results are the basic main element on which the work was done to determine the practical efficiency of application intelligent systems in the learning process. Presented in the paper, research work allows to make several statements which might be formulated as a complex of models and methods capable to undertake the function of complete intellectual complex of existing function of teaching and the process of knowledge control with the minimum participation of teacher and profiling educational institutions.

References

1. Gorbunova, L.G.: On the realization of the rating system in pedagogical high schools. In: Proceedings of 2 International Technical Conference, pp. 105–106. University Education, Penza, Part 1 (1998)
2. Galushkina, A.I.: Scientific. Neyrokompyutery and their Applications, vol. 4, p. 156. M. IPRZhR (2001)
3. Galushkin, A.I.: The theory of neural networks. Neurocomputers and their Applications, vol. 1, p. 221. M. IPRZhR (2000)
4. Golovko, V.A.: Neural networks: learning, organization and applying
5. Barsky, A.B.: Neural networks: recognition, management, decision-making. Financ. Stat. Moscow 30–63 (2004)
6. Bellman, R., Zadeh, L.A.: Decision-Making in Ambiguous Circumstances, Issues Analysis and Decision-Making, pp. 180–199. Springer, Berlin (1976)
7. Bernshteyn, L.S., Bojenyuk, A.V.: Fuzzy Models of Decision Making: Deduction, Induction, Analogy, pp. 78–99. University of Tsure, Taganrog (2001)
8. Bouchon-Meunier, B., Yager, R.R.: Fuzzy Logic and Soft Computing (Advances in Fuzzy Systems: Application and Theory), pp. 84–93, 103–119. World Scientific, Singapore (1995)
9. Nikravesh, M., Zadeh, L.A., Kacprzyk, J.: Soft Computing for Information Processing and Analysis, pp. 93–99 (2005)
10. Nikravesh, M., Aminzadeh, F., Zadeh, L. A.: Soft Computing and Intelligent Data Analysis in Oil Exploration, pp. 273–287 (2003)
11. Jang, J.S.R., Sun, C.-T.: Neuro-fuzzy modeling and control. In: Proceedings of the IEEE, vol. 83, no. 3, pp. 378–406 (1995)
12. Hanss, M.: Applied Fuzzy Arithmetic: An Introduction with Engineering Applications, 1 Edn, pp. 100–116, 139–147. Springer, Berlin (2004)
13. Shahbazova, Sh., Freisleben, B.: A network-based intellectual information system for learning and testing. In: Fourth International Conference on Application of Fuzzy Systems and Soft Computing, pp. 308–313, Siegen, Germany (2000)
14. Shahbazova, S.N.: Development of the knowledge base learning system for distance education. Int. J. Intell. Syst. **27**(4), 343–354 (2012)

15. Shahbazova, S.N.: Application of fuzzy sets for control of student knowledge, ISSN 1683-3511. Appl. Comput. Math. Int. J. **10**(1), 195–208 (2011) (Special Issue on Fuzzy Set Theory and Applications)
16. Shahbazova, S.N.: Applied research in the field of automation of Learning and knowledge control. Studies in Fuzziness and Soft Computing. In: Soft Computing: State of the Art Theory and Novel Application, pp. 223–240. Springer, (June 2012)
17. Zadeh, L.A., Kacprzyk, J.: Fuzzy Logic for the Management of Uncertainty, First Printing edition, pp. 75–84. Wiley-Interscience, New York (1992)
18. Zadeh, L.A., Klir, G.J., Yuan, B.: Fuzzy Sets, Fuzzy Logic, and Fuzzy Systems: Selected Papers by Lotfi A. Zadeh, pp. 60–69. World Scientific, Singapore (1996)
19. Yager, R., Filev, D.: Essentials of Fuzzy Modeling and Control. Wiley, New York (1994)
20. Zadeh, L.A.: A new approach to the analysis of difficulty systems and decision processes. Math. Today Knowl. 23–37 (1974)

Fuzzy Multi-scenario Approach to Decision-Making Support in Human Resource Management

M.H. Mammadova, Z.Q. Jabrayilova and F.R. Mammadzada

Abstract The paper describes the necessity of application of intelligent technologies to support decision making in human resource management (HRM) problems. The specific features of the personnel selection problem are highlighted, immersing the later into a fuzzy environment. Multi-scenario approach is described for solving the problem of employment, taking into account the importance and in equivalence of the indicators, which characterize the eligible candidates for the post, as well as individual character requirements of employers, at a current time. Experiment results for implementing the problem of selection of personnel based on the proposed method for professionals in information technology (IT) are discussed.

Keywords Decisions-making support · Human resource management · Personnel selection problem · Fuzzy environment · Fuzzy multi-criterial model

1 Introduction

In the transition to knowledge-based economy, ensuring effective performance and competitiveness of the organization (enterprises, companies, firms, etc.) requires increased attention to the personnel, i.e., human factor. Employees of the organizations are considered as the main strategic resource, ensuring its performance and achievements of its objectives. According to this concept, the staff becomes one of the main resources of the organization, which is required to be managed appropriately, optimal conditions to be provided for its development, and the necessary funds to be invested in it [1].

The concept basis of personnel management constitutes an increasing role of the worker's individuality, his knowledge of motivational attitudes, and his ability to shape and direct them in accordance with the challenges facing the organization.

M.H. Mammadova (✉) · Z.Q. Jabrayilova · F.R. Mammadzada
Institute of Information Technology of National Academy Science of Azerbaijan,
B.Vahabzadeh Str., 9, Baku C AZ1141, Azerbaijan Republic
e-mail: depart15@iit.ab.az

© Springer International Publishing Switzerland 2016
L.A. Zadeh et al. (eds.), *Recent Developments and New Direction in Soft-Computing Foundations and Applications*, Studies in Fuzziness and Soft Computing 342, DOI 10.1007/978-3-319-32229-2_3

Intelligent capital occupies a special position among other assets and requires specific approaches to the management perspective [2]. Evaluation of intelligent capital of the organization is needed to determine its effectiveness and growth factors, as well as to make decisions on the advisability of investment in this resource.

Objectives of human resource management (HRM) are the basis of personnel policy. The correct solution to these problems, making objective and transparency decisions on HRM, allows the organization to achieve its global goals [3, 4]. In general, today the HRM becomes the strategy of the company or firm. In this case, the funds invested in the development of human resources transform into an investment, not expenditure [5]. The changes, occurred in the labor market, require major changes in the relationship with employees, in the policy of their recruitment, retention, and motivation. In this regard, HRM at the professional level has become a strong modern means used in HR. Fundamentally new attitude toward the personnel as valuable resource of the organization actualizes the importance of developing new conceptual approaches and technologies for HRM. Therefore, in recent years, computer technology is increasingly used for the HRM problem solutions.

Thus, to make more objective decisions regarding personnel planning, selection, recruitment, adaptation, firing, promotion, development, training, and motivation of personnel, the decision-maker (DM) must evaluate and take into account the information in each case that characterizes the applicant, his interests, potential impacts, and results. Essential factor for the quality of personnel management is its assessment using competencies. The problems solved in the field of HRM are complex and varied. They are united by the fact that the finite number of evaluated objects is used as the raw data, and these objects are characterized by a set of diverse features, i.e., these tasks are multi-criterial, and many factors should be taken into account, and many influences, preferences, interests and consequences, and characterizing alternatives should be evaluated [6–8].

Volume, quantitative and qualitative nature, complexity, and contradictions of the information flow to be reached to the DMs, as well as the need to address the interrelationship of numerous factors, dynamic situation created difficulties in decision making on HRM. To overcome these difficulties and consequently more effective HRM of the organization, the application of intelligent decision support technologies seems appropriate [6, 7, 9].

To the number of HRM problems most frequently met in practice, the following problems belong [4, 10]: selection of applicant on a vacant position; compliance of workers to requirements of a workplace, a position; formation of a personnel reserve and planning of vocational advancement, career; selection of people on key positions in operation of business; awarding, compensation of employees, etc.

In this paper, we describe the methodology for personnel selection problem for the vacancy with regard to the importance and nonequivalence of numerous indicators characterizing the alternatives (candidates applying for the position), as well as the requirements of individual character of employers (DM), at the current time.

2 Personnel Selection Problem

Personnel selection often acts as the most important role for controlling the human resource and quality in HRM [11–13]. Effective employee selection is a critical component of a successful organization. Personnel selection is the process of collecting and evaluating information about individuals and choosing those who match the qualifications needed to perform a predefined job in the best way [14]. This process plays a determining role in HRM and is crucial to the success of an organization.

In [15] and [12], authors reviewed the personnel selection studies and found that the several main factors including change in organizations, change in work, change in personnel, change in the society, change of laws, and change in marketing have influenced personnel selection. In literature, there are a number of studies which use heuristic methods for employee selection. A fuzzy MCDM framework based on the concepts of ideal and anti-ideal solutions for the most appropriate candidate is presented in [16]. Also, a fuzzy number ranking method by metric distance for personnel selection problem was proposed in [17] and a personnel selection system based on fuzzy AHP was developed in [18].

In addition, researchers used fuzzy technique for order preference by similarity (TOPSIS) based on the veto threshold for ranking job applicants [9, 19–21]. Recently, owing to the advancements in information technology (IT), researchers have developed decision support systems and expert systems to improve the outcomes of HRM [22, 23].

A model to design an expert system for effective selection and appointment of the job applicants was developed in [24]. Although the applications of expert system or decision support systems on personnel selection and recruitment are increasing [11, 13, 25], however, the research taking into account requirements of the employer in the real time has not been considered in the paper.

Therefore, the goal of personnel selection was to apply valid and effective method to reduce the risks of hiring an unsuitable employee, and increase the opportunities to find an eligible employee who can enhance the productivity of organization [4, 26]. In other words, businesses find eligible employees meet the requirements of organization and occupation from mass of job applications through effective personnel selection methods.

3 Conceptual Model of the Personnel Selection Problem

The problem of personnel selection for the position is classified as semi-structured tasks, which is traditionally reduced to decision making [6, 9, 19–23]. The attitude of the DM and preferences (experience, knowledge, and intuition) of the experts play an important role in the implementation of such tasks. Intelligence support policy of choice (selection of experts), in this case, is defined by a specific manager—DM, experts involved in the evaluation process of alternatives for set of attributes forming the level of satisfaction of alternatives criteria and preference relations for each of them, and the estimating problem of the applicants for the position can be reduced to the adjustment of alternatives in fuzzy initial information.

Before shifting to the methods of candidates' selection, it is important to formalize requirements for the position or workplace of the future employee, based on the development strategy of the organization and characteristics of its corporate culture. Selecting the employee, it is necessary to determine the presence or absence of a candidate's competence, which is needed for the effective performance, i.e., a set of knowledge, skills, abilities, social and personal characteristics, and behavior of employees, defined with the objectives of the organization and set for specific situation. Approach based on competences allows one to link a whole HRM: in recruitment, career planning, assessment of performance, and development in the promising coming years [2]. Selecting the candidates, their competence is assessed and compared with the "portrait of an ideal employee," conveyed by a set of corporate performance at a given workplace [4, 6, 21, 27–29].

Note that the competence of a person is characterized by a number of factors and indicators, and depending on the fields of professional activity, profession, and profile of the organization, these figures have different relative weights of importance [6, 9, 19–21, 28, 30].

Recently, a new trend in the selection of personnel has been observed, which is expressed in individual requirements of the employers for applicants for a certain position, which involves an assessment of the latter one from the standpoint of obligation, desirability, and the lack of demand characterizing the indicators with respect to the proposed position. Hence, the figure, which is mandatory according to the preference of one employer for the purposes and needs of another one, may be desirable or even unnecessary.

This trend is also confirmed by monitoring of supply and demand in the market of IT professionals conducted by the Institute of Information Technologies of the Azerbaijan National Academy of Sciences [31].

Naturally, if a job applicant does not meet at least one indicator listed as obligatory for this specialty by the employer, his chances of getting accepted to the relevant position are equal to zero.

Statistical results of the approach of 72 employers regarding meeting the indicators characterizing education and personal qualities for the specialty of programmer-engineer are presented in Table 1.

Accordingly, as semistructured, the problem of personnel selection is characterized by the following features:

- multi-factorial and multi-criteriality;
- criteria and indicators of qualitative and quantitative nature;
- the need to consider the views in the evaluation process;
- hierarchy rate criteria characterizing evaluated object, expressed in the fact that each top-level individual criterion is based on the aggregation of partial criteria;

Table 1 Results of employers' requirements according to educational and personal qualities criteria for programmer-engineer specialty

Indicators characterizing the employed person	Character of employers' requirements		
	Obligatory (%)	Desirable (%)	Not required (%)
Education:			
Higher education diploma	68.11	25.02	6.87
Higher IT education diploma	30.58	51.43	17.99
Course certificates	5.64	31.97	62.39
Personal qualities			
Performance discipline	75.06	18.07	6.95
Initiative at work	23.63	55.52	20.85
Capability to pass on experience	13.9	56.91	29.19
Team work capability	34.67	29.19	36.14
Analytical thinking	17.99	50.04	31.97

- dependence on employer's requirements that define "portrait of the professional" to occupy particular position, at a real time.

These features "immerse" the task of hiring into a fuzzy environment, i.e., into the "Zade-environment," and cause decision making on the selection of the most suitable candidate for the position in poorly defined fuzzy situation [32, 33].

Thus, an evaluation model referring to fuzzy formalism for the development of an intellectual system supporting decision-making person for realization and reflecting expert knowledge must be proposed (desirability, obligation, and unimportance of criteria indicators).

So, for the solution of evaluation issue in staff management, which requires intelligent support, the following must be known:

- Set of evaluated alternatives:

$$X = \{x_1, x_2, \ldots, x_n\} = \{x_i, i = \overline{1,n}\}.$$

- Set of criteria characterizing alternatives:

$$K = \{K_1, K_2, \ldots, K_m\} = \{K_j, j = \overline{1,m}\}.$$

- Set of evaluable indicators characterizing each criteria:

$$K_j = \{k_{j1}, k_{j2}, \ldots, k_{jT}\} = \{k_{jt}, t = \overline{1,T}\}.$$

- *Y*—value range of each evaluable indicator;
- *T*—set of relationships reflecting desirability, commitment, and insignificance of criteria indicators of employer at the real time;
- *E*—expert group participating in evaluation (decision-making process);

- P—relations in X, K and E sets;
- L—linguistic expressions reflecting the level of relevance and relation of alternatives to criteria indicators;
- W—relative relations in same-group indicators and criteria sets

Listed components of selection are united in below relative-set model:

$$Ms = (X, K, Y, T, E, P, L, W).$$

Solution of evaluation and selection issue based on this model requires development of a relevant method, which refers to solution methods of multi-criteria issues using fuzzy mathematical formalism for this purpose [33, 34].

4 Task Description

The paper proposes an approach that enables to consider the individual requirements of the employers. Thus, we are proposing the approach that enables the selection of the best job applicant among all job applicants considering the individual requirement of the employer regarding meeting the general criteria indicators in order to be hired for specific candidates.

Thus, let us consider that:

$X = \{x_1, x_2, \ldots, x_n\} = \{x_i, i = \overline{1, n}\}$—is a set of job applicants—alternatives the best of which must be selected;

$K = \{K_1, K_2, \ldots, K_m\} = \{K_j, j = \overline{1, m}\}$—is a set of criteria inherent to alternatives and the set is defined by knowledge, capability, and personal qualities of job applicants.

In this case, suitability of alternatives to criteria can be shown in two-dimensional matrix, whereas element of the matrix will be defined by membership functions reflecting the suitability level of x_i alternative to K_j criteria:

$$\varphi_{K_j}(x_i) : X \times K \to [0, 1].$$

Here, $\varphi_{K_j}(x_i)$—reflects the suitability level of x_i alternative to K_j criteria. But these criteria are defined based on multiple indicators of different significance.

That is, $K_j = \{k_{j1}, k_{j2}, \ldots, k_{jT}\} = \{k_{jt}, t = \overline{1, s}\}$.

Let us suppose,

(1) $\left\{\varphi_{k_{j1}}(x_i), \varphi_{k_{j2}}(x_i), \ldots, \varphi_{k_{js}}(x_i)\right\} = \left\{\varphi_{k_{jt}}(x_i), t = \overline{1, s}, j = \overline{1, m}\right\}$—membership function of alternatives to criteria indicators $\left\{k_{jt}, t = \overline{1, s}, j = \overline{1, m}\right\}$ is known (supply base);

(2) Evaluation of the DM regarding obligation (O), desirability (D), and unimportance (U) of meeting $\{k_{jt}, t = \overline{1,s}, j = \overline{1,m}\}$ criteria indicators for occupation of a specific position is known (requirement base).
(3) Result to be obtained based on fuzzy multi-scenario method, $\varphi_K(x_i)$, will express the hiring chance of x_i candidate as a value defined in $[0, 1]$ interval. Depending on this value, experts pre-form following hiring decision options:

- If $\varphi_K(x_i) \in [0, 0.25)$, then this candidate decidedly cannot be hired;
- If $\varphi_K(x_i) \in [0.25, 0.45)$, then hiring of this candidate carries great risk;
- If $\varphi_K(x_i) \in [0.45, 0.62)$, then hiring of this candidate carries a bit of risk;
- If $\varphi_K(x_i) \in [0.62, 0.8)$, then this candidate can be hired;
- If $\varphi_K(x_i) \in [0.8, 1]$, then this candidate is unconditionally hired.

Objective of the issue is to select the best alternative from the supply basis in accordance with demand basis for occupation of a specific vacancy or make a ranked list of alternatives from best to worst: $X : K^* \rightarrow X^*$. Hereby, X—is the set of primary alternatives, K^*—is the set of indicators marked with obligation (O), desirability (D), and unimportance (U), and X^*—is the ranked list of selected alternatives in accordance with demand.

5 Issue Solution

5.1 Modeling of the Demand Basis

Employer's criteria indicators $\{k_{jt}, t = \overline{1,T}, j = \overline{1,m}\}$ for occupation of a specific vacancy are divided into three groups as obligatory (O), desirable (D), and unimportant (U) and form relevant sets: $\{O\}, \{D\}, \{U\}$.

Let us note that,

$$\{O\} \cap \{D\} \cap \{U\} = \varnothing$$

and

$$\{O\} \cup \{D\} \cup \{U\} = \{k_{jt}, t = \overline{1,s}, j = \overline{1,m}\},$$

i.e., these sets do not have a common element, and any $k_{jt} \in K_j \in K$ element can belong to only one of these sets. Following possible situations, scenarios can happen depending on the distribution of $\{k_{jt}, t = \overline{1,s}, j = \overline{1,m}\}$ criteria indicators among $\{O\}, \{D\}, \{U\}$ sets.

Scenario 1. All indicators defining K_j criteria are obligatory: $k_{jt} \in \{O\}, t = \overline{1,s}$;
Scenario 2. A part of indicators defining K_j criteria are obligatory; another part is unimportant: $k_{jt} \in \{O\} \cup \{U\}, t = \overline{1,s}$;
Scenario 3. All indicators defining K_j criteria are desirable: $k_{jt} \in \{D\}, t = \overline{1,s}$;

Scenario 4. A part of indicators defining K_j criteria are desirable; another part is unimportant: $k_{jt} \in \{D\} \cup \{U\}$, $t = \overline{1, s}$;

Scenario 5. A part of indicators defining K_j criteria are obligatory; another part is desirable: $k_{jt} \in \{O\} \cup \{D\}$, $t = \overline{1, s}$;

Scenario 6. A part of indicators defining K_j criteria are obligatory; another part is desirable and a third part is unimportant: $k_{jt} \in \{O\} \cup \{D\} \cup \{U\}$, $t = \overline{1, s}$;

Scenario 7. All indicators defining K_j criteria are unimportant: $k_{jt} \in \{U\}$, $t = \overline{1, s}$.

(Let us note that scenarios 1 and 3 did not emerge during research and scenario 6 was the most common scenario.)

5.2 Evaluation of Alternatives

The following approaches to assess compliance with the required specification alternatives (requests) of the employers in the process of selecting candidates for the vacancy are presented.

Definition of membership function of the alternative to these criteria is realized through a scenario relevant to evaluation of these criteria in the supply basis.

Claim 1 *If a part of indicators defining $K_j = \{k_{jt}, t = \overline{1, s}\}$ criteria (scenario 1, 2, 5, 6) is obligatory and the value of membership function of alternative to at least one of these indicators equals 0, then the membership function of the alternative to the relevant criteria will also equal to 0.*

Claim 2 $K_j = \{k_{jt}, t = \overline{1, s}\}$ *is only defined by desirable (or partly unimportant— scenario) indicators and the value of membership function of alternative to at least one of desirable indicators differs from 0, then the membership function of the alternative to the relevant criteria will also be different from 0. Thus, membership function $K_j, j = \overline{1, m}$ of the alternative depends on distribution of indicators characterizing it among {O}, {D}, {U} sets, scenarios.*

Calculation of membership function of the alternative $K_j = \{k_{jt}, t = \overline{1, s}\}$ to the criteria is based on membership function of the indicators characterizing the criteria and its "curve," i.e., their aggregation based on principal of their importance factor depicted in thus [7, 30, 35], then following are proposed for the calculation of membership function of the alternative to $K_j = \{k_{jt}, t = \overline{1, s}\}$ criteria:

Stage 1. Membership functions are determined for the alternative criteria in accordance with the generated scenario.

1. *Based on Scenario 1,* membership function of the alternative to criteria K_j is calculated using following equation:

$$\varphi_{K_j}(x_i) = \prod_{t=1}^{s} \left[\varphi_{k_{jt}}(x_i) \right]^{w_{jt}} \tag{1}$$

Here, $\varphi_{k_{jt}}(x_i)$—is the membership function of x_i alternative to k_{jt} indicator, and w_{jt}—is the importance factor of k_{jt} indicator. Let us note that $\sum_{t=1}^{s} w_{jt} = 1, t = \overline{1,s}$ condition must be met for criteria indicators.

2. *Based on Scenario 2*: Suppose, g quantity of indicators defining K_j criteria have been evaluated as unimportant and naturally $g < s$. Then, the membership function formula of the alternative to K_j criteria (1) is defined based on s-g quantity of obligatory indicators.

3. *Based on Scenario 3*: Membership function of x_i alternative to K_j criteria is calculated using equation.

$$\varphi_{K_j}(x_i) = \sum_{t=1}^{s} w_{jt} \, \varphi_{k_{jt}}(x_i) \tag{2}$$

4. *Based on Scenario 4*, membership function of x_i alternative to K_j criteria is found only based on formula for indicators included in $\{D\}$ set (2).

5. *Based on Scenario 5*, in order to find the membership function of x_i alternative to K_j criteria, first the difference of membership function of its obligatory indicators from 0 is checked, and if one of them equals zero, then $\varphi_{K_j}(x_i) = 0$ is accepted; otherwise in accordance with formula (2), the value of membership function to K_j criteria is calculated, i.e.,

$$\varphi_{K_j}(x_i) = \begin{cases} 0, & \text{if} \quad \prod_{d=1}^{g} \varphi_{k_{js}}(x_i) = 0 \\ \sum_{t=1}^{s} w_{jt} \varphi_{k_{jt}}(x_i) & \text{if} \quad \prod_{d=1}^{g} \varphi_{k_j}(x_i) \neq 0. \end{cases} \tag{3}$$

Here, $k_{jd} \in \{M\}$, $d = \overline{1,g}$—K_j is the obligatory indicators characterizing K_j criteria and naturally in this case $g < s$.

6. During the solution of the problem based on *Scenario 6*, if S quantity of indicators of K_j is evaluated as unimportant, then it is possible to find the membership function of the alternative to this criterion by carrying out the operational sequence relevant with formula (3) in accordance with s-g quantity of indicators.

7. During the solution of the problem based on *Scenario 7*, during the definition of membership function of the alternative to K (i.e., the value of the job applicant's chance to get the job), its membership function to K_j is not taken into consideration.

Stage 2. Based on the aggregation of membership functions $\left\{ \varphi_{k_j}(x_i), \, j = \overline{1,m} \right\}$ according to the following formula:

Table 2 Defining of membership functions of alternatives X_i, $\{i = \overline{1,n}\}$, to the generalized criterion K

Alternatives	K				
	K_1	...	K_j		K_M
x_1	$\varphi_{K_1}(x_1)$	...	$\varphi_{K_j}(x_1)$	...	$\varphi_{K_M}(x_1)$
...	...	...	...		...
x_1	$\varphi_{K_1}(x_i)$	...	$\varphi_{K_1}(x_i)$	...	$\varphi_{k_{M1}}(x_i)$
...	...	...	...		...
x_n	$\varphi_{K_1}(x_n)$	...	$\varphi_{K_j}(x_n)$	...	$\varphi_{k_{M1}}(x_n)$

$$\varphi_K(x_i), \ i = \overline{1,n}$$

$$\varphi_K(x_i) = \sum_{j=1}^{m} w_j \, \varphi_{K_j}(x_i)$$

is defined $\varphi_k(x_i)$, $i = \overline{1,n}$—membership functions alternatives x_i to generalized criteria K (Table 2). Here, w_j—coefficient of importance of the criteria K_j and $\sum_{j=1}^{m} w_j = 1$.

Stage 3. Obtained results are reviewed based on the rules of 3rd condition and appropriate decision for each alternative.

5.3 Mathematical Formalization of Criteria

A criteria indicator scale is selected in order to determine the membership function— fuzzy value of the alternative criteria indicators—i.e., each criteria indicator is

Table 3 Mathematical formalization of "work experience in specialty"

Quality rating of "work experience in specialty" indicator	Linguistic rating	Fuzzy subset, set in [0, 1] interval
(1) Has three or more years work experience in specialty	Excellent	[0.98–1]
(2) Has 1–3 years work experience in specialty	Good	[0.8–0.97]
(3) Has 6 months to 1 year work experience in specialty	Acceptable	[0.5–0.79]
(4) Has less than half-a-year work experience in specialty	Poor	[0.1–0.49]

divided into rating levels in accordance with quality levels (excellent, good, acceptable, poor, etc.) of the relevant linguistic phrases of the natural language.

After performing of each criteria factor, appropriation of a fuzzy value from the fuzzy set to a linguistic rating level selected for it must be performed (Table 3).

Final—Collective fuzzy value determined by the experts based on individual fuzzy values can be defined in the following ways [36, 37]: (1) by intersection of fuzzy sets; (2) by connection of fuzzy sets; and (3) by making an agreed selection on fuzzy sets.

Based on the last approach, individual evaluation of the "superior" expert with special creativity is considered as the collective value. Such expert must choose such a membership value out of all individual membership values defined by experts as a collective membership value at each point of the possible alternatives space that in general situation, it must differ from remote values in collective and hold a determined "middle" position.

Thus, a "supply basis" is formed by finding a

$$\left\{ \varphi_{k_{j1}}(x_i), \varphi_{k_{j2}}(x_i), \ldots, \varphi_{k_{jT}}(x_i) \right\} = \left\{ \varphi_{k_{jt}}(x_i), t = \overline{1, s}, j = \overline{1, m} \right\}$$

membership function based on how alternatives meet $\left\{ k_{jt}, t = \overline{1, s}, j = \overline{1, m} \right\}$ criteria indicators of alternatives.

5.4 Use of Information About Importance of the Criteria

This point is one of the problems emerging in the solution of personnel management problems, and obtaining of such information gives opportunity to eliminate multi-criterionness and to bring this problem to one-criterion problem. In this case, global criterion is defined as

$$K_Q = \sum_{j=1}^{m} w_j K_j.$$

And here K_j—is criterion characterizing estimated object ($j = 1, 2, \ldots, m$), w_j— is called weight of criterion K_j or importance factor [6]. For importance factor of the criterion, the following condition is foreseen:

$$0 \leq w_j \leq 1; \quad \sum_{j=1}^{m} w_j = 1. \tag{4}$$

The idea of unification is based on the expressions of the person who expresses the opinion about importance of criteria (expert, person who makes a decision) or on determination of appropriate evaluation grade determined to reflect value of considered criterion(in other case, refer to 1–100 point scale) and further normalization within condition (1) of this value. On the basis of the obtained information for today, preparation of methods for determining of criteria importance factors is one of the points the attention is attracted to in the sphere of multi-criterion problems solution [6, 38].

Information about mutual importance, significance of the criteria can be referred by the experts can be:

– expressed by the linguistic expressions representing mutual relative advantage (or weak points) and their pair comparison;
– referred to the establishing of appropriate grade to reflect assessment value of the considered criterion against the background of criteria defining any global factor.

In first case to display mutual relative advantage of the criteria, the linguistic expressions of the type given below are used:

– criterion K_1 has a weak advantage over criterion K_2,
– criterion K_2 has rather more advantage over criterion K_1, etc.

Such linguistic expressions for degree of mutual relative advantage of compared criteria are estimated by 9-point Saati's table (Table 4) [28].

If number of criteria equals n, then by defining of $n - 1$ ratio of pair criteria comparison it is possible to make a matrix of mutual relative relations [28, 39].

After all matrix elements are defined, private vector (w_i^*) is to be found. For this purpose, radical of n-power of matrix line edge (n is the measure of comparison matrix) should be defined and after they are normalized, importance factor w_i of appropriate elements is calculated.

$$
\begin{aligned}
w_i^* &= \sqrt[m]{K_{i1} \times K_{i2} \times \cdots \times K_{im}} \\
w_i &= \frac{w_i^*}{\sum_{i=1}^{m} w_i^*}
\end{aligned}
\tag{5}
$$

It must be noted that importance factors identified by means of formula (5) and condition (4) is being checked up.

Table 4 Defining of relative importance factors of pair comparison on the basis of quality estimations

Importance intensity	Qualitative (linguistic) estimation
1	Criterion K_1 has no advantage over K_2
3	Criterion K_1 has weak advantage over K_2
5	Criterion K_1 has essential advantage over K_2
7	Criterion K_1 has evident advantage over K_2
9	Criterion K_1 has absolute advantage over K_2
2, 4, 6, 8	Intermediate estimations between neighboring estimations

In the second case information about the importance, significance against the background of common criteria reflects value of any criterion.

In such case, it is more advantageous to use method of importance factor on the basis of 10-point system of expert estimation of the criteria [39].

5.5   Detection of Contradictions in the Expressions of Pair Comparison About Criteria Importance

It must be noted that usually in multi-criteria tasks, multiple numbers of criteria and criteria indices lead to the contradictions of expert expressions reflecting their pair comparison made by expert group members.

Thus, before the application of criteria importance factor found by formula (5) in an appropriate way, one of the primary tasks is to identify if contradictory information (expert knowledge) used for their pair comparison is available. For this purpose, maximal private value λ_{max}, consent index (CI), and consent relation (CR) must be calculated.

Calculation of maximal private value λ_{max} is implemented by the pair comparison matrix as follows: Each column of expressions is summarized, then sum of the first one is multiplied to the quantity of the first component of normalized priority vector, and sum of the second column is multiplied with second one; then, all obtained numbers are added, i.e.,

$$\lambda_{max} = \sum_{i=1}^{n} \left(\sum_{j=1}^{n} k_{ij} \times w_i \right).$$

The closer the λ_{max} is to n (n—is a number of compared matrix elements), the more consent the result is.

Decline from consent may be expressed by the value $(\lambda_{max} - n)/(n - 1)$ that will be called CI.

CI is calculated by the following formula: $CI = (\lambda_{max} - n)/(n - 1)$.

If CI is divided into the number appropriate to the chance consent—CC, then we obtain CR.

According to [28], for matrix of the $n = 3$ size, $CC = 0.58$; for matrix of the $n = 4$ size, $CC = 0.90$; for $n = 5$ size, $CC = 1.12$; for $n = 6$ size, $CC = 1.24$ and so on.

CR if identified by the following formula: $CR = CI/CR$.

Consent rate is considered acceptable at $CR \leq 0.1$. If consent rate is higher than 0.1, then expressions should be reconsidered.

6 Applications Fuzzy Multi-scenario Method to Decision Support System for HRM

The Institute of Information Technology of Azerbaijan National Academy of Sciences considered following two problems within the framework of the development of intelligent decision-making support system for HRM:

(1) assessment of employment for personnel bonus;
(2) selection of personnel for vacant positions.

The problem of selection of personnel based on the proposed method is implemented for professionals in IT.

To identify a list of requirements for IT professionals of different occupations and specializations, a survey of employers has been conducted within the monitoring of the IT segment of the labor market [31].

The list of criteria for recruitment as an IT professional set forth by the employers for those wishing to be employed have been determined. Criteria are presented in 6 groups: Criteria are presented as follows:

K_1—age;
K_2—gender;
K_3—education;
K_4—personal qualities;
K_5—professional requirements in IT specialization; and
K_6—additional capabilities.

Each of these criteria is defined by multiple indicators that characterize them.

One of the complication problems during the solution of this issue is the determination of knowledge and capabilities of the job applicant in accordance with professional requirements and determination of his/her suitability level to requirements set forth to occupy this position; that is, above listed are determined through multiple indicators with different importance levels. For instance, it is necessary to determine the level of personal qualities of the job applicant for IT position, such as performance discipline, initiative at work, capability to pass on experience, team work (communication) capability, and analytical thinking, and find their importance coefficient with regard to each other, which requires attraction of experts to the process.

As a result of conducted researches, points reflecting the personal approach to recruitment of IT professionals emerged, which demonstrate themselves in different approaches to requirements set forth by the employer to the job applicant applying for the same position depending on the profile, activity direction, property type (government or non-government, joint, etc.) of the organization.

This point emerges when a requirement indicated as obligatory by one employer for a specific position can be evaluated as desired or even unimportant by another employer. Naturally, if a job applicant does not meet at least one indicator listed as obligatory for this specialty by the employer, his chances of getting accepted to the relevant position are equal to zero (Table 1).

In the first stage, suitability of the job applicant to relevant requirements of the employer on indicators of K_1, K_2, K_3 criteria determined based on documents submitted by the job applicant. In the second stage, evaluation of alternative based on K_4, K_5, K_6 criteria is carried out. Definition of membership function of the alternative to these criteria is realized through a scenario relevant to evaluation of these criteria in the supply basis.

Each of these criteria is defined by multiple indicators that characterize them, for instance, K_4—personal quality criterion is determined based on the below indicators:

k_{41}—performance discipline;
k_{42}—initiative at work;
k_{43}—capability to pass on experience;
k_{44}—team work (communication) capability; and
k_{45}—analytical thinking.

Mathematical formalization of criteria must be carried out in order to find the membership function of $\{k_{jt}, t = \overline{1, s}, j = \overline{1, m}\}$ criteria indicators to alternatives.

K_1, K_2, K_3 are the exact criteria, and relevance of the job applicant to these criteria is determined in a formal order, based on the documentation submitted by the applicant.

An indistinctness and quality characteristic, and support of expert knowledge during the definition of K_4, K_5, K_6 criteria make it necessary to use fuzzy mathematical logic methods that enable to form the linguistic phrases of the natural language [33, 37].

To that effect, it is necessary to develop mathematical formalization of criteria for realization of supply base, and the mechanism of turning the linguistic phrases regarding the level of satisfaction of criteria into a fuzzy value defined in the [0, 1] interval. An experts group of the members of the Scientific Board of the Institute is organized for the mathematical mining of initial information and for determining coefficients of importance of indicators.

On the basis of the expressions said by the expert about theoretical importance of these shown criteria indices, the given in Table 5 relation matrix is created by using

Table 5 Comparision matrix personal quality criteria indicators

	k_{41}	k_{42}	k_{43}	k_{44}	k_{45}	Private vector (w_i^*)	Importance factor (w_i)
k_{41}	1	4	4	0.33	1	1.39	0.22
k_{42}	0,25	1	1	0.2	2	0.63	0.1
k_{43}	0.25	1	1	0.25	0.5	0.57	0.09
k_{44}	3	5	4	1	4	2.99	0.47
k_{45}	1	0.5	2	0.25	1	0.76	0.12
$\sum_{j=1}^{5} k_{4j}$	5.5	11.5	12	2.03	8.5		

relational importance scale displayed in Table 4. While matrix is being compiled, it is referred to its diagonal, symmetric, and transitive features. For instance, evident superiority of criterion index k_{44} over criterion index k_{42}, which is 5, is written in appropriate cell of the matrix, while in diagonally symmetric place cell is 1/5. After matrix has been compiled, importance factors of the criteria are found by means of formula (5).

In next step, the availability of contracting features of used expert expressions is checked. For this purpose, first of all λ_{max} is found.

$$\lambda_{max} = \sum_{i=1}^{5} \left(\sum_{j=1}^{5} k_{ij} \times w_i \right) = 5.41.$$

CI of the used expert expressions is defined.

$$CI = (\lambda_{max} - n)/(n - 1) = 0.102.$$

If we consider CC to be CC = 1.12 for the 5-sized matrix, then we can calculate CR.

$$CR = CI/CR = 0.09.$$

CR was defined to be lower than 0.1, and it means there is no contradiction in the expressions used by the experts about criteria pair comparison and determined importance factor can be used in the realization of the tasks.

7   Conclusion

The proposed technique, which is one of the possible solutions of the problem in personnel selection, takes into account the preferences of employers and enables them to make more reasonable decisions. This paper presents an attempt to provide adequacy of the obtained results to the reality by application of the mentioned methods, as well as proposes approach that allows taking into consideration the individual character requirements of employers, at a current time in decision-making process. This method is successfully applied in various companies to support management decisions on hiring IT personnel.

Thus, fuzzy multi-scenario approach proposed in the article reflects the following points in the process of HRM decision-making process:

– The number of alternatives and criteria is not limited; criteria indicators characterizing the issue are not restricted;
– It gives opportunity to operate over quantity featured criteria as well as over quality featured criteria together with linguistic variables;
– enables to take into consideration the hierarchic nature of criteria;
– enables DM to "maneuver" at a current time in decision-making process.

The proposed method is applied not only for solution of personnel recruitment problems but also for identification of other problems arising in HRM issues. The methods of the system usage are required to be improved in order to take into account the preferences and interests of IT professionals. Currently, a method for making decisions on trade-offs is being developed to deal with the preferences of employers and IT professionals.

References

1. Cole, G.A.: Personnel and Human Resource Management, Thomson Copyright, 511 p (2002)
2. Spencer, L.M., Spencer, S.M.: Competence at Work Models for Superior Performance, 384 p. Wiley India Pvt. Limited (2008)
3. Bazarov, T.Y.: Personnel management, Workshop, 240 p. Unity-Dana, Moscow (2009) (in russian)
4. Makarova, I.K.: Human Resource Management. Five Lessons for effective HR-management, 232 p. Delo, Moscow (2007)
5. Ivantsevich, J.M., Lobanov, A.A.: Human Resources Management, 245 p. Aspect Press, Moscow (2004) (in russian)
6. Larichev, O.I.: Theory and Methods of Decision Making, 296 p. Logos, Moscow (2000) (in russian)
7. Mikoni, S.V.: Multicriteria Selection on the Final Alternative Set, Student Handbook, Lan Publishing, 270 p (2009)
8. Trachtengertz, E.A.: Capabilities and realization of computer decision making support systems, News of Academy of Sciences of Russia. Manag. Theory Syst. (3), 86–113 (2001). http://www.alleng.ru/d/manag/man094.htm (in russian)
9. Mammadova, M.G., Jabrayilova, Z.G.: Application of TOPSIS method in support of decisions made in staff management issues. Comput. Technol. Appl. USA 4(6), 307–316 (2013)
10. Management of organization personnel. Manual. Under edition of Y.M. Kibanov. Infra–M, Moscow (1997) (in russian)
11. Nussbaum, M., Singer, M., Rosas, R., Castillo, M., Flies, E., Lara, R., Sommers, R.: Decision support system for conflict diagnosis in personnel selection. Inf. Manag. 36(1), 55–62 (1999)
12. Robertson, T., Smith, M.: Personnel selection. J. Occup. Organ. Psychol. 74(4), 441–472 (2001)
13. Hooper, RS., Galvin, T.P., Kilmer, R.A., Liebowitz, J.: Use of an expert system in a personnel selection process. Expert Syst. Appl. 14(4), 425–432 (1998)
14. Akhlagh, E.: A rough-set based approach to design an expert system for personnel selection. World Acad. Sci. Eng. Technol. 54, 202–205. http://waset.org/Publications/a-rough-set-based-approach-to-design-an-expert-system-for-personnel-selection/14092 (2011)
15. Borman, W.C., Hanson, M.A., Hedge, J.W.: Personnel selection. Ann. Rev. Psychol. 299–337 (1997)
16. Dursun, M., Karsak, E.: A fuzzy MCDM approach for personnel selection. Expert Syst. Appl. 37, 4324–4330 (2010)
17. Chen, L.S., Cheng, C.H.: Selecting IS personnel use fuzzy GDSS based on metric distance method. Eur. J. Oper. Res. 803–820 (2005)
18. Gungor, Z., Serhadlıoglu, G., Kesen, S.E.: A fuzzy AHP approach to personnel selection problem. Appl. Soft Comput. 9, 641–649 (2009)
19. Kelemenis, A., Askounis, D.: A new TOPSIS-based multi-criteria approach to personnel selections. Expert Syst. Appl. 37, 4999–5008 (2010)

20. Nobari, S.: Design of fuzzy decision support system in employee recruitment. J. Basic Appl. Sci. Res. **1**(11), 1891–1903 (2011)
21. Wang, Y.J., Lee, H.S.: Generalizing TOPSIS for fuzzy multiple-criteria group decision-making. Comput. Math Appl. **53**(11), 1762–1772 (2007)
22. Chen, P.C.: A fuzzy multiple criteria decision making model in employee recruitment. IJCSNS Int. J. Comput. Sci. Netw. Secur. **9**(7), 113–117 (2009)
23. Chien, C.F., Chen, L.F.: Data mining to improve personnel selection and enhance human capital: a case study in high-technology industry. Expert Syst. Appl. **34**(2), 280–290 (2008)
24. Mehrabad, M.S., Brojeny, M.F.: The development of on expert system for effective selection and appointment of the jobs applicants in human resource management. Comput. Ind. Eng. **53**, 306–312 (2007)
25. Larichev, O.I., Sternin, M.: Decision support system of multi-objective problem of assignment. Inf. Syst. Process. **3**, 10–16 (1998)
26. Werner, J.M.: Implications of OCB and contextual performance for human resource management. Hum. Resour. Manag. Rev. **10**(1), 3–24 (2000)
27. Chen, C.T., Lin, C.T., Huang, S.F.: A fuzzy approach for supplier evaluation and selection in supply chain management. Int. J. Prod. Econ. **102**(2), 289–301 (2006)
28. Saaty, T.L.: How to make a decision: the analytic hierarchy process. Eur. J. Oper. Res. **48**, 426–447 (1990)
29. Tai, W.S., Hsu, C.C.: A realistic personnel selection tool based on fuzzy data mining method. In: Proceedings of the 9th Joint Conference on Information Sciences (JCIS), www.atlantis-press.com/php/download_papaer?id=46,9/1/2008
30. Neumann, J.V., Morgenstern, O.: Theory of games and economic behavior, One of Princeton University press is Notable Centenary Titles, 776 p (2007)
31. Mammadova, M.H., Jabrayilova, Z.G., Manafli, M.I.: Monitoring of Demands for Information Technology Specialists, Baku, 199 p (2009)
32. Kofman, A.: Introduction into the Theory of Fuzzy Sets, 432 p. Radio and connection, Moscow (1982) (in russian)
33. Zadeh, L.A.: Fuzzy sets. Inf. Control **8**, 338–353 (1965)
34. Orlovskiy, S.A.: Problems of Decision Making at Fuzzy Initial Information, 208 pp. Nauka, Moscow (1981) (in russian)
35. Sevestyanov, P.V., Dimova, L.G., Kaptur, M., Zenkova, A.V.: Methodology of multi-criterion hierarcic estimation of quality in uncertainty conditions. Inf. Technol. Mosc. (9), 10–13 (2001)
36. Levin V.I.: New generalization of operation above fuzzy sets of News of the Academy of sciences. Theory Control Syst. (1), 143–146 (2001)
37. Zadeh, L.A.: The concept of a linguistic variable and its application to approximate reasoning-I. Inf. Sci. (8), 199–249 (1975)
38. Nogin, V.D.: Decision Making at Muptiple Criteria, St.Petersburg, p. 103 (2007) (in Russion)
39. Jabrayilova, Z.G., Nobari, S.N.: Processing methods of information about the importance of the criteria in the solution of personnel management problems and contradiction detection. Probl. Inf. Technol. Baku (1):57–66

Learning User Intentions in Natural Language Call Routing Systems

Kamil Aida-zade and Samir Rustamov

Abstract The context analysis of customer requests in a natural language call routing problem is investigated in this paper. Understanding of customer intention is one of the most important problems in natural language call routing. The adaptive neuro-fuzzy inference system is examined for solving this problem. This system can be applied to any language call routing domain; that is, there is no lexical or syntactic analysis used in the classification.

1 Introduction

It is impossible to image the modern world without telephone networks. Development of wired and wireless telephone networks has increased the intensity of human life activities. Because getting information of such flight connections from a database or transacting a product order is easy by phone, telephone communication has wide applications. From this point of view, the number of the calls to corporate service centers is currently increasing dramatically. Due to sheer volume, it is necessary to accept, classify, and route these calls automatically.

Call routing is the task of directing callers to the right place in the call center, which could be either the appropriate live agent or an automated service. Call centers typically employ a complex hierarchy of touch-tone or speech-enabled menus to provide self-service using interactive voice response (IVR) which enables skill-based routing. A touch-tone menu implements call routing by having a caller select from a list of options using a touch-tone keypad. If there are more than a few routing destinations, several touch-tone menus are arranged in hierarchical layers. A speech-enabled IVR replaces touch-tone menus with speech-enabled menus,

K. Aida-zade (✉) · S. Rustamov
Institute of Control Systems of Azerbaijan National Academy of Sciences,
Baku, Azerbaijan
e-mail: kamil_aydazade@rambler.ru

S. Rustamov
e-mail: samir.rustamov@gmail.com

© Springer International Publishing Switzerland 2016
L.A. Zadeh et al. (eds.), *Recent Developments and New Direction in Soft-Computing Foundations and Applications*, Studies in Fuzziness and Soft Computing 342, DOI 10.1007/978-3-319-32229-2_4

which allows callers to select an option by either speaking a number ("For…, press or say one") or a keyword ("Say weather, news, stocks, …"). Speech-enabled IVRs are more practicable than standard touch-tone IVRs in many fields, e.g., airline flight information, banking services, and voice portals. Skill-based routing attempts to match the caller to a customer service agent who has the skill set to handle a request.

Touch-tone or speech-enabled IVRs' interfaces are often difficult to use, especially in cases when the number of menu items is large and they are difficult to remember. The other annoying aspect is that callers often cannot determine which touch-tone or voice option best matches their question. Clearly, menu-based systems are inferior to direct contact with a human operator. Recently, however, some companies have attempted to employ natural language call routing systems for call routing.

Natural language call routing (NLCR) lets callers describe the reason for their calls in their own words, instead of presenting them with a closed list of menu options. NLCR directs more callers to the right place, thus saving both caller and agent time.

The understanding of customer intent is one of the most important problems in the NLCR process. We suggest adaptive neuro-fuzzy inference system (ANFIS) as a possible solution to this problem.

2   Related Work

There are some approaches that have been developed for finding user intent in dialogue systems. One of them is the unsupervised learning method: Hidden Topic Markov Modeling [1]. This technique combines two methods of Latent Dirichlet Allocation and HMM in order to learn topics of documents.

The vector-based information retrieval technique was applied for the determination of user intention in NLCR [2]. In the vector-based technique, for every topic in the training corpus, queries are represented as vectors of features (e.g., words) representing the frequencies of terms that occur within them. Then, a distance is computed between the query vector and each topic vector. The topic that is the closest to the query is chosen by the classifier [3, 4].

Topic unigram language modeling is based on counting the number of occurrences of each feature for each topic and involves all words of each topic vocabulary. For each topic, the likelihood of the query given in this topic is calculated, and the topic is chosen that has the maximum likelihood [5, 6].

A Markov Decision Process (MDP) framework is applied for the design of a dialogue agent. The basic assumption in MDPs is that the current state and action of the system determine the next state of the system (Markov property). Partially observable MDPs have been demonstrated that are proper candidates for modeling dialogue agents [7, 8].

In the discriminative term selection method, the discriminative power of the term is measured by measuring the average entropy variation on the topics when the term is present or absent. Each term is assigned a numeric value that indicates its importance [9].

To improve the performance of single classifiers, one can employ automatic relevance feedback, boosting, and discriminative training as mentioned in [10].

Boosting is an iterative method for improving the accuracy of any given learning algorithm. Its basic idea is to build a highly accurate classifier by combining many "weak" or "simple" base classifiers. The algorithm operates by learning a weak rule at each iteration so as to minimize the training error rate [11].

Cache modeling is based on a set of keywords automatically selected for each topic and associated with a unigram statistical distribution. This unigram distribution is continuously compared with the content of a cache memory [12]. The comparison is based on the symmetric Kullback-Leibler divergence which varies in time when new words are considered [13]. A divergence measure is computed between the query and each topic, and the topic selected is the one that has the lowest value.

The BBN Call Director uses a statistical language model for speech recognition and a statistical topic identification system to identify the topic from the call. It uses a multinomial model for keywords and incorporates two different classifiers: Bayesian and log-odds [14, 15].

Neural networks, support vector machines, and radial basis function approaches are also effective in topic identification problems [16].

Wu et al. [17] applied HMM for learning user intention while holding a dialogue in transaction system. They proved that correct identification of user intention seriously improves the dialogue system performance and its decision-making function.

Yet another method applies the theory of fuzzy sets for the understanding of problem [18–27].

Aidazade et al. (including the current author) applied a hybrid fuzzy control and HMM system for the understanding of user intention in the NLCR problem [18, 28]. We developed and improved this model by considering the following operations in the system [28]:

- Instead of a fuzzy control model, we applied an adaptive neuro-fuzzy inference system for our case;
- By means of pruned ICF weighting function, a new feature extraction algorithm is developed.

At the cost of additional variables added within the middle layer of the neural network, ANFIS is able to improve accuracy by a small amount than fuzzy control system. This system can be applied to any language call routing domain; that is, there is no lexical, grammatical, and syntactic analysis used in the understanding process. Our feature extraction algorithm calculates a feature vector based on statistical occurrences of words in the corpus without any lexical knowledge.

3 Automatic Call Routing by a Natural Language Call Router

The goal of NLCR is to understand the caller's request and take one of following appropriate actions:

- Route to the appropriate destination;
- Transfer to a human operator;
- Ask a disambiguation question.

Figure 1 illustrates the architecture of a natural language call router. Callers hear the main routing prompt as an open-ended question, such as "Please tell me, briefly, the reason for your call today." If the caller responds, the response is passed on to speech recognition and natural language understanding (NLU) modules to classify the reason for the call (or topic). Speech recognition transforms the spoken response into a sequence of words. One or a few sentences from the output of speech recognizer are taken as the user request. The language understanding engine then uses different topic identification technologies to determine the reason for the call from the sequence of recognized words. Depending on the caller response, the systems either routes the caller to a customer service agent or to an automated fulfillment system (self-service application) [14, 29].

The most important component in the dialogue management module is the learning of user intention for speech understanding. Assuming the speech recognition module gives us high accuracy, we can work only on the text form of request. We applied supervised machine learning algorithm—ANFIS for solution of this problem. Even though this method is popular in pattern recognition, it has not been thoroughly investigated for user intention in NLCR.

Fig. 1 The general architecture of a natural language call router

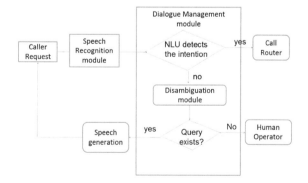

4 Application ANFIS for Learning User Intention in NLCR

The general structure of ANFIS is illustrated in Fig. 2. In response to linguistic statements, the fuzzy interface block provides an input vector to a multilayer neural network. The neural network can be adapted (trained) to yield desired command outputs or decisions [30].

At the first stage, we use a statistical approach for the estimation of the membership function, instead of expert knowledge.

In machine learning-based classification, two disjoint sets of documents are required: a training set and a test set. The training set is used by an automatic classifier to learn the differentiating characteristics of documents, and the test set is used to validate the performance of the automatic classifier. We now describe the data distribution in the corpus (dataset). To build the term list, the following operations are carried out:

- Combine all files from the corpus and make one text file;
- Convert the text to an array of words;
- Sort the array of words: $\begin{pmatrix} A \\ Z \downarrow \end{pmatrix}$;
- Code: $V = \{v_1, \ldots, v_M\}$, where M is the number of different words (terms) in the corpus.

As our target does not use lexical knowledge, we consider every word as one code word. In our algorithm, we do not combine verbs in different tenses, such as present and past, or nouns as singular or plural. Instead, we consider them as different code words.

4.1 Feature Extraction

Feature extraction algorithms are a major part of any machine learning method. We describe such an algorithm which is intuitive, computationally efficient, and does not require either additional human annotation or lexical knowledge.

Below, we describe some of the parameters:

- N is the number of classes (destinations);
- M is the number of different words (terms) in the corpus;

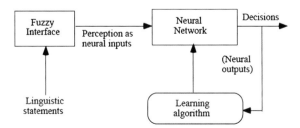

Fig. 2 The structure of ANFIS

- R is the number of observed sequences in the training process;
- $O^r = \left[o_1^r, o_2^r, \ldots, o_{T_r}^r\right]$ are the user requests in the training dataset, where T_r is the length of rth request, $r = 1, 2, \ldots, R$;
- $\mu_{i,j}$ describes the association between ith term (word) and the jth class $i = 1, \ldots, M; j = 1, 2, \ldots, N$;
- $c_{i,j}$ is the number of times ith term occurred in the jth class;
- $t_i = \sum_j c_{i,j}$ denotes the occurrence times of the ith term in the corpus;
- Frequency of the ith term in the jth class

$$\bar{c}_{i,j} = \frac{c_{i,j}}{t_i};$$

- Pruned ICF (inverse class frequency) [31]

$$\mathrm{ICF}_i = \log_2\left(\frac{N}{dN_i}\right),$$

where i is a term and dN_i is the number of classes containing the term i, which is given $\bar{c}_{i,j} > q$, where

$$q = \frac{1}{\delta \cdot N},$$

The value of δ is found empirically for the corpus investigated.

The membership degree of the terms ($\mu_{i,j}$) for appropriate classes can be estimated by the experts or can be calculated by the analytical formulas. Since a main goal is to avoid using human annotation or lexical knowledge, we calculated the membership degree of each term by an analytical formula as follows ($i = 1, \ldots, M; j = 1, 2, \ldots, N$):

$$\mu_{i,j} = \frac{\bar{c}_{i,j} \cdot \mathrm{ICF}_j}{\sum_{v=1}^{N} \bar{c}_{i,v} \cdot \mathrm{ICF}_v}; \tag{1}$$

4.2 Fuzzification Operations

Maximum membership degree is found with respect to the classes for every term of the rth request

$$\bar{\mu}_{i,j}^r = \mu_{i,j}^r,$$

$$j = \arg \max_{1 \le v \le N} \mu_{i,v}^r, \tag{2}$$

$$i = 1, \dots, M.$$

Means of maxima are calculated for all classes:

$$\bar{\bar{\mu}}_j^r = \frac{\sum_{k \in Z_j^r} \bar{\mu}_{k,j}^r}{T_r},$$

$$Z_j^r = \left\{ i : \bar{\mu}_{i,j}^r = \max_{1 \le v \le N} \mu_{i,v}^r \right\}, \tag{3}$$

$$j = 1, \dots, N.$$

We use the center of gravity defuzzification (CoGD) method for the defuzzification operation. The CoGD method avoids the defuzzification ambiguities which may arise when an output degree of membership comes from more than one crisp output value.

We used statistical estimation of membership degree of terms by (1) instead of linguistic statements at the first stage. Then, we applied fuzzy operations (2) and (3). MANN was applied to the output of the fuzzification operation. Outputs of MANN are taken as indexes of classes appropriate to the requests (Fig. 3). MANN is trained by the back-propagation algorithm.

The understanding of a request from an initial incoming call to an information center of an educational company in the Azeri language, and routing according to its intention, was taken as the test problem to be solved.

Calls must be routed to one of the four departments of the company, or be connected to an operator, or be rejected. These departments are as follows: (1) information center, (2) accounting department, (3) test exams center, and (4) service departments.

A total of 357 queries have been taken for the training process and 50 requests for the testing process. Words contained in the user request in human–computer dialogue are taken as observation sequence.

Fig. 3 The structure of MANN in ANFIS

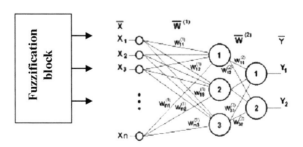

Table 1 Results of ANFIS

Boundary conditions	Correct (%)	Rejection (%)	Error (%)
No restriction	92	0	8
$\Delta_1 = 0, 1; \Delta_2 = 0, 5$	88	8	4
$\Delta_1 = 0, 3; \Delta_2 = 0, 5$	86	10	4
$\Delta_1 = 0, 5; \Delta_2 = 0, 5$	84	14	2

Note that these results are specific to the described test set and could be different in other corpora

For the application of ANFIS to the current test, the membership degree of the words of request is calculated by a fuzzification model in the testing process. By using trained neural networks, parameter estimated the index of class.

We set two boundary conditions for an acceptance decision:

1. $\bar{y}_k \geq \Delta_1$

2. $\bar{y}_k - \tilde{y}_p \geq \Delta_2$

where y_i is the output vector of MANN and

$$\bar{y}_k = \max_{1 \leq i \leq N} y_i, \, k = \arg \max_{1 \leq i \leq N} y_i$$

$$\tilde{y}_p = \max_{1 \leq i \leq k-1; k+1 \leq i \leq N} y_i.$$

Below in Table 1 are shown the results of classification of user requests by ANFIS with different values of Δ_2 and Δ_3.

5   Conclusions

We have described ANFIS classification system structures and applied to detecting user intention in NLCR. A goal of the research was to formulate methods that did not depend on linguistic knowledge and therefore would be applicable to any language. An important component of these methods is the feature extraction process. We focused on analysis of informative features that improve the accuracy of the systems with no language-specific constraints.

It is anticipated that when IF-THEN rules and expert knowledge are inserted into ANFIS, accuracy will improve to a level commensurate with human judgment.

When comparing the current system with others, it is necessary to emphasize that the use of linguistic knowledge does improve accuracy. Since we do not use such knowledge, our results should only be compared with other methods having similar constraints, such as those which use features based on bags of words that are tested on the same dataset. For this reason, we cannot fairly compare our results with others.

References

1. Chinaei, H.R., Chaib-draa, B.: Learning user intentions in spoken dialogue systems. In: ICAART 2009—Proceedings of the International Conference on Agents and Artificial Intelligence, pp. 107–114. Porto, Portugal (2009)
2. Chu-Carroll, J., Carpenter, B.: Vector-based natural language call routing. Comput. Linguist. **25**(3), 361–388 (1999)
3. Jurafsky, D., Martin, J.H.: Speech and Language Processing: An Introduction to Natural Language Processing, Speech Recognition, and Computational Linguistics, 2nd edn. Prentice-Hall, Englewood Cliffs (2009)
4. Kuo, H.-H.J., Lee, C.-H.: A portability study on natural language call steering. In: Proceedings of the Eurospeech-01. Aalborg, Denmark (2001)
5. McDonough, J., Ng, K.: Approaches to topic identification on the switchboard corpus. In: Proceedings of the International Conference on Acoustics, Speech and Signal Processing, pp. 385–388. Yakohamo, Japan (1994)
6. Schwartz, R., Imai, T., Kubala, F., Nguyen, L., Makhoul, J.: A maximum likelihood model for topic classification of broadcast news. In: Proceedings of the European Conference on Speech Communication and Technologie. Rhodes, Greece (1997)
7. Williams, J.D., Poupart, P., Young, S.: Factored partially observable markov decision process for dialogue management. In: 4th IJCAI Workshop on Knowledge and Reasoning in Practical Dialogue Systems. Edinburgh, Scotland (2005)
8. Doshi, F., Roy, N.: Efficient model learning for dialogue management. In: Proceedings of the ACM/IEEE International Conference on Human-robot Interaction (HRI'07), pp. 65–72 (2007)
9. Kuo, H.-K.J., Lee, C.-H., Zitouni, I., Fosler-Luissier, E., Ammicht, E.: Discriminating training for call classification and routing. In: Proceedings of the International Conference on Speech and Language Processing (2002)
10. Zitoni, I., Kuo, H.K.J., Lee, C.H.: Boosting and combination of classifiers for natural language call routing systems. Speech Commun. **41**, 647–661 (2003)
11. Zitouni, I., Kuo, H.-K.J., Lee, C.-H.: Combination of boosting and discriminative training for natural language call steering systems. In: Proceedings of the International Conference on Acoustics, Speech and Signal Processing. Orlando, USA (2002)
12. Kuhn, R., DeMori, R.: A cache-based natural language model for speech recognition. IEEE Trans. Pattern Anal. Mach. Intell. **12**(6), 570–582 (1990)
13. Bigi, B., Brun, A., Haton, J., Smaili, K., Zitouni, I.: Dynamic topic identification: towards combination of methods. Advance in NLP. Tzigov Chark, Bulgaria (2001)
14. Suhm, B., Bers, J., McCarthy, D., Freeman, B., Getty, D., Godfrey, K., Peterson, P.: A comparative study of speech in the call center: natural language call routing vs. touch-tone menus. In: CHI '02 Proceedings of the SIGCHI Conference on Human Factors in Computing Systems, pp. 283–290. NY, USA (2002)
15. Bers, J., Suhm, B., McCarthy, D.: Please tell me briefly the reason of your call understanding natural language call routing. http://bbn.com/resources/pdf/natural-language-call-routing.pdf
16. Joachims, T.: Text categorization with support vector machines: Learning with many relevant features. In: European Conference on Machine Learning, pp. 137–142. Berlin (1998)
17. Wu, C.H., Yan, G., Lin, C.H.: Spoken dialogue system using corpus-based hidden Markov Model. In: The 5th International Conference on Spoken Language Processing, Incorporating the 7th Australian International Speech Science and Technology Conference, vol. 4, pp. 1239–1243. Sydney, Australia, ISCA (1998)
18. Aida-zade, K.R., Rustamov, S.S., Ismayilov, E.A., Aliyeva, N.T.: Using fuzzy set theory for understanding user's intention in human-computer dialogue systems. Trans. ANAS Ser. Phys. Math. Tech. Sci., Baku, vol. XXXI, No 6, pp. 80–90 (2011) (in Azerbaijani)
19. Subasic, P., Huettner, A.: Affect analysis of text using fuzzy semantic typing. Fuzzy systems. FUZZ IEEE 2000. In: International Conference on Fuzzy Systems, vol. 9, issue 4, pp. 483–496 (2001)

20. Salvador, V., Andrade, M., Kawamoto, A.: Fuzzy theory applied on the user modeling in speech interface. In: IADIS International Conference Interfaces and Human Computer Interaction, pp. 201–205 (2007)

21. Rustamov, S.S., Mustafayev, E.E., Clements, M.A.: Sentiment analysis using neuro-fuzzy and hidden markov models of text. In: IEEE SoutheastCon 2013, in press., Jacksonville, USA (2013)

22. Tyson, N., Matula, V.C.: Improved LSI-based natural language call routing using speech recognition confidence scores. In: ICCC 2004, International Conference on Computational Cybernetics, pp. 409–413 (2004)

23. Gorin, A., Parker, B., Sachs, R., Wilpon, J.: How may I help you? Speech Commun. **23**, 113–127 (1997)

24. Haas, J., Hornegger, J., Huber, R., Niemann, H.: Probabilistic semantic analysis of speech. In: DAGM-Symposium, pp. 270–277 (1997)

25. Aida-zade, K.R., Rustamov, S.S., Baxishov, U.C.: The application of hidden Markov model in human-computer dialogue understanding system. Trans. ANAS. Ser. Phys Math. Tech Sci, Baku, vol. XXXII, No 3, pp. 37–46 (2012) (in Azerbaijani)

26. Juang, B.H., Rabiner, L.R.: Hidden Markov models for speech recognition. Technometrics **33** (3), 251–272 (1991)

27. Kaufmann, A., Gupta, M.M.: Introduction to fuzzy arithmetic theory and applications. N: Van Nostrand Reinhold, IEEE Trans. Fuzzy Syst. 483–496 (1991)

28. Aida-zade, K.R., Rustamov, S.S., Mustafayev, E.E., Aliyeva, N.T.: Human-computer dialogue understanding hybrid system. In: International Symposium on Innovations in Intelligent Systems and Applications (INISTA 2012). Trabzon, Turkey (2012)

29. Lee, C.H., Carpenter, B., Chou, W., Chu-Carroll, J., Reichl, W., Saad, A., Zhou, Q.: On natural language call routing. Speech Commun. **31**, 309–320 (2000)

30. Fuller, R.: Neural fuzzy systems (1995)

31. Rustamov, S.S., Clements, M.A.: Sentence-level subjectivity detection using neuro-fuzzy and hidden markov models. In: Proceedings of the 4th Workshop on Computational Approaches to Subjectivity, Sentiment and Social Media Analysis in NAACL-HLT2013, pp. 108–114. Atlanta, USA (2013)

Part II
Aggregation

Bipolarity and Multipolarity in Aggregation Structures

Guy De Tré, Jozo J. Dujmović and Sławomir Zadrożny

Naturally dealing with information processing tasks such as database querying, information retrieval, and decision support requires the adequate handling of bipolarity that might be present in the specification of user preferences in selection criteria. Indeed, past and recent research revealed that users often express their preferences for criteria specifications in a bipolar and sometimes even multipolar way. This means, among others, that different poles of criteria are distinguished, each of them having different semantics and being handled in a different way. For example, a pole of positive (desirable) criteria and a pole of negative (undesirable) criteria might be distinguished. Positive criteria express what the user wants, e.g., a black or blue car, whereas negative criteria express what the user does not want, e.g., a pink car. Likewise, a pole of mandatory criteria, which all have to be satisfied, and a pole of optional criteria, whose satisfaction or dissatisfaction might, respectively, result in a bonus or penalty, can be considered. In this chapter, we study and discuss different manifestations of bipolar and multipolar poles of criteria expressing user preferences from the standpoint of aggregation and try to classify them into two groups: bipolarity based on positive (desirable) and negative (undesirable) poles and bipolarity based on nonuniform input-dependent annihilators. For both groups, we

G. De Tré (✉)
Department of Telecommunications and Information Processing, Ghent University,
Sint-Pietersnieuwstraat 41, 9000 Ghent, Belgium
e-mail: Guy.DeTre@UGent.be

J.J. Dujmović
Department of Computer Science, San Francisco State University,
1600 Holloway Avenue, San Francisco, CA 94132, USA
e-mail: jozo@sfsu.edu

S. Zadrożny
Systems Research Institute, Polish Academy of Science,
Newalska 6, 01-447 Warsaw, Poland
e-mail: zadrozny@ibspan.waw.pl

© Springer International Publishing Switzerland 2016
L.A. Zadeh et al. (eds.), *Recent Developments and New Direction
in Soft-Computing Foundations and Applications*, Studies in Fuzziness
and Soft Computing 342, DOI 10.1007/978-3-319-32229-2_5

49

propose canonical forms of aggregators which allow to aggregate criteria evaluations consistently with human reasoning. Moreover, we study how combinations of different forms of bipolarity can lead to multipolarity in criteria specifications and how aggregators for multipolar criteria evaluations can be constructed. This chapter is an extended and revised version of our contribution entitled 'Bipolarity and Multipolarity in Aggregation Structure' to the 4th World Conference on Soft Computing, May 25–27, 2014, UC Berkeley, CA, USA.

1   Introduction

In the context of information management, the terms *bipolarity* and *multipolarity* are, among others, used to refer to various poles of values or various poles of criteria which have to be interpreted, handled, evaluated, and aggregated in a different way.

When different poles of values are considered, one often distinguishes between positive values, i.e., values which are possible, satisfactory, permitted, desired, or acceptable, and negative values, i.e., values which are impossible, unsatisfactory, not permitted, rejected, undesired, or unacceptable. A typical application is the specification of query preferences where a user specifies for a given criterion which domain values are acceptable (to a certain extent) and which are not. For example, consider the specification of user preferences in the context of selecting the color of a car. For a given user, the positive pole of values might consist of black, dark blue, and dark brown, whereas the negative pole consists of white, yellow, and purple. For other colors, e.g., red, there is some indifference. These colors are neither in the positive, nor in the negative pole, what illustrates the general *heterogeneous* nature of the bipolarity. We refer to bipolarity that relates to a single domain of values as bipolarity that is specified *inside* a single criterion (cf. also [1, 2] where we refer to it as bipolarity at the level of an attribute).

Bipolarity can also be considered at the level of poles of criteria. In such a case, multiple criteria are considered and subdivided into different poles of which the criteria have different semantics and require a different handling during criteria evaluation and aggregation. We refer to bipolarity that relates to multiple criteria as bipolarity that is specified *among* multiple criteria (cf. also [1, 2] where we refer to it as bipolarity at the level of the comprehensive evaluation). The study of this kind of bipolarity is the subject of this chapter. Different approaches to handling and aggregating bipolar criteria have been proposed in the literature. Two of them are the *constraint–wish* approach and the *satisfied–dissatisfied* approach.

In the constraint–wish approach, the criteria in one pole are considered to be constraints, i.e., mandatory criteria, while the criteria in the other pole are rather wishes, i.e., desired criteria. This approach is used among others by Dubois and Prade [3, 4] and by Liétard et al. [5–9].

In the satisfied–dissatisfied approach, a pole of positive criteria, i.e., criteria that should be satisfied, and a pole of negative criteria, which should be dissatisfied, are considered. This approach is used among others by De Tré et al. [10–12] and

Zadrożny et al. [1]. Remark that often, negative criteria might be translated to constraints, while positive criteria might be seen as wishes.

In this chapter, we study bipolarity among multiple criteria from the different perspective of criteria aggregation and aggregation structures. From this point of view, bipolarity is a concept of asymmetric or contrasting logic properties of two inputs, or two groups of inputs. We discuss different interpretations of bipolarity in aggregation structures and try to classify bipolar concepts into two fundamental groups:

- Bipolarity based on positive (desirable) and negative (undesirable) inputs.
- Bipolarity based on nonuniform input-dependent annihilators.

For both groups, we propose canonical forms of aggregators. We discuss how multipolarity results from the combination of bipolarity.

The remainder of the chapter is organized as follows. In Sect. 2, some preliminaries on bipolar criteria are presented, explaining the two approaches that will be discussed in the chapter in more detail. Section 3 deals with aggregators for poles of desirable and undesirable criteria. Aggregation in case of nonuniform input-dependent annihilators is discussed in Sect. 4. Multipolarity is dealt with in Sect. 5. Finally, Sect. 6 contains some conclusions and directions for further research.

2 Preliminaries

Pioneering work in the area of heterogeneous bipolar criteria handling has been done by Lacroix and Lavency in [13], which seems to be the first approach where a distinction has been made between mandatory and desired query criteria. As mentioned earlier, desired and mandatory criteria can be viewed as specifying positive and negative information, respectively. Indeed, the opposite of a mandatory criterion specifies what must be rejected and thus what is considered as being negative with respect to the query result, whereas desired criteria specify what is considered as being positive. In Lacroix and Lavency's approach, the positive criteria are of a somehow lesser importance than negative ones, but this cannot be represented by a simple importance weighting of these criteria. Later on, this idea has been further developed and adapted to be used in 'fuzzy' criterion handling. Existing research approaches on 'fuzzy' bipolar criteria handling can mainly be categorized into two groups, which can be denoted as the *satisfied–dissatisfied* approaches and the *constraint–wish* approaches. Both categories of approaches are briefly introduced in the next two subsections.

2.1 Satisfied–Dissatisfied Approaches

In general, negative and positive criteria may be treated as equally important. Then, we are interested on how to aggregate them under such an assumption. A first assumption that can be made is to consider a 'positive' pole (C^{pos}) and a 'negative' pole (C^{neg}) of elementary criteria. The criteria of each pole are connected using logical connectives [10]. Hereby, only the conjunction operator ($\wedge$), the disjunction operator ($\vee$), and negation operator ($\neg$) can be used. If no brackets are used to enforce priority, then negation has overall priority over conjunction and disjunction and conjunction has priority over disjunction.

The connected positive criteria form a logical expression that expresses what is permitted, whereas the connected negative criteria form another logical expression that expresses what is not permitted. What is neither explicitly permitted, nor forbidden is considered to be unspecified, which could, among others, be due to indifference or hesitation of the user with respect to what is permitted or not, or due to the inability of the user to specify all (un) permitted values within the criteria. This assumption reflects the heterogeneous bipolar characteristic of the approach.

As an example, consider user preferences related to buying a car. The user might look for a recent car, preferably blue or black, which is either a crossover or a berline. She does not wants a white car or a car with no air bags. In this example, the expression corresponding to the positive pole of criteria is as follows:

$$`\text{recent}' \wedge (`\text{blue}' \vee `\text{black}') \wedge (`\text{crossover}' \vee `\text{berline}')$$

whereas the expression corresponding to the negative pole yields

$$`\text{white}' \vee `\text{no airbags}'.$$

In order to keep the heterogeneous characteristics, the approach in [10] proposes to evaluate the logical expressions corresponding with both poles independently of each other and to express criterion satisfaction using *bipolar satisfaction degrees* (BSDs). A BSD is a pair

$$(s, d), \ s, d \in [0, 1] \tag{1}$$

of independent values s and d where s is called the *satisfaction degree* and d is called the *dissatisfaction degree* [11, 12] (cf. also the concept of *evidence couple* [14]).

BSDs can be aggregated in several ways. The standard logical operators for conjunction ($\wedge$), disjunction ($\vee$), and negation ($\neg$) are, respectively,

$$(s_1, d_1) \wedge (s_2, d_2) = (\min(s_1, s_2), \max(d_1, d_2)),$$

$$(s_1, d_1) \vee (s_2, d_2) = (\max(s_1, s_2), \min(d_1, d_2)),$$

$$\neg(s, d) = (d, s).$$

The independent evaluation of the two logical expressions corresponding to the criteria poles C^{pos} and C^{neg} yields a BSD (s, d). The satisfaction degree s is the result of the evaluation of the logical expression of C^{pos}, whereas the dissatisfaction degree d is the result of the evaluation of the logical expression of C^{neg}.

For the aggregation of the satisfaction degrees resulting from the evaluation of the elementary criteria in the logical expression of C^{pos}, the following rule set has been proposed in [10]:

- The conjunction of two satisfaction degrees $s_1, s_2 \in [0, 1]$ is defined by

$$s_1 \wedge s_2 = \min(s_1, s_2).$$

- The disjunction of two satisfaction degrees $s_1, s_2 \in [0, 1]$ is defined by

$$s_1 \vee s_2 = \max(s_1, s_2).$$

For the aggregation of the dissatisfaction degrees resulting from the evaluation of the elementary criteria in the logical expression of C^{neg}, the following rule set has been proposed in [10]:

- Conjunction: The conjunction of two dissatisfaction degrees $d_1, d_2 \in [0, 1]$ is defined by

$$d_1 \wedge d_2 = \max(d_1, d_2).$$

- Disjunction: The disjunction of two dissatisfaction degrees $d_1, d_2 \in [0, 1]$ is defined by

$$d_1 \vee d_2 = \min(d_1, d_2).$$

Both rule sets have been constructed in accordance with the semantics of the standard aggregation operators for BSDs as given above. Besides minimum and maximum, alternative aggregation operators, based on other triangular norms and conorms, can be used if a reinforcement effect is needed or desired.

Consider a three-year-old gray berline car with air bags. This car satisfies the expression corresponding to the positive pole of criteria with a satisfaction degree

$$
\begin{aligned}
s &= s_{\text{recent}} \wedge (s_{\text{blue}} \vee s_{\text{black}}) \wedge (s_{\text{crossover}} \vee s_{\text{berline}}) \\
&= 0.7 \wedge (0 \vee 0) \wedge (0 \vee 1) \\
&= \min(0.7, \max(0, 0), \max(0, 1)) \\
&= 0.
\end{aligned}
$$

Furthermore, the dissatisfaction degree resulting from the evaluation of the expression that corresponds to the negative pole of criteria yields

$$d = d_{\text{white}} \wedge d_{\text{no\_airbags}}$$
$$= 0 \wedge 0$$
$$= \max(0, 0)$$
$$= 0.$$

Hence, the corresponding BSD for the car is (0, 0), which means that the car is neither satisfied, nor dissatisfied. Indeed, because of its gray color, there is hesitation about whether the car would be accepted by the user or not.

Using a nonbipolar search, the car will be rejected because it is neither blue, nor black. This despite the fact that, from a commercial point of view, the car is still interesting because its color has not been explicitly rejected by the user.

2.2 Constraint–Wish Approaches

An alternative kind of bipolar approach in 'fuzzy' criterion specification is to consider two poles of criteria as mandatory criteria (C^{man}) and desired criteria (C^{des}). Bipolarity is thus studied considering queries with preferences as in [13], and the operator 'and possibly' is employed to join two types of conditions.

Reconsidering the car example, the user might specify that she is looking for a car that must have air bags and should not be white. Preferably, the car is recent, is blue or black, and is a crossover or a berline. Under these assumptions, the expression corresponding to the pole of mandatory criteria is as follows:

$$\text{`not white' } \wedge \text{ `airbags'}$$

whereas the expression corresponding to the pole of desired criteria yields

$$\text{`recent' } \wedge (\text{`blue' } \vee \text{ `black') } \wedge (\text{`crossover' } \vee \text{ `berline'}).$$

The independent evaluation of the two logical expressions corresponding to the poles C^{man} and C^{des} yields a pair

$$(c, w), c, w \in [0, 1] \tag{2}$$

where c is the satisfaction degree of the *constraints* in C^{man}, i.e., the required/mandatory conditions, while w is the satisfaction degree of the *wishes* in C^{des}, i.e., the desired conditions which are, in general, not obligatory.

The pair (c, w) may be seen as a special interpretation of the pair (s, d) of (1), where $c = 1 - d$ and $w = s$; i.e., each negative criterion is treated as a complement of a mandatory criterion and the positive criteria are in a sense treated as facultative conditions. Lacroix and Lavency in [13] propose to use the 'And Possibly' aggregator to aggregate c and w, what effectively makes it possible to replace each pair (c, w) with a single value (in [13], originally the set $\{0, 1\}$ is used instead of the interval $[0, 1]$):

$$\text{And Possibly} : [0, 1] \times [0, 1] \rightarrow [0, 1]$$
$$(c, w) \mapsto u \tag{3}$$

The semantics of the 'And Possibly' operator are represented as follows. For a given object o_i whose evaluation yields the pair (c_i, w_i), we consider

$$c_i \text{ And Possibly } w_i = \min(c_i, \text{POSS} \Rightarrow w_i) \tag{4}$$

where

$$\text{POSS} = \max_{j=1,\dots,n} \min(c_j, w_j)$$

and j goes over all the objects $o_j, j = 1, \dots, n$ under evaluation. Thus, the 'And Possibly' operator is context-sensitive and relates the possibility of satisfying both conditions c and w to the actual existence of an object which satisfies both of them.

Lacroix and Lavency proposed the 'And Possibly' operator in the classical logical context; i.e., c and w are binary values. This approach has been adapted to the fuzzy case in [15] via the use of fuzzy logic connectives, as it is already shown in (4). Fuzzy logical connectives may be interpreted in many different ways. In [16], various properties of the 'And Possibly' operator are postulated and it is verified how particular interpretations of fuzzy logical connectives preserve them.

More recently, the semantics of mandatory and desired criteria have been formalized in a slightly different context by means of the so-called 'And If Possible' and 'Or Else' 'fuzzy' connectives [17].

2.2.1 'And if Possible'

With c_1 and c_2 being (elementary) criteria, the expression 'c_1 and if possible c_2' is used to express a weak, nonsymmetric conjunction in the sense that c_2 is less important than c_1. From another perspective, c_1 can be considered as a basic criterion which should be satisfied, whereas the satisfaction of c_2 is only desired, for example, because it is more demanding.

Let γ_c be a fuzzy evaluation function for criterion c, which takes cases from a universe of discourse U as arguments, i.e.,

$$\gamma_c : U \to [0,1]$$
$$x \mapsto \gamma_c(x). \tag{5}$$

This evaluation function evaluates the criterion c for a case or value $x \in U$ and returns a number from the unit interval, where 0 denotes not satisfied at all and 1 represents fully satisfied.

In [17], the semantics of the connective 'And If Possible' have been defined by the following axioms.

D_1: $\gamma_{c_1 \text{ And If Possible } c_2}(x) \geq \min(\gamma_{c_1}(x), \gamma_{c_2}(x))$

D_2: $\gamma_{c_1 \text{ And If Possible } c_2}(x) \leq \gamma_{c_1}(x)$

D_3: $\exists c_1, c_2 : \gamma_{c_1 \text{ And If Possible } c_2}(x) \neq \gamma_{c_2 \text{ And If Possible } c_1}(x)$

D_4: $\gamma_{c_1}(x) \geq \gamma_{c_1'}(x) \Rightarrow \gamma_{c_1 \text{ And If Possible } c_2}(x) \geq \gamma_{c_1' \text{ And If Possible } c_2}(x)$

D_5: $\gamma_{c_2}(x) \geq \gamma_{c_2'}(x) \Rightarrow \gamma_{c_1 \text{ And If Possible } c_2}(x) \geq \gamma_{c_1 \text{ And If Possible } c_2'}(x)$

D_6: $\gamma_{c_1 \text{ And If Possible } c_2}(x) = \gamma_{c_1 \text{ And If Possible } (c_1 \text{ and } c_2)}(x).$

The following definition for $\gamma_{c_1 \text{ And If Possible } c_2}$, satisfying all of the above axioms, has been proposed:

$$\gamma_{c_1 \text{ And If Possible } c_2}(x) = \min(\gamma_{c_1}(x), k\gamma_{c_1}(x) + (1-k)\gamma_{c_2}(x)) \tag{6}$$

where $k \in]0,1[$.

The main difference between the 'And Possibly' operator defined by (4) and the 'And If Possible' operator defined above is the fact that the latter is truth functional while the former is not. Both operators make the notion of 'possibility' operational but in a different way.

In Sect. 4, we show that (6) is a special case of a more general class of aggregation operators introduced by Dujmović in the seventies of the past century.

2.2.2 'Or Else'

The expression 'c_1 Or Else c_2' is used to express a strong, non symmetric disjunction in the sense that c_2 is considered at a less important level as c_1 and therefore is not a full alternative for c_1. From a slightly different perspective, c_1 can be considered as the most demanding criterion which preferably has to be satisfied, but if this is not possible, c_2 is still considered as an acceptable alternative.

In [17], the semantics of the connective 'Or Else' have been defined by the following axioms.

C_1: $\gamma_{c_1 \text{ Or Else } c_2}(x) \leq \max(\gamma_{c_1}(x), \gamma_{c_2}(x))$

C_2: $\gamma_{c_1 \text{ Or Else } c_2}(x) \geq \gamma_{c_1}(x)$

C_3: $\exists c_1, c_2 : \gamma_{c_1 \text{ Or Else } c_2}(x) \neq \gamma_{c_2 \text{ Or Else } c_1}(x)$

C_4: $\gamma_{c_1}(x) \geq \gamma_{c_1'}(x) \Rightarrow \gamma_{c_1 \text{ Or Else } c_2}(x) \geq \gamma_{c_1' \text{ Or Else } c_2}(x)$

C_5: $\gamma_{c_2}(x) \geq \gamma_{c_2'}(x) \Rightarrow \gamma_{c_1 \text{ Or Else } c_2}(x) \geq \gamma_{c_1 \text{ Or Else } c_2'}(x)$

C_6: $\gamma_{c_1 \text{ Or Else } c_2}(x) = \gamma_{c_1 \text{ Or Else}(c_1 \text{ or } c_2)}(x)$

The following definition for $\gamma_{c_1 \text{ Or Else } c_2}$, satisfying all of the above axioms, has been proposed:

$$\gamma_{c_1 \text{ Or Else } c_2}(t) = \max(\gamma_{c_1}(t), k\gamma_{c_1}(t) + (1-k)\gamma_{c_2}(t)) \tag{7}$$

where $k \in]0, 1[$.

In Sect. 4, we show that (7) is a special case of a more general class of aggregation operators introduce by Dujmović.

3 Bipolarity Based on Desirable and Undesirable Criteria

Both the satisfied–dissatisfied approach and the constraint–wish approach can also be studied from the perspective of aggregation and aggregation structures. Such a study is the scientific contribution of this chapter. In this section, the satisfied–dissatisfied approach is handled. The study of the constraint–wish approach is detailed in Sect. 4. For both approaches, canonical forms of aggregators are presented.

3.1 General Considerations

Considering criteria evaluation in the context of database querying or decision support, we can observe that all evaluation problems consist of defining a set of elementary criteria for a group of selected attributes and then evaluating their contribution to the overall (output) suitability z of an evaluated system. Often each elementary criterion is defined for a single attribute. In some cases, we can identify two poles of criteria: criteria expressing what is desirable and criteria expressing what is undesirable. For example, a home buyer may consider the proximity to job location to be a desirable attribute and the proximity to a noisy or polluted airport to be an undesirable attribute. Similarly, a car buyer can consider the strong power of engine to be a desirable attribute and the high cost of fuel and maintenance to be undesirable attributes. Infrequently, the same attribute can be used to express what is desirable (from some standpoint or in a given interval of values) and what is undesirable (from another standpoint, or in another interval of values). In such rare cases, both poles contain a criterion that is defined for the same attribute. For example, a car buyer may consider a blue or black car color desirable and a white car color undesirable. In what follows, we will briefly call an attribute desirable (resp. undesirable) if its corresponding criterion expresses what is desirable (resp. undesirable).

For the sake of simplicity, assume that elementary criteria may be expressed only on the level of individual attributes. With the understanding that a denotes an attribute, dom_a denotes its associated domain of allowed values, and $[a]$ denotes the actual value of a under consideration, a desirable and an undesirable pole of attributes can be specified as follows.

The *pole of desirable criteria* includes m elementary criteria $g_i : dom_{a_i} \rightarrow [0, 1]$, $i = 1, \ldots, m$ for m attributes $a_1, \ldots, a_m$, hereafter called the desirable attributes, that are evaluated so that the attribute suitability degrees $x_i = g_i([a_i])$ satisfy the condition $\partial z/\partial x_i \geq 0$, $i = 1, \ldots, n$, with z being the overall, aggregated (output) suitability.

The *pole of undesirable criteria* includes n elementary criteria $h_i : dom_{b_i} \rightarrow [0, 1]$, $i = 1, \ldots, n$ for n attributes $b_1, \ldots, b_n$, hereafter called the undesirable attributes, that are evaluated so that the attribute unsuitability degrees $y_i = h_i([b_i])$ satisfy the condition $\partial z/\partial y_i \leq 0$, $i = 1, \ldots, k$.

3.2 Canonical Aggregation Structures

In the area of logic aggregation structures, some structures have an easily justifiable regular form. Such structures are called *canonical structures*. A survey of frequently used canonical aggregation structures can be found in [18]. Bipolarity based on desirable/undesirable criteria can be modeled using a bipolar canonical LSP aggregation structure as shown in Fig. 1. This aggregation structure is discussed below. *Logic Scoring of Preference* (LSP) is a decision support technique for the specification, evaluation, and aggregation of multiple criteria [19, 20]. Aggregation in LSP is done by using an *aggregation structure*. This structure is specifically designed for the decision process under consideration and has to reflect the human decision-making process as adequate as possible. The basic components of the aggregation structure are the simple LSP aggregators, which act as logical connectives. Simple LSP aggregators can on their turn be combined into compound aggregators.

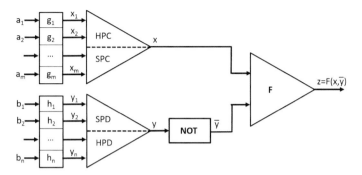

Fig. 1 The canonical aggregation structure for bipolar criteria over desirable/undesirable attributes

3.2.1 LSP Aggregators

The formal basis for LSP aggregators is the so-called generalized conjunction/disjunction (GCD) function which can be expressed by

$$
M(x_1, \ldots, x_n; W_1, \ldots, W_n; r) =
\begin{cases}
(\sum_{i=1}^{n} W_i x_i^r)^{1/r}, & \text{if } 0 < |r| < +\infty \\
\prod_{i=1}^{n} x_i^{W_i}, & \text{if } r = 0 \\
x_1 \wedge \ldots \wedge x_n, & \text{if } r = -\infty \\
x_1 \vee \ldots \vee x_n, & \text{if } r = +\infty.
\end{cases}
\tag{8}
$$

The values $x_i \in [0, 1]$, $1 \leq i \leq n$, are the input suitability degrees (hereby, 0 and 1, respectively, denote 'not suitable at all' and 'completely suitable'). The given (or precomputed) static weights, W_i, $1 \leq i \leq n$ determine the relative importance of the inputs and have to sum up to 1. Furthermore, the given (or precomputed) exponent $r \in [-\infty, +\infty]$ determines the logical properties of the aggregator. Special cases of exponent values are as follows:

- $+\infty$ corresponding to full disjunction,
- $-\infty$ corresponding to full conjunction, and
- 1 corresponding to weighted average.

The other exponent values allow to model other aggregators, ranging continuously from full conjunction to full disjunction, and can be computed from a desired value of orness (ω), i.e., an index expressing how 'close' the aggregator should be to the regular disjunction operator in its behavior. The following numeric approximation for r can be used [20]:

$$r = \frac{0.25 + 1.89425x + 1.7044x^2 + 1.47532x^3 - 1.42532x^4}{\omega(1-\omega)} \tag{9}$$

where

$$x = \omega - 1/2 \text{ and } 0 < \omega < 1.$$

The andness (α) is obtained as the complement of the orness, i.e.,

$$\alpha = 1 - \omega.$$

Andness is hence an index expressing how 'close' the aggregator should be to the regular conjunction operator in its behavior.

For $\omega > 0.5$, we have disjunction $\omega = 1$ corresponds with $r = +\infty$ and is called *full disjunction*. For $0.75 < \omega < 1$, a *hard partial disjunction* (HPD) operator is obtained, whereas $0.5 < \omega < 0.75$ yields a *soft partial disjunction* (SPD) operator. So, $\omega = 0.75$ can be considered as corresponding with a *neutral partial disjunction* operator.

Likewise, for $\alpha > 0.5$, we have conjunction $\alpha = 1$ corresponds with $r = -\infty$ and is called *full conjunction*. For $0.75 < \alpha < 1$, a *hard partial conjunction* (HPC) operator is obtained, whereas $0.5 < \alpha < 0.75$ yields a *soft partial conjunction* (SPC) operator. Also here, $\alpha = 0.75$ can be considered as corresponding with a *neutral partial conjunction* operator.

If $\alpha = \omega = 0.5$, the neutral (*weighted*) *arithmetic mean* operator is obtained. This corresponds with the case where $r = 1$.

3.2.2 Aggregators for Desirable and Undesirable Criteria

Aggregating for Desirable Criteria

Now, reconsider the canonical LSP criterion structure shown in Fig. 1. For the group of desirable attributes $a_1, \ldots, a_m$, we have that the higher their corresponding suitability degrees are, the better the situation is. We now assume that this group of attributes can be separated in two subgroups: *mandatory* and *nonmandatory* attributes.

Mandatory attributes are aggregated using an aggregation structure based on HPC. By definition, HPC is an aggregator that has the annihilator 0 supported (uniformly) by all inputs; that is, if *any* of the input suitability degrees is 0, then the output must also be 0.

Nonmandatory desired attributes are aggregated by an aggregation structure that is based on SPC. SPC is a conjunctive aggregator that does not support the annihilator 0; only one positive input is sufficient to secure a positive output.

If all mandatory inputs are aggregated using a HPC structure that yields the suitability x_{man} and all nonmandatory (desired or optional) inputs are aggregated using a SPC structure that yields the suitability x_{nman}, then the aggregated suitability x can be obtained by using a conjunctive partial absorption (CPA) operator $\unrhd$ [21], i.e., $x = x_{man} \unrhd x_{nman}$. CPA is further discussed in Sect. 4.

Aggregation for Undesirable Criteria

For the group of undesirable attributes $b_1, \ldots, b_n$, we have that the higher their corresponding suitability degrees are, the worse the situation is. We can also subdivide this group of attributes into two groups: *sufficient* and *nonsufficient* attributes.

Sufficient undesired attributes are aggregated using an aggregation structure based on HPD. By definition, HPD is an aggregator that has the annihilator 1 supported (uniformly) by all inputs; that is, if *any* of the input unsuitability degrees is 1, then the output must also be 1. This makes each input in such a group sufficient to point out that the evaluated object is unacceptable.

Nonsufficient undesired attributes are aggregated by an aggregation structure that is based on SPD. SPD is a disjunctive aggregator that does not support the annihilator 1; only one input that is less than 1 is sufficient to result in an output that is less than 1.

If all sufficient undesired inputs are aggregated using a HPD structure that yields the unsuitability y_{suf} and all nonsufficient undesired inputs are aggregated using a SPD structure that yields the unsuitability y_{nsuf}, then the aggregated unsuitability y can be obtained by using a disjunctive partial absorption (DPA) operator $\overline{\rhd}$ [21], i.e., $y = y_{suf} \overline{\rhd} y_{nsuf}$. DPA is further discussed in Sect. 4.

Computing Overall Suitability

The aggregation of the desirable attribute scores (suitability degrees) generates the overall suitability score x, whereas the independent aggregation of the undesirable attribute scores (unsuitability degrees) generates the independent overall unsuitability degree y as shown in Fig. 1. The degrees x and y have similar semantics as the satisfaction degree s and dissatisfaction degree d of the BSD (s, d) that is obtained in the satisfied–dissatisfied approach presented in [10] (cf. Equation 1).

However, in the canonical aggregation structure given in Fig. 1, it is proposed to aggregate x and y in order to obtain an overall suitability degree z. The motivation for this is that for some applications, such an overall suitability degree might be required.

Basic Approach. Let us now consider a process of aggregating the suitability x and the unsuitability y. A simplified verbal criterion could be 'we simultaneously want x and not y.' In an extreme Boolean case this criterion yields the material nonimplication or the abjunction function $z = x \wedge \bar{y}$. The abjunction function is a non-idempotent aggregator which can also be denoted as a negated implication: $x \wedge \bar{y} = \overline{x \rightarrow y}$.

General Approach. Not surprisingly, in a general case, we can apply a partial abjunction function based on GCD. To convert the unsuitability score y to a corresponding suitability score $\bar{y}$, we need a negation operator that preferably supports involution, e.g.,

$$\bar{y} = \frac{1 - y}{1 + p \cdot y} \tag{10}$$

where $p > -1$.

The final stage of the canonical aggregation structure is the aggregation of suitability of desirable attributes x and unsuitability of undesirable attributes y creating the output overall suitability

$$z = F(x, \bar{y}). \tag{11}$$

The output aggregation function F can take several forms, not necessarily referring to the strict abjunction mentioned above. In particular, F can be a HPC operator, i.e., $z = Wx \Delta \overline{W}\bar{y}]] >, 0 < W < 1, \overline{W} = 1 - W, \bar{y} = y - 1$. This aggregator is a weighted partial abjunction of x and y. In some cases, it can be justifiable to use SPC. On the other hand, a special interpretation of desirable and undesirable attributes may be assumed, e.g., the one related to the constraint–wish approach. Then, F can be an asymmetric aggregator, where x is a mandatory input and $\bar{y}$ is an optional input (i.e., a CPA [21]).

The canonical aggregation structure from Fig. 1 can be transformed if the elementary criteria of undesired attributes can be 'negated' as exemplified in Fig. 2. In this example if the distance from the airport is undesirable, then it is equivalent to use either the unsuitability criterion or its complement, the suitability criterion.

If all undesired attributes are evaluated using suitability criteria, then we can replace the nonidempotent canonical aggregation structure from Fig. 1 with the idempotent canonical aggregation structure shown in Fig. 3.

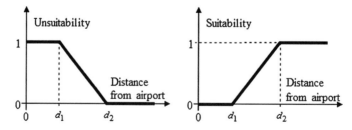

Fig. 2 Suitability and unsuitability criteria for the distance of home from an airport

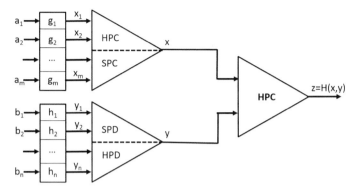

Fig. 3 Symmetric canonical aggregation structure

4 Bipolarity Based on Nonuniform Input-Dependent Annihilators

4.1 General Considerations

The interpretations of constraint–wish approaches of bipolarity are based on asymmetric aggregation operators. Among these, the *conjunctive partial absorption* $\unrhd$ (CPA) and the *disjunctive partial absorption* $\overline{\rhd}$ (DPA) seem especially interesting. The concept of asymmetric aggregators and the CPA can be considered as the oldest and perhaps most frequently used form of bipolarity introduced in 1974 [22]. Both the CPA and the DPA were then studied in [19]. A detailed quantitative analysis and synthesis of PA aggregators were presented in [21].

4.2 Canonical Aggregation Structures

The basic canonical forms of CPA and DPA aggregators have two inputs: x and y, and their internal organization is presented in Figs. 4 and 5.

In these figures, the symbols ∇, $\varDelta$, and $\ominus$, respectively, denote partial disjunction, partial conjunction, and the neutrality (arithmetic mean). More specifically, $\overline{\nabla}$ denotes HPD and $\overline{\varDelta}$ denotes HPC, whereas W_1 and W_2 are precomputed weights (see below).

4.2.1 Conjunctive Partial Absorption

In the case of CPA $\unrhd$ (Fig. 4), the input x is mandatory and it supports the annihilator 0, i.e.,

Fig. 4 Two basic versions of conjunctive partial absorption

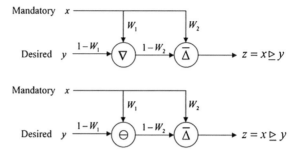

Fig. 5 Two basic versions of disjunctive partial absorption

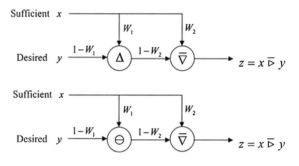

$$\forall y > 0 : 0 \trianglerighteq y = 0.$$

As opposed to that, the optional input y does not support the annihilator 0, i.e.,

$$\forall x > 0 : 0 < x \trianglerighteq 0 < x.$$

The CPA $\trianglerighteq$ (Fig. 4) aggregates a mandatory input x and a nonmandatory (desired or optional) input y, as follows:

$$x \trianglerighteq y = W_2 x \overline{\overline{\varDelta}} (1 - W_2)[W_1 x \tilde{\nabla} (1 - W_1)y] \tag{12}$$

where $\overline{\overline{\varDelta}} \in \{\wedge, \overline{\varDelta}\}$ and $\tilde{\nabla} \in \{\vee, \underline{\nabla}, \overline{\nabla}, \ominus\}$. Here, $\overline{\varDelta}$ denotes hard partial conjunction, $\underline{\nabla}$ represents soft partial disjunction, and $\overline{\nabla}$ denotes hard partial disjunction.

The weights W_1 and W_2 in equation Eq. (12) are computed so as to reflect as adequately as possible the impact of the mean penalty P and mean reward R percentages provided by the user. Hereby, the underlying semantics of P and R are defined by the following border conditions for the CPA $\trianglerighteq$ [21]:

$$\forall 0 < x \leq 1 : x \trianglerighteq 0 = x(1 - p), 0 \leq p < 1 \tag{13}$$

(hence, if the optional condition is not satisfied at all, then criterion satisfaction is decreased with a penalty of p).

$$\forall 0 < x < 1 : x \underline{\triangleright} 1 = x(1+r), 0 \le r < 1/x - 1 \tag{14}$$

(hence if the optional condition is fully satisfied, then criterion satisfaction is increased with a reward of r). Note that p and r can be zero. The values P and R are (approximately) the mean values of p and r and usually expressed as percentages. Users select desired values of P and R and use them to compute the corresponding weights W_1 and W_2. More details on this computation can be found in [21].

Let us note that Eq. (6) is a special case of Eq. (12) that is obtained if $\overline{\overline{\Delta}} = \wedge$ and $\overline{\nabla} = \ominus$. The use of (hard) partial conjunction in Eq. (12) enables the use of a reward.

4.2.2 Disjunctive Partial Absorption

In the case of DPA $\overline{\triangleright}$ (Fig. 5), the input x is sufficient and it supports the annihilator 1, i.e.,

$$\forall y < 1 : 1 \overline{\triangleright} y = 1.$$

As opposed to that, the optional input y does not support the annihilator 1, i.e.,

$$\forall x < 1 : x < x \overline{\triangleright} 1 < 1.$$

The DPA $\overline{\triangleright}$ (Fig. 5) aggregates a sufficient input x and a nonsufficient (desired or optional) input y, as follows:

$$x \overline{\triangleright} y = W_2 x \overline{\overline{\nabla}}(1 - W_2)[W_1 x \widetilde{\Delta}(1 - W_1)y] \tag{15}$$

where $\overline{\overline{\nabla}} \in \{\vee, \overline{\nabla}\}$ and $\widetilde{\Delta} \in \{\wedge, \underline{\Delta}, \overline{\Delta}, \ominus\}$. Here, $\overline{\nabla}$ denotes hard partial disjunction, $\underline{\Delta}$ represents soft partial conjunction, and $\overline{\Delta}$ denotes the hard partial conjunction.

Like with CPA, the weights W_1 and W_2 in equation Eq. (15) are computed so as to reflect as adequately as possible the impact of the mean penalty P and mean reward R percentages provided by the user. The underlying semantics of P and R are defined by the dual counterparts of the border conditions for the CPA, given in the previous subsection (see [21] for more details on this).

Note that Eq. (7) is a special case of Eq. (15), which is obtained if $\overline{\overline{\nabla}} = \vee$ and $\widetilde{\Delta} = \ominus$. The use of (hard) partial disjunction in Eq. (15) enables the use of a penalty.

4.3 Comparison with the 'And if Possible' and 'Or Else' Operators

Studying the fundamental properties of CPA and DPA and comparing them with those of the 'And If Possible' and 'Or Else' operators presented in Sect. 2, we obtain the following.

The fundamental properties of CPA are as follows:

1. $$\forall 0 \le y \le 1 : 0 \underline{\triangleright} y = 0$$

2. $$\forall 0 < x \le 1 : 0 < x \underline{\triangleright} 0 \le x$$

3. $$\forall 0 < x < 1 : x \le x \underline{\triangleright} 1 < 1$$

Comparing these properties with the axioms of the 'And If Possible' operator reveals that axiom $[C_2]$ is in conflict with the third fundamental property of CPA. In fact, axiom $[C_2]$ permits penalizing and prevents rewarding. For example, in a criterion for evaluating the quality of available parking facilities at a home, let the mandatory requirement be the availability of a private garage for one or ideally for two cars. Let the optional requirement be the availability of a quality street parking. Now, if the mandatory requirement is partially or perfectly satisfied, but the street parking is not available, the function (6) will penalize such a home. However, if a mandatory requirement is partially satisfied (there is a garage for one car), and there is a perfect street parking, the function (6) does not permit to compensate imperfections of the home parking with the quality street parking, and this may be inconsistent with human reasoning.

The fundamental properties of DPA are as follows:

1. $$\forall 0 \le y \le 1 : 1 \overline{\triangleright} y = 1$$

2. $$\forall 0 < x < 1 : 0 < x \overline{\triangleright} 0 \le x$$

3. $$\forall 0 < x < 1 : x \le x \overline{\triangleright} 1 < 1$$

Comparing these properties with the axioms of the 'Or Else' operator reveals that axiom $[D_2]$ is in conflict with the third fundamental property of DPA. In fact, based on axiom $[D_2]$, the aggregator (7) permits only rewards and no penalty, which is frequently inconsistent with human reasoning. Consider, for example, two car descriptions where a sufficient criterion 'low fuel consumption' is perfectly satisfied for both. If a desired condition 'air bags' is only satisfied for the first car and there is no penalty facility available, then it would not be possible to distinguish between the overall satisfaction of both cars. However, humans would naturally assign a penalty to the second car and prefer the first one.

The main advantage of the CPA and DPA operators is that they enable the use of both a reward and a penalty. As illustrated above, both rewards and penalties are required if we want to adequately reflect human reasoning.

5 Multipolarity

It is useful to note that the canonical aggregation structures in both Figs. 1 and 3 use 'bipolarity inside bipolarity' (or a kind of 'multipolarity') identifying four categories of attributes:

1. mandatory–desired,
2. optional–desired,
3. mandatory–undesired, and
4. optional–undesired.

Thus, bipolarity, tripolarity [23] (mandatory/desired/optional, shown in Fig. 6), and generally 'multipolarity' reflect situations where we have two or more logically dissimilar clusters of attributes, and each cluster contains attributes that have a similar logical impact on the overall suitability score.

More recently, multipolarity has also been studied in the context of constraint–wish approaches by Liétard et al. [24]. Essentially, the 'And If Possible' and 'Or Else' operators are hereby generalized to expressions of the form

$$c_n \text{ And If Possible } c_{n-1} \text{ And If Possible} \ldots \text{ And If Possible } c_1 \qquad (16)$$

and

$$c_1 \text{ Or Else } c_2 \text{ Or Else} \ldots \text{ Or Else } c_n \qquad (17)$$

where the underlying assumption is that c_{i+1} is a relaxation of c_i, for $i = 1, \ldots, n-1$. Hence, for multipolarity with respect to the 'And If Possible' aggregator, the semantics are to satisfy a condition and, if possible, to satisfy a more restrictive variant of this condition. Likewise, for mulipolarity with respect to the 'Or Else' aggregator, the semantics are to satisfy a condition, or else, to satisfy a relaxation of this condition.

By connecting multiple canonical CPA aggregators, a similar kind of multipolarity as the one that is obtained from the generalization of the 'And If Possible'

Fig. 6 Tripolarity realized using nested bipolarity: $S = x'(y'z)$

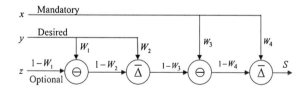

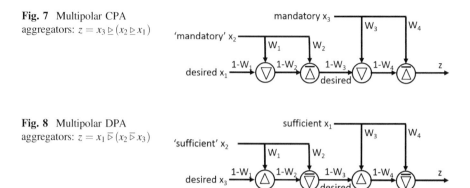

Fig. 7 Multipolar CPA aggregators: $z = x_3 \trianglerighteq (x_2 \trianglerighteq x_1)$

Fig. 8 Multipolar DPA aggregators: $z = x_1 \overline{\triangleright} (x_2 \overline{\triangleright} x_3)$

aggregator can be modeled. This is illustrated in Fig. 7 for the case where $n = 3$. Consider n inputs x_i, $i = 1, \ldots, n$, where for $i = 1, \ldots, n - 1$, x_{i+1} is obtained from a criterion which is a relaxation of the criterion from which x_i is obtained. Then, the following multipolar CPA aggregator can be constructed

$$z = x_n \trianglerighteq (x_{n-1} \trianglerighteq (\ldots) \trianglerighteq x_1). \tag{18}$$

Hereby, the most stringent input x_n is considered to be a mandatory input, whereas satisfaction of the remainder of the expression is considered to be desired. In this desired part, the first input x_{n-1} is considered to be mandatory, whereas the remainder of the expression is desired and so on. Thus, in order to have a satisfaction of the desired input $x_{n-1} \trianglerighteq (x_{n-2} \trianglerighteq (\ldots) \trianglerighteq x_1)$, x_{n-1} must be satisfied, whereas the satisfaction of the remainder of the expression is desired, and so on.

Likewise, by connecting multiple canonical DPA aggregators, a similar kind of multipolarity as the one that is obtained from the generalization of the 'Or Else' aggregator can be modeled. This is illustrated in Fig. 8 for the case where $n = 3$. Reconsider n inputs x_i, $i = 1, \ldots, n$, where for $i = 1, \ldots, n - 1$, x_{i+1} is obtained from a criterion which is a relaxation of the criterion from which x_i is obtained. Then, the following multipolar DPA aggregator can be constructed

$$z = x_1 \overline{\triangleright} (x_2 \overline{\triangleright} (\ldots) \overline{\triangleright} x_n). \tag{19}$$

Hereby, the least stringent input x_1 is considered to be a sufficient input, whereas satisfaction of the remainder of the expression is considered to be desired. In this desired part, the first input x_2 is considered to be sufficient, whereas the remainder of the expression is desired and so on. Thus, in order to have a satisfaction of the desired input $x_2 \overline{\triangleright} (x_3 \overline{\triangleright} (\ldots) \overline{\triangleright} x_n)$, it is sufficient that x_2 is satisfied, whereas the satisfaction of the remainder of the expression is desired, and so on.

6 Conclusions

In this chapter, we discussed and studied bipolarity among multiple criteria from the standpoint of aggregation and aggregation structures. Two main approaches are distinguished: bipolarity based on desirable and undesirable criteria and bipolarity based on nonuniform input-independent annihilators. For both approaches, canonical forms of aggregators are proposed.

Bipolarity based on desirable and undesirable criteria is obtained by considering different poles of criteria which all have their own different semantics: a pole of 'positive' criteria that express what the user will accept and a pole of 'negative' criteria that denote what the user will reject.

Bipolarity based on nonuniform input-dependent annihilators is studied from the standpoint of conjunctive and disjunctive partial absorption. Herewith, it is shown that the 'And If Possible' and 'Or Else' operators of the well-known 'constraint–wish' approach are in fact special cases of partial absorption and can be generalized by canonical forms of conjunctive and disjunctive partial absorption operators.

Furthermore, we explored some issues of multipolarity. Multipolarity naturally occurs when mandatory and nonmandatory criteria are distinguished within the 'positive' and 'negative' criteria poles. A special kind of such multipolarity is the so-called tripolarity where mandatory, desired, and optional criteria are distinguished.

Another kind of multipolarity, studied in the context of the 'constraint–wish' approach is obtained by connecting a list of multiple criteria by multiple 'And If Possible' operators (or by multiple 'Or Else' operators). Each criterion in the list then has to be a relaxation of the previous criterion in the list. This kind of multipolarity can be generalized by connecting multiple canonical partial absorption aggregators.

In our future research, we plan to further investigate the different aspects and concepts of multipolarity.

References

1. Zadrożny, S., De Tré, G., Kacprzyk, J.: Remarks on various aspects of bipolarity in database querying. In: Proceedings of the 2010 International Workshop on Database and Expert Systems Applications Proceedings (DEXA'10), Bilbao, Spain, pp. 323–327 (2010)
2. Zadrożny, S., Kacprzyk, J.: Bipolarity in database querying: Various aspects and interpretations. In: Pivert, O., Zadrożny, S. (eds.) Flexible Approaches in Data, Information and Knowledge Management, pp. 71–91. Springer, Heidelberg (2013)
3. Dubois, D., Prade, H.: Handling bipolar queries in fuzzy information processing. In: Galindo, J. (ed.) Handbook of research on fuzzy information processing in databases, pp. 97–114. IGI Global, New York (2008)
4. Dubois, D., Prade, H.: An introduction to bipolar representations of information and preference. Int. J. Intell. Syst. **23**, 866–877 (2008)

5. Bosc, P., Pivert, O., Mokhtari, AM, Liétard, L.: Extending relational algebra to handle bipolarity. In: Proceedings of SAC'10, Sierre, Switzerland, pp. 1718–1722 (2010)
6. Liétard, L., Rocacher, D.: On the definition of extended norms and co-norms to aggregate fuzzy bipolar conditions. In: Proceedings of the 2009 IFSA/EUSFLAT Conference, Lisbon, Portugal, pp. 513–518 (2009)
7. Liétard, L., Tamani, N., Rocacher, D.: Linguistic Quantifiers and Bipolarity. In: Proceedings of the 2011 IFSA World Congress and the 2011 AFSS International Conference, Surabaya and Bali Island, Indonesia (2011)
8. Liétard, L., Pivert, O., Rocacher, D.: On a graded inclusion of bipolar fuzzy relations. In: Proceedings of the 8th EUSFLAT Conference, Milan, Italy (2013)
9. Tamani, N., Liétard, L., Rocacher, D.: Bipolarity and the relational division. In: Proceedings of the 7th EUSFLAT Conference, Aix-les-Bains, France (2011)
10. De Tré, G., Zadrożny, S., Matthé, T., Kacprzyk, J., Bronselaer, A.: Dealing with positive and negative query criteria in fuzzy database querying : bipolar satisfaction degrees. Lect. Notes Comput. Sci. **5822**, 593–604 (2009)
11. Matthé, T., De Tré, G.: Bipolar query satisfaction using satisfaction and dissatisfaction degrees: bipolar satisfaction degrees. In: Proceedings of the ACM Symposium on Applied Computing (ACM SAC'09), Honolulu, Hawaii, pp. 1699–1703 (2009)
12. Matthé, T., De Tré, G., Zadrożny, S., Kacprzyk, J., Bronselaer, A.: Bipolar database querying using bipolar satisfaction degrees. Int. J. Intell. Syst. **26**(10), 890–910 (2011)
13. Lacroix, M., Lavency, P.: Preferences: Putting more knowledge into queries. In: Proceedings of the VLDB'87 Conference, Brighton, UK, pp. 217–225 (1987)
14. Rodriíguez, J.T., Camilo, A.F., Montero F.J.: On the semantics of bipolarity and fuzziness. In: Melo-Pinto, P. et al. (eds.) Eurofuse 2011: Workshop on Fuzzy Methods for Knowledge-Based Systems, pp. 193–205. Springer, Heidelberg (2012)
15. Zadrożny, S., Kacprzyk, J.: Bipolar queries and queries with preferences. In: Proceedings of the DEXA Conference (DEXA'06), Krakow, Poland, pp. 415–419 (2006)
16. Zadrożny, S., Kacprzyk, J.: Bipolar queries: An aggregation operator focused perspective. Fuzzy Sets Syst. **196**, 69–81 (2012)
17. Bosc, P., Pivert, O.: On four noncommutative fuzzy connectives and their axiomatization. Fuzzy Sets Syst. **202**, 42–60 (2012)
18. Dujmović, J.J., De Tré, G.: Multicriteria methods and logic aggregation in suitability maps. Int. J. Intell. Syst. **26**(10), 971–1001 (2011)
19. Dujmović, J.J.: Extended continuous logic and the theory of complex criteria. J. Univ. Belgrade **537**, 197–216 (1975)
20. Dujmović, J.J.: Preference logic for system evaluation. IEEE Trans. Fuzzy Syst. **15**(6), 1082–1099 (2007)
21. Dujmović, J.J.: Partial absorption function. J Univ Belgrade **659**, 156–163 (1979)
22. Dujmović, J.J.: New results in the development of the mixed averaging by levels method for system evaluation (In Serbo-Croatian). In: Proceedings of the Informatica Conference, Bled, Yugoslavia, paper 4.36 (1974)
23. Su, S.Y.W., Dujmović, J.J., Batory, D.S., Navathe, S.B., Elnicki, R.: A cost-benefit decision model: analysis, comparison, and selection of data management systems. ACM Trans. Database Syst. **12**(3), 472–520 (1987)
24. Liétard, L., Hadjali, A., Rocacher, D.: Towards a gradual QCL model for database querying. In: Information Processing and Management of Uncertainty in Knowledge-Based Systems, Communications in Computer and Information Science, vol. 444, pp. 130–139. Springer, New York (2014)

Choquet Integral with Interval Type 2 Sugeno Measures as an Integration Method for Modular Neural Networks

Gabriela E. Martínez, Olivia Mendoza, Juan R. Castro, Patricia Melin and Oscar Castillo

Abstract In this paper, a new method for response integration, based on the Choquet integral with interval type 2 Sugeno measures, is presented. Type 1 and interval type 2 fuzzy systems for edge detection based on the Sobel and morphological gradient are used, which is a preprocessing system applied to the training data for better performance in the modular neural network. Fuzzy Sugeno measures are represented by an interval type 2 fuzzy system. The Choquet integral is used as a method to integrate the outputs of the modules of the modular neural networks (MNN). A database of faces was used to perform the preprocessing, the training, and the combination of information sources of the MNN.

1 Introduction

An integration method is a mechanism which takes input as a number n of data and combines them to form a value representative of the information, methods exist which combine information from different sources which can be aggregation operators as arithmetic mean, geometric mean, ordered weighted averaging (OWA) [1], and inter alia.

Artificial neural networks were introduced by W.S. McCullogh and W. Pitts in 1943 [2] and can be used in a variety of applications; however, there are problems that cannot be processed in a single network either because of their complexity or

G.E. Martínez (✉) · O. Mendoza · J.R. Castro
Faculty of Chemical Sciences and Engineering Autonomous,
University of Baja California, Tijuana, Mexico
e-mail: gabriela.martinez14@uabc.edu.mx

P. Melin · O. Castillo
Division of Graduate Studies and Research, Tijuana Institute of Technology,
Tijuana, Mexico
e-mail: pmelin@tectijuana.mx

O. Castillo
e-mail: ocastillo@tectijuana.mx

© Springer International Publishing Switzerland 2016   71
L.A. Zadeh et al. (eds.), *Recent Developments and New Direction
in Soft-Computing Foundations and Applications*, Studies in Fuzziness
and Soft Computing 342, DOI 10.1007/978-3-319-32229-2_6

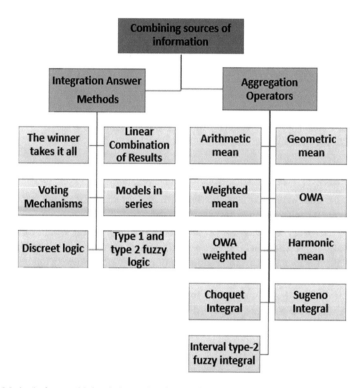

Fig. 1 Methods for combining information from different sources

the amount of information they have; in these cases, a modular neural network (MNN) is used so that each of the modules is responsible for a sub-task or a part of the problem [3–7]. Therefore, it is necessary to have a mechanism to integrate information from different modules to provide the general solution to the problem.

In a MNN, it is common to use some methods such as fuzzy logic type 1 and type 2 [8–10], the fuzzy Sugeno integral [11], interval type 2 fuzzy logic Sugeno integral [12], a probabilistic sum integrator [13], a Bayesian learning method [14], among others, as shown in Fig. 1.

The Choquet integral is an aggregation operator, which has been successfully used in various applications [15–17]. In this paper, the fuzzy Sugeno measures are represented by an interval type 2 fuzzy system in combination with the Choquet integral.

This paper is organized as follows: Sect. 2 shows the concepts of fuzzy measures, interval type 2 fuzzy measures, and Choquet integral which is the technique that was applied for the combination of the several information sources. Section 3 shows edge detection based on Sobel and morphological gradient with interval type 2 fuzzy system. Section 4 shows the modular neural network proposal, and in Sect. 5, the simulation results are shown. Finally, Sect. 6 shows the Conclusions.

2 Fuzzy Measures and Choquet Integral

Initially, Michio Sugeno defined the concept of "fuzzy measures and fuzzy integral" in 1974 [18]. A fuzzy measure is a non-negative monotone function of defined values in "classical sets." Currently, when referring to this topic, the term "fuzzy measures" has been replaced by the term "monotonic measures," "non-additive measures," or "generalized measures" [19–21]. When fuzzy measures are defined on fuzzy sets, we speak about monotonous fuzzified measures [21].

2.1 Fuzzy Measure

If $x = \{x_1, x_2, \ldots, x_n\}$ is a finite set, a fuzzy measure μ with respect to the data set X is a function $\mu : 2^x \rightarrow [0, 1]$ that must satisfy the following conditions:

(1) $\mu(X) = 1$; $\mu(\emptyset) = 0$
(2) If $A \subseteq B$, then $\mu(A) \leq \mu(B)$

where in the second condition, A and B are subsets of X.

A fuzzy measure is a Sugeno measure or λ-fuzzy, if it satisfies condition (1) of addition for some $\lambda > -1$.

$$\mu(A \cup B) = \mu(A) + \mu(B) + \lambda\mu(A)\mu(B) \tag{1}$$

where λ can be calculated with (2):

$$f(\lambda) = \left\{ \prod_{i=1}^{n} (1 + M_i(x_i)\lambda) \right\} - (1 + \lambda) \tag{2}$$

The value of the λ parameter is determined by the conditions of the Theorem 1.

Theorem 1 Let $\mu(\{x\}) < 1$ for each $x \in X$ and let $\mu(\{x\}) > 0$ for at least two elements of X, then (2) determines a unique parameter λ in the following way:

If $\sum_{x \in X} \mu(\{x\}) < 1$, then is in the interval $(0, \infty)$.
If $\sum_{x \in X} \mu(\{x\}) = 1$, then $\lambda = 0$; That is the unique root of the equation.
If $\sum_{x \in X} \mu(\{x\}) > 1$, then λ is in the interval $(-1, 0)$.

The fuzzy measure represents the importance or relevance of the sources when computing the aggregation [22], and the method to calculate Sugeno measures is performed recursively using (3) and (4).

$$\mu(A_1) = \mu(M_1) \tag{3}$$

$$\mu(A_i) = \mu(A_{(i-1)}) + \mu(M_i) + \left(\lambda\mu(M_i) * \mu(A_{(i-1)})\right) \tag{4}$$

where A_i represents the fuzzy measure and M_i represents the fuzzy density determined by an expert, where $1 < M_i \le \ldots \le n$ should be permuted with respect to the descending order of their respective $\mu(A_i)$.

2.2 Fuzzy Measures for Interval Type 2 Fuzzy Sets

For the estimation of the fuzzy densities of each information source, we take the maximum value of each X_i, where an interval of uncertainty is added.

You need to add an uncertainty footprint or FOU which will create an interval based on the fuzzy density. Equation (5) can be used to approximate the center of the interval for each fuzzy density, and Eqs. (6) and (7) are used to estimate left and right values of the interval for each fuzzy density. Note that the domain for $\mu_L(x_i)$ and $\mu_U(x_i)$ is given in Theorem 1 [23].

Calculation of the fuzzy densities:

$$\mu_c(x_i) = \max(X_i) \tag{5}$$

$$\mu_L(x_i) = \begin{cases} \mu_c(x_i) - \text{FOU}_\mu/2; & \text{if } \mu_c(x_i) > \text{FOU}_\mu/2 \\ 0.0001 & \text{otherwise} \end{cases} \tag{6}$$

$$\mu_U(x_i) = \begin{cases} \mu_c(x_i) + \text{FOU}_\mu/2; & \text{if } \mu_c(x_i) < (1 - \text{FOU}_\mu/2) \\ 0.9999 & \text{otherwise} \end{cases} \tag{7}$$

Calculating the parameters λ_L and λ_U for each side of the interval with (8) and (9)

$$\lambda_L + 1 = \prod_{i=1}^{n}(1 + \lambda_L\mu_L(\{x_i\})) \tag{8}$$

$$\lambda_U + 1 = \prod_{i=1}^{n}(1 + \lambda_U\mu_U(\{x_i\})) \tag{9}$$

Once the λ_U, λ_L are obtained, parameters can be calculated fuzzy measures left and right by extending the recursive formulas (10, 11) (12, 13):

$$\mu_L(A_1) = \mu_L(x_1) \tag{10}$$

$$\mu_L(A_i) = \mu_L(x_i) + \mu_L(A_{i-1}) + \lambda_L\mu_L(x_i)\mu_L(A_{i-1}) \tag{11}$$

$$\mu_U(A_1) = \mu_U(x_1) \tag{12}$$

$$\mu_U(A_i) = \mu_U(x_i) + \mu_U(A_{i-1}) + \lambda_U \mu_U(x_i)\mu_U(A_{i-1}) \tag{13}$$

There are two types of integral that performed the calculation of Sugeno measures: the integral of Sugeno and Choquet Integral.

2.3 Choquet Integral

The Choquet integral can be calculated using (14) or an equivalent expression (15)

$$\text{Choquet} = \sum_{i=1}^{n} \left\{ \left[A_i - A_{(i-1)} \right] * D_i \right\} \tag{14}$$

with $A_0 = 0$,
or also

$$\text{Choquet} = \sum_{i=1}^{n} A_i * \left\{ \left[D_i - D_{(i+1)} \right] \right\} \tag{15}$$

with $D_{(n+1)} = 0$,
where A_i represents the fuzzy measurement associated with data D_i.

3 Edge Detection

Edge detection can be defined as a method consisting of identifying changes that exist in the light intensity, which can be used to determine certain properties or characteristics of the objects in the image.

We used the ORL Database of Faces [24] to perform the training of the modular neural network, which has images of 40 people with 10 samples of each individual. To each of the images was applied to a preprocessing by making use of Sobel edge detector and morphological gradient with type 1 and type 2 fuzzy logic system [25] in order to highlight features, some of the images can be displayed in Fig. 5b, d.

3.1 The Morphological Gradient

To perform the method of morphological gradient, we need to calculate every one of the four gradients as commonly done in the traditional method using (16–20), see

Fig. 2 Calculation of the
gradient in the four directions

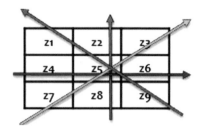

Fig. 2; however, the addition of the gradients is performed by a type 1 and type 2 fuzzy system [25]; in Fig. 3, the membership functions of the fuzzy system are shown, and the resulting image can be viewed in Fig. 5b, d.

$$D1 = \sqrt{(z5 - z2)^2 + (z5 - z8)^2} \qquad (16)$$

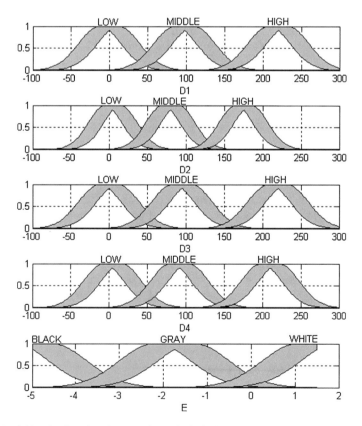

Fig. 3 Variables for the edge detector of morphological gradient the type 2

$$D2 = \sqrt{(z5 - z4)^2 + (z5 - z6)^2} \tag{17}$$

$$D3 = \sqrt{(z5 - z1)^2 + (z5 - z9)^2} \tag{18}$$

$$D4 = \sqrt{(z5 - z7)^2 + (z5 - z3)^2} \tag{19}$$

$$G = D1 + D2 + D3 + D4 \tag{20}$$

3.2 Sobel

The Sobel operator is applied to a digital image in gray scale is a pair of 3×3 convolution masks, one estimating the gradient in the x-direction (columns) (21) and the other estimating the gradient in the y-direction (rows) (22) [26].

$$\text{sobel}_x = \begin{bmatrix} -1 & 0 & 1 \\ -2 & 0 & 2 \\ -1 & 0 & 1 \end{bmatrix} \tag{21}$$

$$\text{sobel}_y = \begin{bmatrix} 1 & 2 & 1 \\ 0 & 0 & 0 \\ -1 & -2 & -1 \end{bmatrix} \tag{22}$$

If we have $I_{m,n}$ as a matrix of m rows and r columns, where the original image is stored, then g_x and g_y are matrices having the same dimensions as I, which at each element contains the horizontal and vertical derivative approximations and are calculated by (23) and (24) [25].

$$g_x = \sum_{i=1}^{i=3} \sum_{j=1}^{j=4} \text{Sobel}_{x,ij} * I_{r+i-2,c+j-2} \quad \begin{array}{l} \text{for} = 1, 2, \ldots, m \\ \text{for} = 1, 2, \ldots, n \end{array} \tag{23}$$

$$g_y = \sum_{i=1}^{i=3} \sum_{j=1}^{j=4} \text{Sobel}_{y,ij} * I_{r+i-2,c+j-2} \quad \begin{array}{l} \text{for} = 1, 2, \ldots, m \\ \text{for} = 1, 2, \ldots, n \end{array} \tag{24}$$

In the Sobel method, the gradient magnitude g is calculated by (25).

$$g = \sqrt{g_x^2 + g_y^2} \tag{25}$$

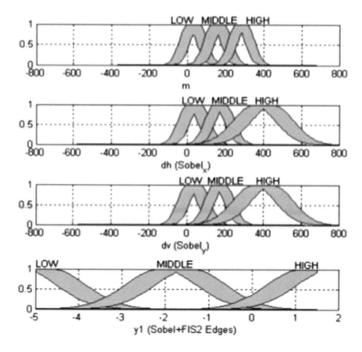

Fig. 4 Variables for the edge detector with the type 2 fuzzy Sobel

For the type 1 and type 2 fuzzy inference systems, 3 inputs can be used, in which 2 of them are the gradients with respect to the x-axis and y-axis, calculated with (23) and (24), which we call DH and DV, respectively. The third variable m is the image after the application of a low-pass filter hMF in (26); this filter allows to detect image pixels belonging to regions of the input where the mean gray level is lower. These regions are proportionally more affected by noise, which it is supposed to be uniformly distributed over the whole image [26]. The membership functions of interval type 2 system are shown in Fig. 4.

$$hMF = \frac{1}{25} * \begin{bmatrix} 1 & 1 & 1 & 1 & 1 \\ 1 & 1 & 1 & 1 & 1 \\ 1 & 1 & 1 & 1 & 1 \\ 1 & 1 & 1 & 1 & 1 \\ 1 & 1 & 1 & 1 & 1 \end{bmatrix} \qquad (26)$$

After applying the type 1 and type 2 edge detector with Sobel Method, the resulting image can be viewed in Fig. 5c, e.

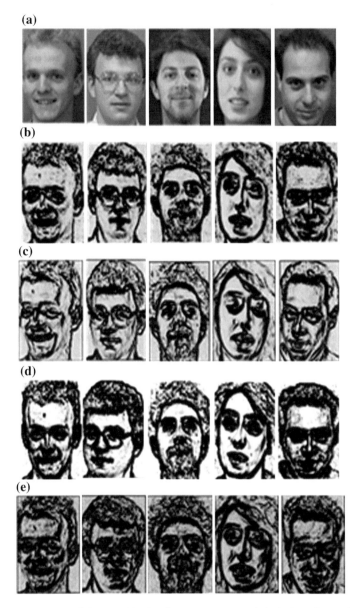

Fig. 5 **a** Original image, **b** image with edge detector type 1 morphological gradient, **c** image with edge detector type 1 Sobel, **d** image with interval type 2 morphological gradient, **e** image with edge detector interval type 2 Sobel

4 Modular Neural Networks

We trained a MNN of 3 modules with the data set of ORL. To each image, a methodology of edge detector was applied as described in Sect. 3 and then the image was divided into three horizontal sections, each of which was used as training data in each of the modules, as shown in Fig. 6.

The integration of the modules of the MNN was performed with the Choquet integral (14) and (15). In Table 1, we can appreciate the data distribution of the database.

4.1 Training Parameters

Training method: gradient descendent with momentum and adaptive learning rate back-propagation (Traingdx).
Each module of the MNN has two hidden layers [200 200].
Error goal: 0.00001.
Epochs: 500.

In Table 2, the distribution of the training data in the MNN is shown; 70 % of data are used for training, 15 % for validation, and the other 15 % for testing.

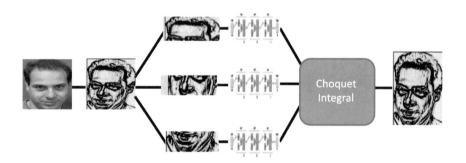

Fig. 6 Proposed architecture of the modular neural network

Table 1 Procedure performed in the experiment

Database	People quantity	Samples per people	Size of training set 80 %	Size of test set 20 %
ORL	40	10	320	80

Table 2 Distribution of the training data

Training 70 %	Validation 15 %	Test 15 %
224	48	48

Table 3 Procedure performed in the experiment

1. Define the database of images
2. Define the edge detector
3. Detect the edges of each image
4. Add the edges to the train set
5. Divide the images in three parts
6. Calculate the recognition rate using the k-fold cross-validation method
(a) Calculate the indices for k-folds
(b) Train the modular neural network $k − 1$ times for each training fold
(c) Simulate the modular neural network with the k test fold
7. Calculate the mean of rate for all the k-folds using Choquet integral with interval type 2 fuzzy measures

4.2 The Experiment Consists of a Modular Neural Network Recognition System and Choquet Integral for the Modules Fusion

The experiment consist on apply a preprocessing at the images through an edge detector on database of faces ORL to obtain a data set with which we train a modular neural network, this with the purpose of compare the recognition rate obtained using the k-fold cross validation method [27], see Table 3.

In the experiments, we performed 27 tests in each simulation of the trainings with each edge detector making variations in fuzzy densities and performing the calculation of the parameter λ_U **and** λ_L with the bisection method.

In Table 4, the parameters shown are used for the integration of information; the first column shows the tests number performed, the second shows the diffuse density associated with each information source—in this case each one of the 3 modules—and the third and fourth columns show the value of lambda for the upper λ_U and lower λ_L intervals calculated from the fuzzy densities.

5 Simulation Results

In Table 5, are shown an example of the results obtained for face recognition. For this case, the preprocessing of the image is done with the interval type-2 morphological gradient edge detector, and the aggregation method used for the MNN is the interval type-2 Choquet integral.

Table 4 Parameters of the fuzzy densities, λ_U and λ_L

Test	Fuzzy densities			λ_U	λ_L
1	0.1	0.1	0.1	9.76E−18	5.04E−17
2	0.1	0.1	0.5	1.40E−16	0.00E+00
3	0.1	0.1	0.9	1.60E−16	−1.00E+00
4	0.1	0.5	0.1	1.40E−16	0
5	0.1	0.5	0.5	1.236	−0.7307
6	0.1	0.5	0.9	−0.6262	−1.00E+00
7	0.1	0.9	0.1	−8.35E−18	−0.9998
8	0.1	0.9	0.5	−0.6262	−1
9	0.1	0.9	0.9	−9.38E−01	−1
10	0.5	0.1	0.1	1.40E-16	0
11	0.5	0.1	0.5	1.236	−0.7307
12	0.5	0.1	0.9	−0.6262	−1
13	0.5	0.5	0.1	1.236	−0.7307
14	0.5	0.5	0.5	−0.4428	−0.9043
15	0.5	0.5	0.9	−0.8715	−1
16	0.5	0.9	0.1	−0.6262	−1
17	0.5	0.9	0.5	−0.8715	−1
18	0.5	0.9	0.9	−0.9691	−1
19	0.9	0.1	0.1	−8.35E−18	−1.00E+00
20	0.9	0.1	0.5	−0.6262	−1
21	0.9	0.1	0.9	−0.6262	−1
22	0.9	0.5	0.1	−0.6262	−1
23	0.9	0.5	0.5	−8.72E−01	−1
24	0.9	0.5	0.9	−0.9691	−1
25	0.9	0.9	0.1	−0.9376	−1
26	0.9	0.9	0.5	−0.9691	−1
27	0.9	0.9	0.9	−0.9911	−1

In Table 6, the percentage of recognition of the Choquet integral with each interval edge detector is displayed, and a higher percentage of recognized with the usage of the type 2 edge detector with the gradient morphological was obtained with a 0.9549 %.

In Table 7, the percentage of recognition of the Choquet integral using interval fuzzy measures presents a small increase with respect to Choquet integral without interval fuzzy measures.

Table 5 Results obtained in the experiment

Test	Training data			Test data		
	Mean rate	Std rate	Max rate	Mean rate	Std rate	Max rate
1	1	0	1	0.865	0.006	0.875
2	1	0	1	0.863	0.015	0.888
3	1	0	1	0.858	0.014	0.875
4	1	0	1	0.868	0.014	0.888
5	1	0	1	0.863	0.023	0.9
6	1	0	1	0.86	0.016	0.888
7	1	0	1	0.868	0.014	0.888
8	1	0	1	0.868	0.021	0.9
9	1	0	1	0.865	0.014	0.888
10	1	0	1	0.86	0.011	0.875
11	1	0	1	0.855	0.014	0.875
12	1	0	1	0.853	0.011	0.863
13	1	0	1	0.865	0.011	0.875
14	1	0	1	0.858	0.019	0.888
15	1	0	1	0.858	0.011	0.875
16	1	0	1	0.87	0.014	0.888
17	1	0	1	0.87	0.014	0.888
18	1	0	1	0.865	0.011	0.875
19	1	0	1	0.86	0.011	0.875
20	1	0	1	0.863	0.009	0.875
21	1	0	1	0.86	0.014	0.875
22	1	0	1	0.86	0.006	0.863
23	1	0	1	0.86	0.011	0.875
24	1	0	1	0.86	0.014	0.875
25	1	0	1	0.865	0.014	0.888
26	1	0	1	0.868	0.014	0.888
27	1	0	1	0.865	0.011	0.875
	1	**0**	**1**	**0.863**	**0.013**	**0.881**

Table 6 Results with the Choquet integral using type 1 and type 2 edge detector

Method	mean_rate	std_rate	max_rate
T1-Sobel	0.93125	0.0385	0.925
T1-Morphological gradient	0.94	0.0677	0.975
T2-Sobel	0.9431	0.0193	0.9625
T2-Morphological gradient	0.9549	0.0482	0.9625

Table 7 Results with the Choquet integral using type 2 edge detector with interval fuzzy measures

Uncertainty footprint	mean_rate	std_rate	max_rate
FOU 0.1	0.955	0	0.925
FOU 0.2	0.957	0.04825	0.9625

6 Conclusions

The use of Choquet integral as an integration method of responses of a modular neural network applied to face recognition has yielded favorable results when performing the aggregation process of the preprocessed images with the detectors type 1 and type 2 of Sobel edges and morphological gradient; the use of the Sugeno measure by intervals allowed increase of the percentage of data recognition; however, it is still necessary to use a method that optimizes the value of the Sugeno measure assigned to each source of information, and also how to calculate the interval because these were designated arbitrarily. Future work could be considering the optimization of the proposed method, as in [28–30].

7 Future Works

Although good results were obtained by applying the Choquet integral as an aggregation operator of the MNN, more testing is needed on other benchmark databases to verify results obtained also to find another way to generate an interval of uncertainty among the data, fuzzy measures, value of lambda, and fuzzy densities.

Acknowledgment We are grateful to the MyDCI program of the Division of Graduate Studies and Research, UABC, Tijuana Institute of Technology and for the financial support provided by our sponsor CONACYT contract grant number: 189350.

References

1. Zhou, L-G., Chen, H-Y., Merigó, J.M., Anna, M.: Uncertain generalized aggregation operators. Expert Syst. Appl. **39**, 1105–1117 (2012)
2. McCulloh, W.S., Pitts, W.: A logical calculus of the ideas immanent in nervous activity. Bull. Math. Biophys. **5**, 115–133 (1943)
3. Alves, V.M.O., Cavalcanti, G.D.C.: A Nonexclusive Task Decomposition Method for Modular Neural Networks. IEEE, New York (2010) (978-1-4244-8126-2/10)
4. Bo, Y-C., Qiao, J-F., Yang, G.: A Modular Neural Networks Ensembling Method Based on Fuzzy Decision-Making. IEEE, New York (2011) (978-1-4244-8039-5/11)
5. Vazirani, H., Kala, R., Shukla, A., Tiwari, R.: Diagnosis of Breast Cancer by Modular Neural Network. IEEE, New York (2010) (978-1-4244-5540-9/10)
6. Turchenko, I., Kochan, V., Sachenko, A.: Recognition of Multi-sensor Output Signal Using Modular Neural Networks Approach. TCSET, Lviv-Slavsko, Ukraine (2006)

7. Liu, Y., Yao, X.: Evolving Modular Neural Networks Which Generalise Well. IEEE, New York (1997) (0-7803-3949-5/97)
8. Hidalgo, D.: Fuzzy inference systems type 1 and type 2 as integration methods in neural networks for multimodal biometrics and me-optimization by means of genetic algorithms, Master Thesis, Tijuana Institute of Technology (2008)
9. Sánchez D., Melin P.: Modular neural network with fuzzy integration and its optimization using genetic algorithms for human recognition based on iris, ear and voice biometrics. Soft Comput. Recognit. Based Biom. 85–102 (2010)
10. Sánchez, D., Melin, P., Castillo, O., Valdez, F.: Modular neural networks optimization with hierarchical genetic algorithms with fuzzy response integration for pattern recognition. MICAI, pp. 247–258 (2012)
11. Melin, P., Gonzalez, C., Bravo, D., Gonzalez, F., Martínez, G.: Modular neural networks and fuzzy sugeno integral for pattern recognition: the case of human face and fingerprint. In: Hybrid Intelligent Systems: Design and Analysis. Springer, Heidelberg, Germany (2007)
12. Melin, P., Mendoza, O., Castillo O.: Face recognition with an improved interval type-2 fuzzy logic sugeno integral and modular neural networks. IEEE Trans. Syst. Man Cybernet. Part A Syst. Hum. **41**(5) (2011)
13. Meena, Y., Arya, K.V., Kala, R.: Classification using redundant mapping in modular neural networks. In: Second World Congress on Nature and Biologically Inspired Computing, Dec 15–17, in Kitakyushu, Fukuoka, Japan (2010)
14. Wang, P., Xua, L., Zhou, S-M., Fan, Z., Li, Y., Feng, S.: A novel Bayesian learning method for information aggregation in modular neural networks. Expert Syst. Appl. **37**, 1071–1074 (2010)
15. Kwak, K.-C., Pedrycz, W.: Face recognition: a study in information fusion using fuzzy integral. Pattern Recogn. Lett. **26**, 719–733 (2005)
16. Timonin, M.: Robust optimization of the Choquet integral. Fuzzy Sets Syst. **213**, 27–46 (2013)
17. Yanga, W., Chena, Z.: New aggregation operators based on the Choquet integral and 2-tuple linguistic information. Expert Syst. Appl. **39**(3), 2662–2668 (2012)
18. Sugeno, M.: Theory of fuzzy integrals and its applications. Thesis Doctoral, Tokyo Institute of Technology, Tokyo, Japan (1974)
19. Murofushi, T., Sugeno, M.: Fuzzy Measures and Fuzzy Integrals. Department of Computational Intelligence and Systems Science, Tokyo Institute of Technology, Yokohama, Japan (2000)
20. Song, J., Li, J.: Lebesgue theorems in non-additive measure theory. Fuzzy Sets Syst. **149**(3), 543–548 (2005)
21. Wang, Z., Klir, G.: Generalized Measure Theory. Springer, New York (2009)
22. Torra, V., Narukawa, Y.: The interpretation of fuzzy integrals and their application to fuzzy systems. Int. J. Approx. Reason. **41**, 43–58 (2006)
23. Mendoza, O., Melin, P.: Extension of the Sugeno Integral with Interval Type-2 Fuzzy Logic. Fuzzy Information Processing Society, NAFIPS (2008)
24. Database ORL Face. Cambridge University Computer Laboratory. http://www.cl.cam.ac.uk/research/dtg/attarchive/facedatabase.html (2012)
25. Mendoza, O., Melin, P., Castillo, O., Castro, J.: Comparison of fuzzy edge detectors based on the image recognition rate as performance index calculated with neural networks. Soft Comput. Recognit. Biom. Stud. Comput. Intell. **312**, 389–399 (2010)
26. Mendoza, O., Melin, P., Licea, G.: A hybrid approach for image recognition combining type-2 fuzzy logic, modular neural networks and the Sugeno integral. Inf. Sci. Int. J. **179**(13), 2078–2101 (2009)
27. Mendoza, O., Melin, P.: Quantitative evaluation of fuzzy edge detectors applied to neural networks or image recognition. Adv. Res. Dev. Digit. Syst. 324–335 (2011)
28. Sánchez D., Melin P.: Multi-objective hierarchical genetic algorithm for modular granular neural network optimization. Soft Comput. Appl. Optim. Control Recognit. 157–185 (2013)

29. Sánchez, D., Melin, P., Castillo, O., Valdez, F.: Modular granular neural networks optimization with multi-objective hierarchical genetic algorithm for human recognition based on iris biometric. IEEE Congr. Evolut. Compu. 772–778 (2013)
30. Sánchez, D., Melin, P.: Optimization of modular granular neural networks using hierarchical genetic algorithms for human recognition using the ear biometric measure. Eng. Appl. AI **27**, 41–56 (2014)

Part III
Decision-Making

Fuzzy Logic Ideas Can Help in Explaining Kahneman and Tversky's Empirical Decision Weights

Joe Lorkowski and Vladik Kreinovich

Abstract Analyzing how people actually make decisions, the Nobelist Daniel Kahneman and his co-author Amos Tversky found out that instead of maximizing the expected gain, people maximize a weighted gain, with weights determined by the corresponding probabilities. The corresponding empirical weights can be explained qualitatively, but quantitatively, these weights remain largely unexplained. In this paper, we show that with a surprisingly high accuracy, these weights can be explained by fuzzy logic ideas.

1 Empirical Decision Weights: Formulation of the Problem

Decisions are important. One of the main objectives of science and engineering is to help people make decisions. For example, we try to predict weather, so that people will be able to dress properly (and take an umbrella if needed), so that if a hurricane is coming, people can evacuate. We analyze quantum effects in semi-conductors so that engineers can design better computer chips. We analyze diseases so that medical doctors can help select the best treatment, etc.

In complex situations, people need help in making their decisions. In simple situations, an average person can easily make a decision. For example, if the weather forecast predicts rain, one should take an umbrella with him/her, otherwise one should not.

In more complex situations, however, even when we know all the possible consequences of each action, it is not easy to make a decision. For example, in medicine, many treatments come with side effects: A surgery can sometimes result

J. Lorkowski (✉) · V. Kreinovich
Department of Computer Science, University of Texas at El Paso,
El Paso, TX 79968, USA
e-mail: lorkowski@computer.org

V. Kreinovich
e-mail: vladik@utep.edu

© Springer International Publishing Switzerland 2016   89
L.A. Zadeh et al. (eds.), *Recent Developments and New Direction
in Soft-Computing Foundations and Applications*, Studies in Fuzziness
and Soft Computing 342, DOI 10.1007/978-3-319-32229-2_7

in a patient's death; immune system suppression can result in an infectious disease, etc. In such situations, it is not easy to compare different actions, and even skilled experts would appreciate computer-based help.

To help people make decisions, we need to analyze how people make decisions. One of the difficulties in designing computer-based systems which would help people make decisions is that to make such systems successful, we need to know what exactly people want when they make decisions. Often, people cannot explain in precise terms why exactly they have selected this or that alternative.

In such situations, we need to analyze how people actually make decisions, and then try to come up with formal descriptions which fit the observed behavior.

Experiments start with decision making under full information. To analyze how people make decisions, researchers start with the simplest situations, in which we have the full information about the following:

- we know all possible outcomes $o_1, \ldots, o_n$ of all possible actions;
- we know the exact value u_i (e.g., monetary) of each outcome o_i; and
- for each action a and to each outcome i, we know the probability $p_i(a)$ of this outcome.

Seemingly reasonable behavior. The outcome of each action a is not deterministic. For the same action, we may get different outcomes u_i with different probabilities $p_i(a)$. However, usually similar situations are repeated again and again.

If we repeat a similar situation several times, then the average expected gain of selecting an action a becomes close to the mathematical expectation of the gain, i.e., to the value

$$u(a) \stackrel{\text{def}}{=} \sum_{i=1}^{n} p_i(a) \cdot u_i.$$

Thus, we expect that a decision maker selects the action a for which this expected value $u(a)$ is the largest. In the first crude approximation, this is how people actually make decisions. But if we want a more precise description of human behavior, we—somewhat surprisingly—have to modify this formula.

How people actually make decisions is somewhat different. In their famous experiments, the Nobelist Daniel Kahneman and his co-author Amos Tversky found out that a much more accurate description of human decision making can be obtained if we assume that, instead of maximizing the expected gain, people maximize a weighted gain, with weights determined by the corresponding probabilities; see, e.g., [1] and references therein.

In other words, people select the action a for which the weighted gain

$$w(a) \stackrel{\text{def}}{=} \sum_{i} w_i(a) \cdot u_i$$

attains the largest possible value, where $w_i(a) = f(p_i(a))$ for an appropriate function $f(x)$.

This empirical transformation $f(x)$ from probabilities to weights takes the following form:

Probability	0	1	2	5	10	20	50	80	90	95	98	99	100
Weight	0	5.5	8.1	13.2	18.6	26.1	42.1	60.1	71.2	79.3	87.1	91.2	100

How can we explain this empirical transformation? There are qualitative explanations for this phenomenon, but not a quantitative one.

In this paper, we propose a quantitative explanation based on the fuzzy logic ideas.

2 Fuzzy-Motivated Idea: Considering "Distinguishable" Probabilities

Main idea. The main idea behind our explanation is based on the fact that when people make decisions, they do not estimate probabilities as numbers from the interval [0, 1] and do not process them. If a person is asked about the probability of a certain event, in many cases, the answer will not come as an exact number, it will most probably come as an imprecise ("fuzzy") word, like "low," "high," and "medium"; see, e.g., [2, 4, 6].

In other words, instead of using all *infinitely* many possible real numbers from the interval [0, 1], we only use *finitely* many possible values—i.e., in effect, we estimate the probability on a finite scale. The reason for this discretization is that if the two probability values are close to each other, intuitively, we do not feel the difference. For example, there is a clear difference between 10 % chances of rain or 50 % chances of rain, but we do not think that anyone can feel the difference between 50 % and 51 % chances. So, the discrete scale is formed by probabilities which are distinguishable from each other. Let us show how this idea can be formalized.

Comment. In this formalization, we will follow ideas first outlined in [3].

How to formalize when probabilities are distinguishable. Probability of an event is estimated, from observations, as the frequency with which this event occurs. For example, if out of 100 days of observation, rain occurred in 40 of these days, then we estimate the probability of rain as 40 %. In general, if out of n observations, the event was observed in m of them, we estimate the probability as the ratio

$$\frac{m}{n}.$$

This ratio is, in general, different from the actual (unknown) probability. For example, if we take a fair coin, for which the probability of head is exactly 50 %, and flip it 100 times, we may get 50 heads, but we may also get 47 heads, 52 heads, etc.

It is known (see, e.g., [5]) that the expected value of the frequency is equal to p and that the standard deviation of this frequency is equal to

$$\sigma = \sqrt{\frac{p \cdot (1 - p)}{n}}.$$

It is also known that, due to the central limit theorem, for large n, the distribution of frequency is very close to the normal distribution (with the corresponding mean p and standard deviation σ).

For normal distribution, we know that with a high certainty, all the values are located within 2–3 standard deviations from the mean, i.e., in our case, within the interval

$$(p - k_0 \cdot \sigma, p + k_0 \cdot \sigma),$$

where $k_0 = 2$ or $k_0 = 3$: For example, for $k_0 = 3$, this is true with confidence 99.9 %. We can thus say that the two values of probability p and p' are (definitely) distinguishable if the corresponding intervals of possible values of frequency do not intersect, and thus, we can distinguish between these two probabilities just by observing the corresponding frequencies.

In precise terms, the probabilities $p < p'$ are distinguishable if

$$(p - k_0 \cdot \sigma, p + k_0 \cdot \sigma) \cap (p' - k_0 \cdot \sigma', p + k_0 \cdot \sigma') = \emptyset,$$

where

$$\sigma' \stackrel{\text{def}}{=} \sqrt{\frac{p' \cdot (1 - p')}{n}},$$

i.e., if $p' - k_0 \cdot \sigma' \geq p + k_0 \cdot \sigma$. The smaller p', the smaller the difference $p' - k_0 \cdot \sigma'$. Thus, for a given probability p, the next distinguishable value p' is the one for which

$$p' - k_0 \cdot \sigma' = p + k_0 \cdot \sigma.$$

When n is large, these values p and p' are close to each other; therefore, $\sigma' \approx \sigma$. Substituting an approximate value σ instead of σ' into the above equality, we conclude that

$$p' \approx p + 2k_0 \cdot \sigma = p + 2k_0 \cdot \sqrt{\frac{p \cdot (1 - p)}{n}}.$$

If the value p corresponds to the ith level out of m, i.e., in fuzzy terms to the truth value,

$$\frac{i}{m}$$

then the next value p' corresponds to the $(i + 1)$-st level, i.e., to the truth value

$$\frac{i+1}{m}.$$

Let $g(p)$ denote the fuzzy truth value corresponding to the probability p. Then,

$$g(p) = \frac{i}{m} \text{ and } g(p') \frac{i+1}{m}.$$

Since the values p and p' are close, the difference $p' - p$ is small, and therefore, we can expand the expression $g(p') = g(p + (p' - p))$ in Taylor series and keep only linear terms in this expansion:,

$$g(p') \approx g(p) + (p' - p) \cdot g'(p)$$
$$g'(p) = \frac{dg}{dp}$$

where $g'(p)$ denotes the derivative of the function $g(p)$. Thus,

$$g(p') - g(p) = \frac{1}{m} = (p' - p) \cdot g'(p).$$

Substituting the known expression for $p' - p$ into this formula, we conclude that

$$\frac{1}{m} = 2k_0 \cdot \sqrt{\frac{p \cdot (1 - p)}{n}} \cdot g'(p).$$

This can be rewritten as

$$g'(p) \cdot \sqrt{p \cdot (1 - p)} = \text{const}$$

for some constant, and thus,

$$g'(p) = \text{const} \cdot \frac{1}{\sqrt{p \cdot (1 - p)}}.$$

Integrating this expression and taking into account that $p = 0$ corresponds to the lowest (0th) level—i.e., that $g(0) = 0$—we conclude that

$$g(p) = \text{const} \cdot \int_0^p \frac{dq}{\sqrt{q \cdot (1-q)}}.$$

This integral can be easily computed if we introduce a new variable t for which $q = \sin^2(t)$. In this case,

$$dq = 2 \cdot \sin(t) \cdot \cos(t) \cdot dt,$$
$$1 - p = 1 - \sin^2(t) = \cos^2(t) \text{ and therefore,}$$
$$\sqrt{p \cdot (1-p)} = \sqrt{\sin^2(t) \cdot \cos^2(t)} = \sin(t) \cdot \cos(t).$$

The lower bound $q = 0$ corresponds to $t = 0$ and the upper bound $q = p$ corresponds to the value t_0 for which $\sin^2(t_0) = p$, i.e., $\sin(t_0) = \sqrt{p}$ and $t_0 = \arcsin(\sqrt{p})$. Therefore,

$$g(p) = \text{const} \cdot \int_0^p \frac{dq}{\sqrt{q \cdot (1-q)}}$$
$$= \text{const} \cdot \int_0^{t_0} \frac{2 \cdot \sin(t) \cdot \cos(t) \cdot dt}{\sin(t) \cdot \cos(t)} \int_0^{t_0} 2 \cdot dt$$
$$= 2 \cdot \text{const} \cdot t_0.$$

We know how t_0 depends on p, so we get

$$g(p) = 2 \cdot \text{const} \cdot \arcsin(\sqrt{p}).$$

We can determine the constant from the condition that the largest possible probability value $p = 1$ should correspond to the right-most point $g(p) = 1$. From the condition that $g(1) = 1$, taking into account that

$$\arcsin(\sqrt{1}) = \arcsin(1) = \frac{\pi}{2},$$

we conclude that

$$g(p) = \frac{2}{\pi} \cdot \arcsin(\sqrt{p}). \tag{1}$$

Description of the resulting discretization. For a scale from 0 to some number m, the value $g(m)$ is equal to the ratio

$$\frac{i}{m}.$$

So, $i = m \cdot g(p)$.

Thus, the desired discretization means that to each probability p, we assign the value $i \approx m \cdot g(p)$ on the scale from 0 to m, where $g(p)$ is described by the above formula.

3 Distinguishable Probabilities Can Explain Empirical Decision Weights

How do we select weights? If we need to select finitely many weights from the interval $[0, 1]$, then it is natural to select weights which are equally distributed on this interval, i.e., weights

$$0, \frac{1}{m}, \frac{2}{m}, \ldots, \frac{m-1}{m}, 1. \tag{2}$$

How to assign weights to probabilities: idea. We have m, a finite list of distinguishable probabilities $0 = p_0 < p_1 < \ldots < p_m = 1$. These probabilities correspond to degree

$$g(p_i) = \frac{i}{m}, \tag{3}$$

where $g(p)$ is determined by the formula (1). We need to assign, to each of these probabilities, an appropriate weight from the above list (2).

The larger the probability, the more weight we should assign to the corresponding outcome. Thus, we arrive at the following assignment of weights to probabilities:

- to the value $p_0 = 0$, we assign the smallest possible weight $w_0 = 0$;
- to the next value p_1, we assign the next weight

$$w_1 = \frac{1}{m};$$

- to the next value p_2, we assign the next weight

$$w_2 = \frac{1}{m};$$

- …
- to the value $p_m - 1$, we assign the weight

$$w_{m-1} = \frac{m-1}{m};$$

- finally, to the value $p_m = 1$, we assign the weight

$$w_m = 1.$$

In general, to the value p_i, we assign the weight

$$w_i = \frac{i}{m}.$$

By comparing this assignment with the formula (3), we conclude that to each value p_i, we assign the value $w_i = g(p_i)$.

How to assign weights to probabilities: result. Our arguments show that to each probability $p \in [0, 1]$, we assign the weight $g(p)$, where the function $g(p)$ is determined by the formula (1).

Comparing our weights with empirical weights: first try. Let us compare the probabilities p_i, Kahneman's empirical weights $\tilde{w}_i$, and the weight $w_i = g(p_i)$ computed by using the formula (1):

p_i	0	1	2	5	10	20	50	80	90	95	98	99	100
$\tilde{w}_i$	0	5.5	8.1	13.2	18.6	26.1	42.1	60.1	71.2	79.3	87.1	91.2	100
$w_i = g(p_i)$	0	6.4	9.0	14.4	20.5	29.5	50.0	70.5	79.5	85.6	91.0	93.6	100

The estimates $w_i = g(p_i)$ are closer to the observed weights $\tilde{w}_i$ than the original probabilities, but the relation does not seem very impressive.

We will show that the fit is much better than it seems at first glance. At first glance, the above direct comparison between the observed weights $\tilde{w}_i$ and the estimated weights $w_i = g(p_i)$ seems to make perfect sense. However, let us look deeper.

The weights come from the fact that users maximize the weighted gain $w(a) = \sum w_i(a) \cdot u_i$. It is easy to observe that if we multiply all the weights by the same positive constant $\lambda > 0$, i.e., consider the weights $w_i'(a) = \lambda \cdot w_i(a)$, then for each action, the resulting value of the weighted gain will also increase by the same factor:

$$w_i' = \lambda \cdot w_i$$

The relation between the weighted gains of two actions a and a' does not change if we simply multiply both gains by a positive constant:

- if $w_i(a) < w_i(a')$, then, multiplying both sides of this inequality by λ, we get $w_i'(a) < w_i'(a')$;
- if $w_i(a) = w_i(a')$, then, multiplying both sides of this equality by λ, we get $w_i'(a) = w_i'(a')$;
- if $w_i(a) > w_i(a')$, then, multiplying both sides of this inequality by λ, we get $w_i'(a) > w_i'(a')$.

All we observe is which of the two actions a person selects. Since multiplying all the weights by a constant does not change the selection, this means that based on the selection, we cannot uniquely determine the weights: An empirical selection which is consistent with the weights w_i is equally consistent with the weights $w_i' = \lambda \cdot w_i$.

This fact can be used to *normalize* the empirical weights, i.e., to multiply them by a constant so as to satisfy some additional condition.

In [1], to normalize the weights, the authors use the requirement that the weight corresponding to probability 1 should be equal to 1. Since for $p = 1$, the estimated weight $g(1)$ is also equal to 1, we get a perfect match for $p = 1$—but a rather lousy march for probabilities intermediate between 0 and 1.

Instead of this normalization, we can select λ so as to get the best match "on average".

How to improve the fit: details. A natural idea is to select λ from the least squares method, i.e., select λ for which the relative mean squares difference

$$\sum_i \left(\frac{\lambda \cdot w_i - \tilde{w}_i}{w_i} \right)^2$$

is the least possible. Differentiating this expression with respect to λ and equating the derivative to 0, we conclude that

$$\sum_i \left(\lambda - \frac{\tilde{w}_i}{w_i} \right) = 0,$$

i.e., that

$$\lambda = \frac{1}{m} \cdot \sum_i \frac{\tilde{w}_i}{w_i}.$$

Resulting match. For the above values, this formula leads to $\lambda = 0.910$.

p_i	0	1	2	5	10	20	50	80	90	95	98	99	100
$\tilde{w}_i$	0	5.5	8.1	13.2	18.6	26.1	42.1	60.1	71.2	79.3	87.1	91.2	100
$\tilde{w}'_i = \lambda \cdot g(p_i)$	0	5.8	8.2	13.1	18.7	26.8	45.5	64.2	72.3	77.9	82.8	87.4	91.0

The resulting values are indeed much closer to the empirical weights.

For most probabilities p_i, the difference between the fuzzy-motivated weights w'_i and the empirical weights w_i is so small that it is below the accuracy with which the empirical weights can be obtained from the experiment.

Conclusion. Fuzzy-motivated ideas indeed explain Kahneman and Tversky's empirical decision weights.

Acknowledgment This work was supported in part by the National Science Foundation grants HRD-0734825 and HRD-1242122 (CyberShARE Center of Excellence) and DUE-0926721.

References

1. Kahneman, D.: Thinking, Fast and Slow. Farrar, Straus, and Giroux, New York (2011)
2. Klir, G., Yuan, B.: Fuzzy Sets and Fuzzy Logic. Prentice Hall, Upper Saddle River, New Jersey (1995)
3. Nguyen, H.T., Kreinovich, V., Lea, B.: How to combine probabilistic and fuzzy uncertainties in fuzzy control. In: Proceedings of the Second International Workshop on Industrial Applications of Fuzzy Control and Intelligent Systems, College Station, pp. 117–121, 2–4 Dec 1992
4. Nguyen, H.T., Walker, E.A.: A First Course in Fuzzy Logic. Chapman & Hall/CRC, Boca Raton, Florida (2006)
5. Sheskin, D.J.: Handbook of Parametric and Nonparametric Statistical Procedures. Chapman & Hall/CRC, Boca Raton, Florida (2011)
6. Zadeh, L.A.: Fuzzy sets. Inf. Control **8**, 338–353 (1965)

A Fuzzy Multiagent Approach
for Integrated Product Life Cycle
Environment

V.V. Taratukhin, Y.V. Yadgarova and E.Y. Skachko

Abstract One of the main questions in product life cycle management is how to create the comprehensive framework for autonomous, intelligent decision-making which integrates business and scheduling data. The key problem is to simulate human-like decision-making process to provide agile manufacturing process. Multiagent technologies play a key role in this problem and form an integration platform between human and manufacturing. Also developing distributed control system based on multiagent technologies encounters difficulties, due to ambiguous, vague, or missing information. In the area of intelligent manufacturing systems, there are a number of fuzzy scheduling models presented (Subramaniam et al. in Int J Adv Manuf Technol 16(10): 759–764, [1]; Srinoi et al. in Int J Prod Res 44 (11): 1–21, [2]). These frameworks only deal with manufacturing processes. The research presents multiagent framework that integrates design, manufacturing, and control process in fuzzy area using resource-based approach to agent's interaction. This model is applicable to work with various manufacturing agents in conjunction with different design agents and control systems.

V.V. Taratukhin (✉)
ERCIS Competence Center, University of Münster, Leonardo-Campus 3,
48149 Münster, Germany
e-mail: victor.taratoukhine@ercis.uni-muenster.de

Y.V. Yadgarova · E.Y. Skachko
Department of Computer Systems and Complex Automation, Bauman Moscow
State Technical University, 2-ya Baumanskaya, 5, Moscow, Russia
e-mail: y.v.yadgarova@gmail.com

E.Y. Skachko
e-mail: ekaterina_skachko@inbox.ru

L.A. Zadeh et al. (eds.), *Recent Developments and New Direction
in Soft-Computing Foundations and Applications*, Studies in Fuzziness
and Soft Computing 342, DOI 10.1007/978-3-319-32229-2_8

1   Introduction

On the one hand, developing a distributed control system that can operate with ambiguous and vague information is still an open problem, due to it's complexity. But with rapid progress of computer capacity, new earlier unavailable technologies and solutions are taken into account.

On the other hand, the high-value products and infrastructure require engineering services, such as maintenance support throughout the life cycle. The future product life cycle management system should provide loosely coupled integration with design process (productive maintenance support on the stage of design, control of the design mistakes, and feedback) [3].

Smart industry, a key trend in manufacturing, is the ability of machines and devices to be self-organized, to communicate independently with each other, and to provide agile and adaptive design and manufacturing environment. One of the main ideas proposed in this paper is the new approach of communication and integration of the engineering design, manufacturing and planning systems' fuzzy self-oriented entities.

This paper proposes a new way of organizing manufacturing/control and design scene by including fuzzy logic approach in design and control subsystems. Unlike the traditional fuzzy multiagent scheduling systems, such system includes integration with design process agents to solve design process problems and fuzzy resource–goal approach to interact with agents.

2   Agent-Based Reconfigurable Model

The design principles of large-scale systems composed of heterogeneous working in real-time components require unified theory of real-time operations that includes existing results and novel solutions. Also, the cross-layer design is reached, specifying that each device should be designed based on hardware, operating system, middleware, sensing, actuation, as well as communication as a whole. One of the most appropriate technologies for creation such system is multiagent systems. The main benefits of multiagent systems are their decentralization and simplicity of development of agents.

In this paper, general three-layer architecture of multiagent system with design (D-), manufacturing (M-), and control (C-) agents including fuzzy logic approach of M-agent is presented. Resource–goal network approach is used for interaction of agents on M-layer, controlling and distributing resources among the agents.

Concept framework can be presented formally as follows:

$$CF = MP_1, \ldots, MP_t, \ldots, MP_n,$$

MP$_t$ is the t-manufacturing part, $t = 1, 2, \ldots, n$.

$$AP = DA_1, \ldots, DA_i, \ldots, DA_m, \ldots, MA_1, \ldots, MA_f, \ldots, MA_k, \ldots, CA_1, \ldots, CA_j, \ldots, CA_n.$$

2.1 Agent Types

For defining this framework, three types of agents were defined:
DA$_i$ is the ith Design Agent (D-agent), $i = 1, 2, \ldots, m$.
MA$_f$ is the fth Manufacturing Agent (M-agent), $f = 1, 2, \ldots, k$.
CA$_j$ is the jth Control Agent (C-agent), $j = 1, 2, \ldots, n$.

Each DA$_i$ consists of five elements: FB—facts base, which includes information about geometric characteristics of the part and material type; KB—knowledge base; K—corrector block, which adapts knowledge base, as a result of communications with any other agents; I—inference engine; and LI—local interface mechanism.

Manufacturing agent MA$_f$ represents execution unit of the manufacturing system and consists of actuator and fuzzy multiagent control system. Fuzzy inference engine in each MA$_f$ is responsible for routing and scheduling mechanism. Communication with each other implements with resource–goal fuzzy network approach.

Each CA$_j$ consists of the following: MB—metaknowledge base; knowledge base of control agent; inference engine; corrector block; and local and global interface mechanism (GI).

On this layer, agents are using FIPA-compliant communication languages (ACL); global and local interface translates messages from external ACL to internal description and from internal description to external representation. K—corrector is realized for internal adaptation of knowledge and fact bases that will be described later in this report.

Each M-agent operates as an independent entity and processes associated manufacturing parts.

2.2 Resource–Goal Network Approach

Resource–goal network [4] as formalism for the MAS specification integrated to manufacturing environment is weighted directed graph:

$$\text{NET}_{RG} = (A, C, K, \text{RES}, G, W, t)$$

where set of vertex is a set of agents A, set of arcs C consist of goal links C_G and resource links C_{RES}, $C = C_G \cup C_{RES}$, $C_G \cap C_{RES} = \emptyset$, K—types of agents A, RES—set of resources in MAS, G—set of goals, W—set of conduction for resources and goals, and t—set of discrete time moments.

Every agent as a vertex of resource–goal network $a_i \epsilon A$ characterized by type of agent $k_j \epsilon K$ and amount of resources $RES(a_i) \epsilon RES$. Every vertex $c_{ij} \epsilon C$ is characterized by conductivity (capacity) $w_{ij} \epsilon W$. Conductivity of agent defines the maximum value of resource that one agent can transfer to another during one time interval. For resource–goal network is defined weight matrix along arcs $W_c = \left[w_c \left(a_i a_j \right) \right]$ and weight matrix along resources $W_{res} = \left[w_{res} \left(a_i a_j \right) \right]$.

Decision-making of the agent about the use of resources or relocating it in this network is based on negotiations between agent and network characteristics at the moment.

3 Design Agent: Inference

To include fuzzy logic approach in MA-set fuzzy resource–goal routing among the manufacturing agents (MA) within order processing restrictions is implemented. Reconfigurable model works with several similar machines and tools that control the agent are able to vary in order to achieve the optimal route.

On the other hand, another restriction of the model is mismatching that can occur within manufacturing process. Design is multidisciplinary process and involves several stages such as conceptual design, basic structural design, detail design, production design, manufacturing processes analysis, and documentation. Design agent (DA) must detect and resolve two main types of mismatches:

- mismatches of integration
- concurrency mismatches [5–11].

Mismatches of integration are assembly mismatches. These types of mismatches detect geometric, material characteristics of the part for checking assembly possibility.

According to [12], "Concurrent engineering is getting the right people together at the right time to identify and resolve design problems. Concurrent engineering is designing for assembly, availability, cost, customer satisfaction, maintainability, manageability, manufacturability, operability, performance, quality, risk, safety, schedule, social acceptability, and all other attributes of the product."

Concurrency mismatches are controlled by the design agents. Such type of agent stores knowledge base, facts base, and inference engine. There is information about sequences part's structure parameters. Details of D-agents are presented in [13].

4 Manufacturing Agent: Fuzzy Resource–Goal Network Approach

Configuration of the manufacturing system consists of warehouse (W), robocara (R), two milling machines (MM_1, MM_2), two turning lathe (TL_1, TL_2), two industrial robots (M_1, M_2) and control system (C). Parts require a different set of operations $\{op_1, op_2, op_3, ...\}$ and can be routed for several different ways according to processing route and weight coefficient of link between agents.

Resource–goal network as a model of agent's interaction is based on matrix of goals and matrix of resources. To interact with other agents and route parts in different stages of manufacturing process, agents must exchange information about their resources, goals, and state of the environment. However, the problems occur when information about agent's resources or state of the environment is incomplete, ambiguous, and vague. One way of solving this problem is using fuzzy logic approach to present resource's state of agent. Unlike the classical logic propositions in fuzzy logic are not limited to one of the two values True, $False_1$ (or 0, 1) [14].

Every two agents a_i and a_j in resource–goal network has **Goal conductivity** $w_c(a_ia_j)$ and **Resource conductivity** $w_{res}(a_ia_j)$.

To include fuzzy logic approach in resource–goal formalism is defined fuzzy variables [15]:

Fuzzy resource conductivity indicates capacity of the vertex in the goal–resource network. Physical sense of this variable is the maximum value of resource that one agent can transfer to another during one time interval.

$$FRC_{ij} = \text{minimum, low, middle, large, max}$$

If FRC_{ij} of two agents is known, it is possible to define **Weight matrix** along resources in manufacturing system.

For example, fuzzy weight matrix between three actuators can be defined as:

$$\begin{pmatrix} \text{low} & \text{middle} & \text{low} \\ \text{middle} & \text{high} & \text{high} \\ \text{low} & \text{middle} & \text{low} \end{pmatrix}$$

Fuzzy resource amount

$$FRC = \text{low, middle, large}$$

Fuzzy workload

$$FW = \text{low, middle, large}$$

Fuzzy workload of the MA_f indicates quantity of parts requested for this agent. Quantitative representation of this value is queue size of the MA_f.

Generally, fuzzy sets depend on the type of the manufacturing: Resource conductivity of industrial robot and milling tool is different. Also workload level of each actuator is different. In this work, it is considered that parts group by similar resource requirements.

As the fuzzy output variable is defined **Fuzzy Requirement** variable represented by sets:

$$FR_{ij} = low, average, high, critical$$

Fuzzy Requirement variable shows demand of resource that MA_i should transfer to MA_j.

For every pair of manufacturing agents MA_f and MA_j, conditional fuzzy propositions were presented in this form :

$$\text{If } FRC_{ij} \text{ is } * \text{ and } FW_i \text{ is } * \text{ and } FRA_j \text{ is } * \text{ then } FR_{ij} \text{ is } *$$

The set of fuzzy rules consists of 27 (3 * 3 * 3) rules. This set is constructed for every FR_{ij} variable. The simple subset for two manufacturing agents is presented on Fig. 1 (Tables 1 and 2).

Fig. 1 Fuzzy sets for FW_i variable

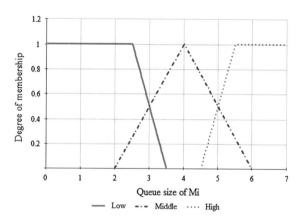

Table 1 Fuzzy variables

Linguistic variable	Fuzzy set
FRC_{ij}	Minimum, low, middle, large, maximum
FW_i	Low, middle, large
FRA_i	Low, middle, large
FR_{ij}	Low, average, high, critical

Table 2 Fuzzy rules for FR_{ij} variable

FR_{ij}	FW_i	FRA_j	FRC_{ij}
Average	Low	High	Middle
Average	Low	Middle	Large
Critical	High	High	Minimum
High	High	Middle	Middle
Low	High	Low	Middle

Resource demands information required to solve problems:

- Routing parts to actuators with satisfied resource demands and high value of workload;
- Normalize amount of resources for MA_f.

Final routing between actuators provided by CA_i combined fuzzy inference engine and depends on manufacturing process, collisions during processing, and resource demands of the MA_f.

5   Control Agent: Inference

Control agent stores metaknowledge values about technological environment such as map of the production system, distance between machines, stages of the design process, and overall design time. Also control agent stores actual knowledge base (amount, state and location of the parts, and resource demands of MA_f) and overall state of environment.

One of the main functions of the control agent is final routing parts through manufacturing line. Based on manufacturing agent's workload, resource demands, state and design agent's information, a combined inference engine of this agent can be defined (Fig. 2).

Knowledge base in control agent is responsible for order processing and routing parts between manufacturing agents. Generally, order processing in this case is strictly defined by technologist, but in case of several same machine tools, control agent must route part on the least loaded machine tool with low demand of resources. So, order processing is the restriction of our model.

According to mismatches restrictions in design agent, intermediate route has been produced. Then on the final step, selected routings are the input to the control agent that produces string route of the part based on its metaknowledge base and state of the part.

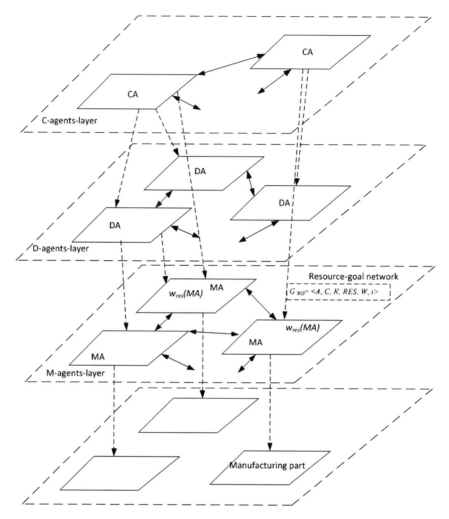

Fig. 2 Multiagent framework architecture

6 Prototype Implementation

In this research, a prototype of the control system which is capable of integrating with hardware devices is proposed. Such subsystem provides flexible configuration of each device. Rules of communication, ontology, knowledge base, and facts base are configured separately by an engineer of corresponding domain.

The initial multiagent system consists of different types of agents, group of agents, and objects. In the designed subsystem, the universal circuit board can be embedded in any device, controlled by a UNIX class operating system which contains an agent platform.

As a manufacturing agent, an implementation with RFID readers to store real-time information about the product is used. Such information (destination, state, and location of the part) influences the control agent inference engine.

6.1 Hardware Overview

Prototype of the control system consists of a single board computer, RFID reader, and actuator. The scheme of the control system is presented on the Fig. 3.

The single board computer holds UNIX OS on board, provides the software for controlling the agent, and power supply to the whole system.

The RFID reader is connected by means of SPI (serial peripheral interface) via development board. This interface proves to be the most suitable for the kind of task, which the constructed device (the agent) fulfills. The interface can handle data speeds up to 10 Mbit/s. It works in full-duplex mode.

It also allows arbitrary choice of message size, content, and purpose. SPI is preferable for energy consumption factor as well, because in the current design, the device's peripherals do not use any outer power supply. Devices communicate in master/slave mode, where the master device initiates the data frame. In the present design, the microcontroller of the single board computer plays a role of master device.

The exemplary executive device is Mitsubishi industrial robot (RV-1A series). It is a compact 6-axis robot-manipulator, which gives excellent options for building a flexible manufacturing system. You can see the device in Fig. 4. The robots controller supports RS-232 interface, so TxD and RxD pins of the single board computer are used to connect it.

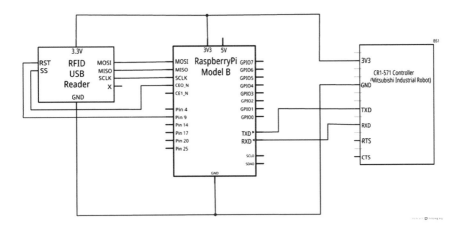

Fig. 3 Schema of the control/routing device

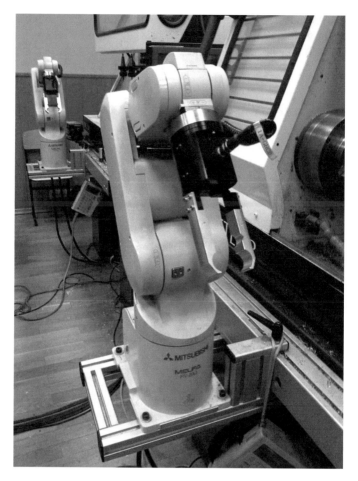

Fig. 4 Mitsubishi industrial robot

6.2 Software Overview

After hardware connection, there is a need to set a convenient software interface of communicating with RFID. The reader requires interface with registers and its native instructions. From the point of high-level code, it is inconvenient. Another level of abstraction for communicating with RFID is needed.

In order to control and to interact with RFID reader, a third-party Python library is used. Python allows importing GPIO and SPI wrapper. It hides low-level abstractions and provides quite a convenient interface of communication with RFID.

It operates 64 registers of the RFID device and uses native instructions, common for SPI interaction.

The functions of software are as follows:

- Reading part's data and store current information of it's state;
- Updating data during processing;
- Realization of resource–goal formalism;
- Communicating with other MA_f and CA_j.

7 Conclusions

In this paper, a new multiagent framework with fuzzy logic approach was developed. This framework includes distributed control system based on multiagent technologies which allows solving problems related with ambiguous or missing information. The research presents approach that integrates design, manufacturing, and control process using resource-based fuzzy network to agent's interaction. Prototype of the system is presented. This model is applicable to work with various manufacturing agents in conjunction with different design agents and control systems.

References

1. Subramaniam, V. et al.: Job shop scheduling with dynamic fuzzy selection of dispatching rules. Int. J. Adv. Manuf. Technol. **16**(10), 759–764 (2000)
2. Srinoi, P., Shayan, E., Ghotb, F.: Scheduling of flexible manufacturing systems using fuzzy logic. Int. J. Prod. Res. **44**(11), 1–21 (2002)
3. Taratoukhine, V., Bechkoum, K.: Towards a consistent distributed design: a multi-agent approach. In: Proceedings. IEEE International Conference on Information Visualization, pp. 384–389 (1999)
4. Dyundyukov, V., Tarassov, V. Goal-resource networks and their application to agents communication and co-ordination in virtual enterprises. Manuf. Model. Manage. Control **7**(1), 347–352 (2013)
5. Kandel, A.: Fuzzy Mathematical Techniques with Applications (1986)
6. D'Ambrosio, J., Darr, T., Birmingham, W.: Hierarchical concurrent engineering in a multiagent framework. Concurrent Eng. **4**(1), 47–57 (1996)
7. Szczerbicki, E.: Distributed multiagent system for concurrency evaluation: conceptual architecture. Syst. Anal. Model. Simulation **16**(4), 277–291 (1994)
8. Pham, D.T., Dimov, S.S.: An approach to concurrent engineering. Proc. Inst. Mech. Eng. Part B J. Eng. Manuf. **212**(1), 13–27 (1998)
9. Tong, G., Fitzgerald, B.: Concurrent engineering battleground. World Class Design Manuf. **1**(3), 45–48 (1994)
10. Danesh, M. R., Jin, Y.: An agent-based decision network for concurrent engineering design. Concurrent Eng. **9**(1), 37–47 (2001)
11. Sun, J., Zhang, Y.F., Nee, A.Y.C.: Agent-based product design and planning for distributed concurrent engineering. In: Proceedings of Robotics and Automation. ICRA'00. IEEE International Conference on, IEEE, vol. 4, pp. 3101–3106, 2000

12. Dean, E.B., Unal, R.: Elements of designing for cost. In: Proceedings of AIAA 1992 Aerospace Design Conference, 1992
13. Tarchinskaya, E., Taratoukhine, V., Matzner, M.: Cloud-based engineering design and manufacturing: state of the art. Manuf. Model. Manage. Control **7**(1), 335–340 (2013)
14. Prasad, B.: Concurrent Engineering Fundamentals, vol. 1. Prentice Hall, Englewood Cliffs, NJ (1996)
15. Zadeh, L. A.: The Concept of a Linguistic Variable and Its Application to Approximate Reasoning, pp. 1–10. Springer, US (1974)

Part IV
Image Processing and Pattern Recognition

Fuzzy Information Measure for Improving HDR Imaging

Annamária R. Várkonyi-Kóczy, Sándor Hancsicska and József Bukor

Abstract Digital image processing can often improve the quality of visual sensing of images and real-world scenes. Recently, high dynamic range (HDR) imaging techniques have become more and more popular in the field. Both classical and soft computing–based methods proved to be advantages in revealing the non-visible parts of images and realistic scenes. However, extracting as much details as possible is not always enough because the sensing capability of the human eye depends on many other factors and the visual quality is not always proportional to the rate of accurate reproduction of the scene. In this paper, a new fuzzy information measure is introduced by which the quality of HDR images can be improved and both the amount of visible details and the quality of sensing can be increased.

Keywords Information enhancement · Image quality improvement · Fuzzy image processing · High dynamic range imaging · Tone mapping

A.R. Várkonyi-Kóczy (✉)
Institute of Mechatronics and Vehicle Engineering, Óbuda University,
Budapest, Hungary
e-mail: varkonyi-koczy@uni-obuda.hu

A.R. Várkonyi-Kóczy · S. Hancsicska
Integrated Intelligent Systems Japanese-Hungarian Laboratory,
Óbuda University, Budapest, Hungary
e-mail: hancsisandor@gmail.com

A.R. Várkonyi-Kóczy · J. Bukor
Department of Mathematics and Informatics, J. Selye University,
Komarno, Slovakia
e-mail: bukorj@selyeuni.sk

1 Introduction

It is well known that the (high dynamic) range of illumination of real-world scenes can be much wider than that captured by a human eye. The situation becomes even worse if an ordinary display transmits between the scene and the eye. The range of light in the real world spans 10 orders of magnitude, a single scene's captured luminance values may have as much as 4 orders of magnitude difference, while a typical CRT can only display 2 orders of magnitude. Any medium results in only a limited stimulus in the perception [1].

Another problem follows from the adaptation features of the light perception. The human eye is able to adapt itself to the temporal lighting conditions and captures the 4 orders of magnitude in the actual range of illumination; however, it takes time. When the lighting conditions change, e.g., we enter from a light room to a dark one, and we do not see anything first. After 1 min, the eye starts to adapt to the new conditions. This adaptation procedure finishes only after approximately 20 min.

The opposite direction is less noticeable. When we switch on the lamp in a dark room or we leave a tunnel and enter in the bright sunshine, we start to see well (~80 %) much earlier, after approximately 2 s [2].

The wideness and possible sudden changes in the range of illumination may cause that a part of the information becomes non-displayable and/or non-visible in the highly illuminated and less illuminated regions, i.e. are lost for the observer and for data storage. I.e., the high dynamic range (HDR) of illumination may cause serious distortions and problems in the view and further processing of real-world scenes and digital images.

Digital signal processing techniques can often improve the visual quality of photographs and sceneries [3]. Noise filtering, image information enhancement methods, image sharpeners, etc. all aim to enhance and preserve visual information while suppressing noise. HDR imaging techniques transform the HDR of illumination into the visible/displayable range of illumination while trying to preserve the quality of the scenery and as much details as possible.

In this paper, a new tone mapping function–based algorithm is introduced which applies a locally tuned nonlinear mapping. The compression of the regions basically depends on the amount of visual information in the region. The method assigns bigger low dynamic range (LDR) regions to those HDR regions which contain more details taking into account that the human visual sensing is limited in terms of the minimum distance of which details can be distinguished. On the other hand, the technique offers a way to keep parts invariant if their importance makes it necessary or if the region contains correct image data. By this, there is a possibility for compressing regions where unimportant or sparse information is stored and for extending the important or dense regions. The amount of information is determined by applying a new fuzzy measure. The complexity of the method is low, i.e., it allows also real-time processing.

The paper is organized as follows: In Sect. 2, HDR imaging is summarized. Section 3 is devoted to the proposed new fuzzy information measure. In Sect. 4, an adaptive tone mapping function–based HDR technique is introduced which takes into account the amount of details in the different luminance regions during the compression of the scale, while Sect. 5 presents examples and comparisons with other techniques. Finally, Sect. 6 is devoted to the conclusions.

2 High Dynamic Range Imaging

HDR imaging covers image processing techniques that aim to transform (compress) a wide (wider than that can be captured) dynamic range of illumination (luminosity) into the visible range in such a way that preserves (enhances) local features. A non-HDR display or camera can show/take pictures with limited exposure range which may result in the loss of important details. HDR processing can offer a solution and can also capture the information hided in the dark and extremely bright areas.

HDR techniques can be divided into two main groups according to their focus of processing: We can differentiate between local and global methods. In the first group, we find techniques such as anchor-based algorithms (see, e.g., [4–6]), image sensors (and special computer renderings) (see, e.g., [7–9]), and multiple exposure time synthesization (see, e.g., [10–12]), while tone mapping function–based algorithms (see, e.g., [13–15]) belongs to the second group.

For comparison of the techniques, see also [16].

2.1 Anchoring Theory

In the following, for simplicity, let us speak about grayscale problems. The perception of the luminosity depends basically on two factors and can be determined as the product of the reflectance at the corresponding object point and the intensity of the illumination at that point. On the other hand, the amount of light projected to the eyes (luminance) depends on several factors: the illumination that strikes visible surfaces, the proportion of light reflected from the surface, and the amount of light absorbed, reflected, or deflected by the prevailing atmospheric conditions [17]. If a visual system only made a single measurement of luminance, acting as a photometer, then there would be no way to distinguish a white surface in dim light from a black surface in bright light.

Humans are usually able to tie their perception to the absolute light intensity scale more or less correctly because we possess a skill called lightness constancy. However, a gray patch appears brighter when viewed against a dark background and darker when viewed against a bright background. Further, there is a tendency of the highest luminance of a scene or image to appear white and a tendency of the largest area also to appear white [1]. (This effect is known as simultaneous contrast.) In many cases, we are more certain about relative lightnesses than their specific (absolute) lightness values. Thus, lightness values cannot easily be tied to absolute luminance values because there is no systematic relationship between absolute luminance and surface reflectance. We need an anchoring rule which associates at least one shade in the scene to the perceptible absolute black to white scale. By this, we become able to derive the specific tones from the relative luminance values.

The anchor can be put to any of the luminance points. For example, Gilchrist et al. [4] anchor the average luminance value (average luminance rule: the average luminance in the visual field is perceived as middle gray), while Rudd and Zemach [18] tie the highest luminance (highest luminance rule: the value of white is assigned to the highest luminance in the display).

2.2   Local Anchoring

In case of complex images, the anchoring rule is applied in an indirect way. First, the image is decomposed into segments (so-called frameworks) by using some kind of clustering, e.g., the k-means algorithm [19]. Then, the anchoring rule is used separately on the individual frameworks. Roughly, the steps are as follows:

The k-means clustering algorithm is used to find the centroids of the segments. After finding them, those which are 'very close' (closer than a given threshold) are merged. During the merging, the centroids are weighted by their areas (i.e., the area of the corresponding framework).

Besides developing as much details of the images as possible, we also want to keep the 'character' of the image. Because of this, we apply a so-called articulation factor associated with each framework which shows the 'significance' of the frameworks in the determination of the output. Thus, the next step can be the determination of the articulation factors and the modification of the intensity values by these articulation factors. By this, frameworks with wider intensity range will have a greater (more characteristic) contribution to the final luminance values of the pixels.

The most important step is the determination of the anchors. They are estimated separately from framework to framework. During this procedure, one of the rules, e.g., the highest luminance rule, can be applied, however not directly. The local anchor can be estimated after removing a small amount, let us say 5 % of all pixels in the framework's area that has the highest luminance, as the highest luminance of the rest of the pixels [5]. (This is important because of the self-luminosity [20]; the 5 % is an experimental factor.)

Finally, the global output lightness values of the pixels are determined. This means the merging of the frameworks where the lightness modification of the pixels is evaluated. Since the borders of the frameworks are not ambiguous and over-lapping, they can be handled as fuzzy sets as well (i.e., to each pixel, a fuzzy membership value is assigned to define the membership of the pixels belonging to the frameworks). An appropriate fuzzy rule base can be built for the merging, and the resulting lightness modification value corresponds to the weighted sum of the local lightness values, where the membership values serve as weighting factors. This fuzzy decision making usually leads to better results.

2.3 Global Tone Mapping

The main idea of global tone mapping lies in the definition of a simple, monotonous, low-complexity mapping function, which maps the wide range of luminance values into a displayable one. Monotonicity is required to ensure that the mapping will keep the relativity of the luminance, i.e., lighter regions will remain lighter while darker regions darker. The boundedness of the function guarantees that any HDR region can be transformed to any visible one. Typical examples are bounded linear and saturating functions like the logarithmic mapping (see, e.g., [21]).

2.4 Image Synthesization

Image synthesization belongs to the gradient-based approaches. This method uses several images made with different exposures of the same scene. During the processing, the images are divided into regions. The method selects the most informative subimage, i.e., that contains the highest density of information within the region, for each local image region. The chosen image parts (taken with different exposures) are merged together into one image in such a way that all the involved data are preserved in it. The final output image is generated by smoothing the temporal output. The aim of this step is to eliminate the sharp transitions, which unavoidably arise at the borders of the regions. For more details, see, e.g., [10–12] and [22].

3 Fuzzy Information Measure for the Determination of the Level of Details

An important question in image processing and data storage is as follows: What carries the information in the images? Although there are many possible replies to this question considering statistical elements or the histogram of the luminance values, a simple, new, low complexity one can be the amount of the intensity changes in the image. Let us consider that the most characteristic about the objects are their boundaries (which usually can be extracted as edges, representing significant intensity changes). This means that the level of details in an image is directly proportional (and thus can be measured and represented) by summing up the intensity changes, e.g., in case of RGB representation, the sum of gradient magnitudes of the RGB components.

The sum of gradient magnitudes, i.e., the amount of the intensity changes, can be measured also by the edge points. In [23], a fuzzy edge detection method is described which offers a way to qualify edges according to their fuzzy edge-ness (how strong edges they are). By this, the evaluation of the sum of intensity changes can be turned to a task where the edge points weighted by their fuzzy edgeness are summed. This technique is especially advantageous in case of preprocessed (e.g., edge-detected) images.

The complexities of these two approaches are significantly lower than those of the other ones, and as we demonstrate below, they provide good results.

The amount of information in an image is strongly related to the number and complexity of the objects in the image. The boundary edges of the objects carry the primary information about the object's shape. Thus, image content information can be represented by the characteristic features, such as corners and edges in the image, i.e., the number of characteristic pixels is proportional to the amount of information in the image.

Let $I^R(x, y)$, $I^G(x, y)$, and $I^B(x, y)$ be the R, G, and B intensity components of the pixel at location $[x, y]$ in the image to be processed. Let us consider the group of neighboring pixels which belong to a 3×3 window centered on $[x, y]$. For calculating the gradients of the intensity functions in horizontal ΔI_x and vertical ΔI_y directions at position $[x, y]$, the intensity differences of the RGB components between the neighboring pixels are considered. For simplicity, we show the expressions for only one (let us say the R) component (the same has to be evaluated in case of the other two, G and B components, and in case of grayscale images):

$$\begin{aligned} \Delta I_x^R &= \left| I^R(x+1, y) - I^R(x, y) \right| \\ \Delta I_y^R &= \left| I^R(x, y-1) - I^R(x, y) \right| \end{aligned} \tag{1}$$

For further processing, the maximum of the estimated gradient values should be chosen, which solves as the input of the normalized linear mapping function P defined as follows:

$$P(v) = \frac{v}{I_{max}}. \tag{2}$$

Here, I_{max} stands for the maximum intensity value. (For 8-bit RGB scales, it equals 255.) Let $\mathbf{R}$ be a rectangular image region of width rw and height rh, with upper left corner at position $[x_r, y_r]$. The R component level of the detail inside of the region $\mathbf{R}$ is defined as

$$M_D^R(\mathbf{R}) = \sum_{i=0}^{rw} \sum_{j=0}^{hw} P\left(\max(\Delta I_x^R(x_r+i, y_r+j), \Delta I_y^R(x_r+i, y_r+j)) \right) \tag{3}$$

The sum of the three, R, G, and B components gives the total level of detail in region $\mathbf{R}$.

Fig. 1 Fuzzy membership function m_{LA} of 'edgeness.' $L - 1$ equals the maximum intensity value, and p and q are tuning parameters

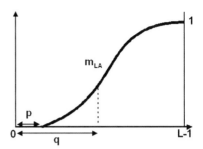

$$M_D(\mathbf{R}) = M_D^R(\mathbf{R}) + M_D^G(\mathbf{R}) + M_D^B(\mathbf{R}) \tag{4}$$

As higher the calculated M_D value is as detailed the analyzed region is.

To follow the other method, i.e., to evaluate the fuzzy edges, first, the edges have to be extracted. A good performance fuzzy edge detection method proposed by Russo in [23] can be summarized, as follows. (For simplicity, the expressions are given for grayscale images.)

As previously described, the group (3×3 window) of the neighboring pixels is considered around the processed pixel at location $[x, y]$. The output of the edge detector is yielded by

$$I_{x,y}^P = (L - 1)\max\{m_{LA}(\Delta v_1), m_{LA}(\Delta v_2)\}$$
$$\Delta v_1 = |I(x - 1, y) - I(x, y)| \tag{5}$$
$$\Delta v_2 = |I(x, y - 1) - I(x, y)|$$

where $I_{x,y}^P$ denotes the pixel luminance in the edge-detected image and m_{LA} stands for the used membership function (see Fig. 1). $I(x, y)$, $I(x - 1, y)$, and $I(x, y - 1)$ correspond to the luminance values of the processed pixel (at location $[x, y]$) and its left and upper neighbors. $L - 1$ equals the maximum luminance value (e.g., 255).

After edge detection, the fuzzy information level of the image can be evaluated by simply summing the membership values of the edge pixels.

4 The New Adaptive Fuzzy Information Level Based on Tone Mapping Function

The fuzzy level of details (FLD) can be used to scale the transformation between the visible and HDR light intensity ranges according to the amount of information in the different intensity ranges.

As starting, any mapping function corresponding to the requirements described in Sect. 2.3 can be used. As example, consider the logarithmic mapping shown in Fig. 2.

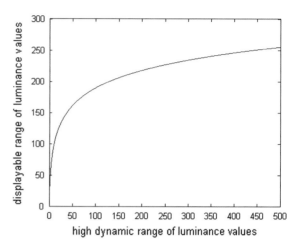

Fig. 2 Example for the most frequently used, logarithmic tone mapping function

This simple and easily evaluable mapping function can be combined with a nonlinear scaling of the vertical axis (the axis of displayable luminance values), thus extending the mapping possibilities. The nonlinear vertical axis on the one hand enables to keep the image data, which should not be modified, invariable, while on the other hand the high or less illuminated areas can be corrected without having any influence on the areas containing correct image data. The mapping function makes possible to compress regions where unimportant or sparse information is stored, thus offering a way to keep wider parts of the displayable or viewable domain for the important (dense) regions. The importance of the regions can be estimated easily and automatically by the FLD described in Sect. 3. The magnitude of 'strong' intensity changes (e.g., density of the edges) within the region which is characteristic of the density of the represented amount of (seen or hided) information in this region. The displayable luminance region can be allocated proportionally to this measure, i.e., if we have an important (dense) high dynamic region, we can modify the originally corresponding region (assigned according to the log function) proportional to the characteristic information dense measure. The mapping will keep the relativity of the luminance, i.e., lighter regions will remain lighter while darker regions darker. Figure 3 illustrates a possible mapping function and a simple nonlinear vertical axis of the displayable luminance values. The nonlinearity of the vertical axis is influenced by a set of linear functions. By changing the linear functions, the nonlinear characteristics of the vertical axis can also be modified.

The mapping function in Fig. 3 has the form

$$
\begin{aligned}
&\text{If} \quad 0 \leq \log(L_\mathrm{w}) \leq a \quad \text{then} \quad L_\mathrm{d} = \log(L_\mathrm{w})/\cos \alpha_1 \\
&\text{If} \quad a \leq \log(L_\mathrm{w}) \leq b \quad \text{then} \quad L_\mathrm{d} = \log(L_\mathrm{w})/\cos \alpha_2 \\
&\text{If} \quad b \leq \log(L_\mathrm{w}) \leq c \quad \text{then} \quad L_\mathrm{d} = \log(L_\mathrm{w})/\cos \alpha_3
\end{aligned}
$$

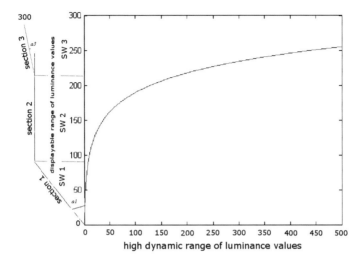

Fig. 3 Example of the proposed FLD adaptive function: the logarithmic mapping combined with nonlinear (or piecewise linear) vertical axis

where

$$a = SW1; b = (SW1 + SW2); c = (SW1 + SW2 + SW3)$$

and   (6)

$$SW1/\cos\alpha_1 + SW2/\cos\alpha_2 + SW3/\cos\alpha_3 = L_{\mathrm{dmax}}$$

Here, L_w denotes a wide range of luminance values and L_d stands for the corresponding displayable luminance value (luminance value in the resulted output image, with upper limit L_{dmax} (i.e., after determining the adaptively modified widths of the sections, the total length of the transformed range has to be scaled to the total width of the displayable range). SWi represents the width of the ith region on the logarithmic scale, and α_i is the angle between the side of the ith section of the axis and the original vertical axis. The α_i angles are tuned according to the FLDs in the regions.

Remark The proposed fuzzy level of details can advantageously be used to measure the information densities in the local image regions of multiple exposure images, as well.

5 Examples

To illustrate the effectiveness of the proposed technique, two examples are shown. The first example shows the HDR image of a track with a part brightly enlightened (see Fig. 4), while in Fig. 5 the image is transformed using the new technique. Figure 6 presents the image improved by anchoring-based algorithm using three frameworks, and Fig. 7 illustrates the image processed by logarithmic mapping.

Fig. 4 Example 1: Original image of an brightly enlightened track

Fig. 5 Example 1: Improved image of the track processed by the new method

In the second example, the processing of the image of a satellite can be followed. In Fig. 8, the original image can be seen where a part of the scene is highly illuminated and the details can badly be recognized. Figure 9 shows the output after using the new adaptive fuzzy information level–based tone mapping function. For comparison, we include also the image improved by simple logarithmic tone mapping (see Fig. 10).

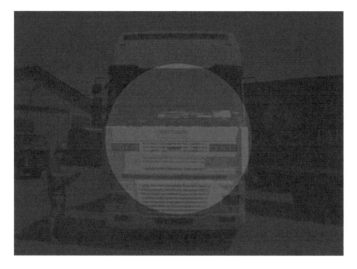

Fig. 6 Example 1: Image of the track improved by anchoring-based algorithm

Fig. 7 Example 1: Image of the track improved by non-adaptive logarithmic tone mapping

As conclusion, we can state that the introduced method well enhances the majority of the previously hided details. The technique seems to perform much better than the simple anchoring algorithm and usually extracts more details, especially in the bright regions, than the simple logarithmic mapping. According to our experiences, after processing, approximately 90 % of the hided details become visible.

The complexity of the introduced information measuring method is proportional to the number of pixels in the image, i.e., it does not explode even in case of huge images.

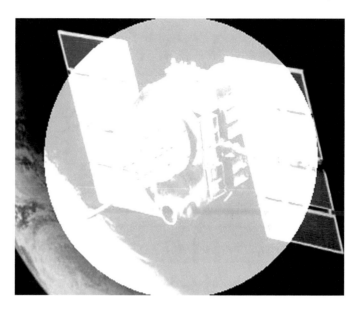

Fig. 8 Example 2: Original image of a highly illuminated satellite

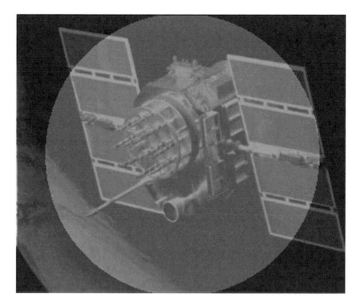

Fig. 9 Example 2: Improved image of the satellite processed by the new method

In our future work, we will work on an improved method where the distortion of the logarithmic mapping (i.e., that a wide intensity section is assigned to the same size region in the dark intensity range while a narrow zone in the bright intensity

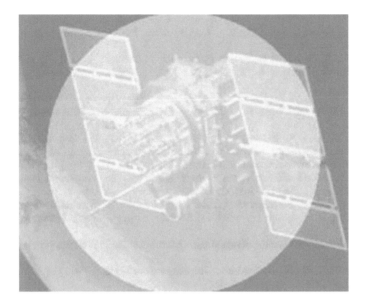

Fig. 10 Example 2: Image of the satellite improved by non-adaptive logarithmic tone mapping

range) will be better compensated and the transformation based on the information level will be better matched to the starting tone mapping function.

6 Conclusions

The HDR of illumination may cause serious distortions and problems in the view and further processing of digital images. Important information can be hided in the highly or extreme lowly illuminated parts. This paper deals with the reproduction of such images and introduces a new fuzzy information measure which may help in improving the previously existing global tone reproduction and image synthesization methods to adaptively develop the hardly or non-viewable features and content of the images.

Acknowledgment This work has partially been sponsored by the Hungarian National Scientific Fund under contract OTKA 105846 and the Research & Development Operational Program for the project "Modernization and Improvement of Technical Infrastructure for Research and Development of J. Selye University in the Fields of Nanotechnology and Intelligent Space", ITMS 26210120042, co-funded by the European Regional Development Fund.

References

1. Adelson, E.H.: Lightness perception and lightness illusions. In: The Cognitive Neurosciences, 2nd edn, pp. 339351. MIT Press, Cambridge, MA (2000)
2. Palmer, S.: Chapter 3.3, Surface-based color processing. In: Vision Science: Photons to Phenomenology. The MIT Press, Cambridge (1999)

3. Russo, F.: Design of fuzzy relation-based image sharpeners. In: New Advances in Intelligent Signal Processing. Ser. Studies in Computational Intelligence, vol. 372, pp. 115–131. Springer, Berlin (2011)
4. Gilchrist, A., Kossyfidis, C., Bonato, F., Agostini, T., Cataliotti, J., Li, X., Spehar, B., Annan, V., Economou, E.: An anchoring theory of lightness perception. Psych. Rev. **106**(4), 795–834 (1999)
5. Krawczyk, G., Myszkowski, K., Seidel, H.P.: Lightness perception in tone reproduction for high dynamic range images. In: 26th Annual Conference EUROGRAPHICS, vol. 24(3), pp. 635–645 (2005)
6. Beeler, T., Hahn, F., Bradley, D., Beckel, B., Beardsley, P., Gotsman, C., Sumner, R.W., Gross, M.: High-quality passive facial performance capture using anchor frames. J. ACM Trans. Graph. (TOG), Proceedings of ACM SIGGRAPH 2011, 30(4), paper 75 (July 2011)
7. Kawahito, S.: An ultimate dynamic range imaging device from star light to sun light. Inter-Academia **2005**(1), 105–113 (2005)
8. Wang, X., Wolfs, B., Meynants, G., Bogaerts, J.: An 89 dB dynamic range CMOS image sensor with dual transfer gate pixel. In: International Image Sensor Workshop, paper R36 (2011)
9. Darmont, A.: High Dynamic Range Imaging: Sensors and Architectures, 1st edn. SPIE Press (2012)
10. Rubinstein, R., Brook, A.: Fusion of differently exposed images. Final Project Report, Israel Institute of Technology, pp. 14 (2004)
11. Várkonyi-Kóczy, A.R., Rövid, A., Hashimoto, T.: Gradient based synthesized multiple exposure time color HDR image. IEEE Trans. Instrum. Meas. **57**(8), 1779–1785 (2008)
12. Vonikakis, V., Andreadis, I.: Fast automatic compensation of under/over-exposed image regions. In: Advances in Image and Video Technology: Second Pacific Rim Symposium (PSIVT), 510 (Dec 2007)
13. Qiu, G., Guan, J., Duan, J., Chen, M.: Tone mapping for HDR image using optimization—A new closed form solution. In: 18th International Conference on Pattern Recognition, vol. 1, pp. 996–999 (2006)
14. Kovesi, P.: Phase preserving tone mapping of non-photographic high dynamic range images. In: Digital Image Computing: Techniques and Applications, pp. 1–8 (2012)
15. Sajeena, A., Sunitha Beevi, K.: An Improved HDR Image Processing Using Fast Global Tone Mapping. Int. J. Res. Eng. Technol. **2**(12), 298–302 (2013)
16. Pourazad, M.T., Mai, Z., Nasiopoulos, P.: Effect of global and local brightness-change on the quality of 3D visual perception. Int. J. Adv. Telecommun. **5**(1 and 2), 101–110 (2012)
17. Adelson, E.H., Pentland, A.P.: The perception of shading and reflectance. In: Knill, D., Richards, W. (eds.) Perception as Bayesian Inference, pp. 409–423. Cambridge University Press, New York (1996)
18. Rudd, M.E., Zemach, I.K.: The highest luminance anchoring rule in lightness perception. J. Vision **3**(9), 56a (2003)
19. Theiler, J., Gisler, G.: A contiguity-enhanced k-means clustering algorithm for unsupervised multispectral image segmentation. SPIE **3159**, 108–118 (1997)
20. Fleming, R.W., Dror, R.O., Adelson, E.H.: Real-world illumination and the perception of surface reflectance properties. J. Vision **3**, 347–368 (2003)
21. Drago, F., Myszkowski, K., Annen, T., Chiba, N.: Adaptive logarithmic mapping for displaying high contrast scenes. Eurographics 2003, Comput. Graph. Forum **22**(3), 419426 (2003)
22. Salvi, G., Sharma, P., Raman, S.: Efficient Image Retargeting for High Dynamic Range Scenes. arxiv:1305.4544 [cs.CV] (2013)
23. Russo, F.: Recent advances in fuzzy techniques for image enhancement. IEEE Trans. Instrum. Meas. **47**(6), 1428–1434 (1998)

Optimization of Type-1 and Type-2 Fuzzy Systems Applied to Pattern Recognition

Daniela Sánchez, Patricia Melin and Oscar Castillo

Abstract In this paper, a new method of fuzzy inference system optimization using a hierarchical genetic algorithm (HGA) is proposed. The fuzzy inference system is used to combine the different responses of modular neural networks (MMNs). In this case, the MMNs are used to perform the human recognition using 4 biometric measures: face, iris, ear, and voice. The main idea is the optimization of some parameters of a fuzzy inference system such as the type of fuzzy logic (FL), type of system, number of membership functions in each input, type of membership functions in each variable, their parameters, and the consequences of the fuzzy rules.

1 Introduction

The recognition of people is of great importance, since it allows us to have a greater control about when a person has access to certain information; biometrics accurately identifies each individual and distinguishes one from another [1]. The achieved results indicate that recognition-based biometry techniques are much more precise and accurate than the traditional techniques [2].

There is a great diversity of techniques to solve complex problems, such as fuzzy logic (FL), neural networks (NNs), and genetic algorithms (GAs). These techniques are complementary; thus, they may be used in combination. These systems are computational systems that integrate different intelligent techniques. Hybrid intelligent systems allow the representation and manipulation of different types and

D. Sánchez (✉) · P. Melin · O. Castillo
Tijuana Institute of Technology, Tijuana, Mexico
e-mail: danielasanchez.itt@hotmail.com

P. Melin
e-mail: pmelin@tectijuana.mx

O. Castillo
e-mail: ocastillo@tectijuana.mx

© Springer International Publishing Switzerland 2016
L.A. Zadeh et al. (eds.), *Recent Developments and New Direction
in Soft-Computing Foundations and Applications*, Studies in Fuzziness
and Soft Computing 342, DOI 10.1007/978-3-319-32229-2_10

forms of data and knowledge which may come from various sources [3]. There are many works where these kinds of systems have been implemented [4–8], and they have demonstrated that the integration of different intelligent techniques provides good results. In this paper, modular neural network (MNNs), FL, and hierarchical genetic algorithm (HGA) are combined to improve the performance that each technique has separately.

This paper is organized as follows: The basic concepts used in this work are presented in Sect. 2. Section 3 contains the proposed method. Section 4 presents experimental results, and in Sect. 5, the conclusions of this work are presented.

2 Basic Concepts

2.1 Modular Neural Networks

Neural networks (NNs) can be used to extract patterns and detect trends that are too complex to be noticed by either humans or other computer techniques [9, 10].

The results of the different applications involving MNNs lead to the general evidence that the use of MMNs implies a significant learning improvement comparatively to a single NN and especially to the backpropagation NN, because the concept of modularity is an extension of the principle of divide and conquer: The problem should be divided into smaller subproblems that are solved by experts (modules), and their partial solutions should be integrated to produce a final solution. Each NN works independently in its own domain, and each of the NNs is build and trained for a specific task [11–13].

2.2 Type-2 Fuzzy Logic

FL is a useful tool for modeling complex systems and deriving useful fuzzy relations or rules [14]. The concept of a type-2 fuzzy set was introduced by Zadeh (1975) as an extension of the concept of an ordinary fuzzy set (henceforth called a "type-1 fuzzy set"). A type-2 fuzzy set is characterized by a fuzzy membership function, i.e., the membership grade for each element of this set is a fuzzy set in [0, 1], unlike a type-1 set where the membership grade is a crisp number in [0, 1]. Such sets can be used in situations where there is uncertainty about the membership grades themselves, e.g., an uncertainty in the shape of the membership function or in some of its parameters. Consider the transition from ordinary sets to fuzzy sets. When the membership cannot be determined of an element in a set as 0 or 1, fuzzy sets of type-1 are used. Similarly, when the situation is so fuzzy that there is a trouble in determining the membership grade even as a crisp number in [0, 1], fuzzy sets of type-2 are used [15, 16].

2.3 Hierarchical Genetic Algorithm

A GA is an optimization and search technique based on the principles of genetics and natural selection where the fittest individuals survive [17–19]. GAs are non-deterministic methods that employ crossover and mutation operators for deriving offspring. GAs work by maintaining a constant-sized population of candidate solutions known as individuals (chromosomes) [20–22].

Introduced in [23], a HGA is a type of GA. Its structure is more flexible than the conventional GA. The basic idea under HGA is that for some complex systems, which cannot be easily represented, this type of GA can be a better choice. The complicated chromosomes may provide a good new way to solve the problem [24, 25].

3 Proposed Method

The proposed method combines the MMN responses using FL as a technique for the integration. In this paper, 4 biometric measures are used (each biometric measure is a MMN), and the combination of responses is performed by a fuzzy integrator. The number of inputs in the fuzzy integrator will depend on the number of biometric measures. In this work, the fuzzy integrator has 4 inputs (one for each biometric measure). An example of this fuzzy integrator is shown in Fig. 1.

3.1 Description of the Hierarchical Genetic Algorithm for Fuzzy Integrators Optimization

The proposed HGA performs the optimization of fuzzy inference system. The main idea is to obtain a fuzzy integrator with good parameters because if this occurs, the response combination of MNNs will have a good effectiveness. For this reason, some parameters of a fuzzy inference system are optimized, such as type of FL, type of system, number of membership functions in each inputs, type of membership functions in each variable, their parameters, and the consequences of the fuzzy rules. The chromosome of the proposed HGA is shown in Fig. 2.

The optimization of the consequences is described below. First, all the possible rules are generated, for example, the maximum number of membership functions in

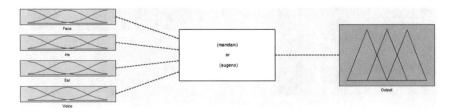

Fig. 1 Example of the fuzzy integrator

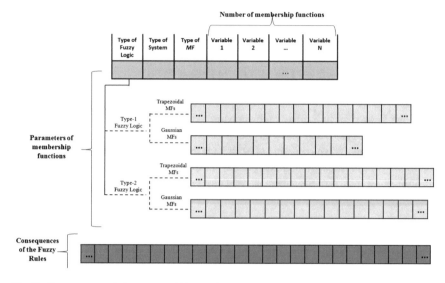

Fig. 2 Chromosome of the proposed hierarchical genetic algorithm

each variable is 2, and the number of inputs is equal to 4, that means the total of possible rules is 16. The consequences are taken from the established genes. When, for example, the number of membership functions for each input variable is 2 (for face variable), 2 (for iris variable), 2 (for ear variable) and 2 (for voice variable), the number of rules is 8. The possible rules for this combination are taken with their respective consequences. An example is shown in Fig. 3. The parameters used for the HGA are shown in Table 1.

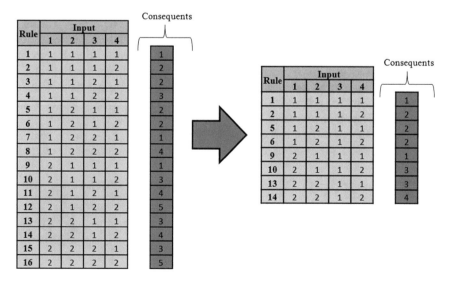

Fig. 3 Example of selection of the rules

Table 1 Parameters of the HGA

Genetic operator	Value
Population size	10
Maximum number of generations	100
Selection	Roulette wheel
Selection rate	0.85
Crossover	Single point
Crossover rate	0.9
Mutation	bga
Mutation rate	0.01

Fig. 4 Examples of the face images from CASIA database

3.2 Databases

Only 77 persons of each database were taken to perform the recognition. The databases used are described below:

(1) *Face*: The database of face from the Institute of Automation of the Chinese Academy of Sciences [26] was used. It contains 5 images per person. The image dimensions are 640×480 with BMP format. An example of the database is shown in Fig. 4.

(2) *Iris*: The database of human iris from the Institute of Automation of the Chinese Academy of Sciences [27] was used. It contains 14 images (7 for each eye) per person. The image dimensions are 320×280 with JPEG format. An example of the database is shown in Fig. 5.

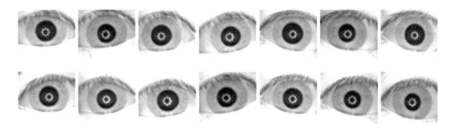

Fig. 5 Examples of the human iris images from CASIA database

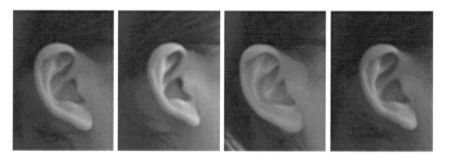

Fig. 6 Examples of Ear Recognition Laboratory from the University of Science and Technology Beijing (USTB)

(3) *Ear*: The database of ear from the University of Science and Technology of Beijing [28] was used. It contains four images per person; the image dimensions are 300×400 pixels, and the format is BMP. An example of the database is shown in Fig. 6.

(4) *Voice*: In the case of voice, the database was obtained from the students of Tijuana Institute of Technology, and it consists of 10 voice samples in WAV format. The word that they said in Spanish was "ACCESAR."

4 Experimental Results

The proposed method is used to perform the human recognition based on face, iris, ear, and voice.

4.1 Non-optimized Modular Neural Networks

Ten training sessions for each biometric measure were performed. The architecture of each training was randomly established. The results obtained in each training for each biometric measure are presented in Table 2.

4.2 Non-optimized Fuzzy Integration

Different training sessions presented above were combined, and 5 cases were considered. Two non-optimized fuzzy integrators were used to compare with the optimization. The first fuzzy integrator is a Type-1 fuzzy inference system, and the second fuzzy integrator is a Type-2 fuzzy inference system; both fuzzy integrators

Table 2 The results of each biometric measure

Training	Face (%)	Iris (%)	Ear (%)	Voice (%)
1	87.01	79.10	94.80	87.79
2	85.71	81.81	77.92	91.77
3	52.92	96.10	96.10	90.16
4	45.77	82.57	79.22	91.88
5	60.17	94.37	97.40	91.23
6	37.01	90.90	57.14	93.18
7	47.18	63.20	81.81	90.04
8	70.77	84.95	90.90	89.93
9	68.83	92.72	82.46	92.85
10	70.65	98.26	67.53	86.36

Fig. 7 Example variable of type-1 fuzzy integrator

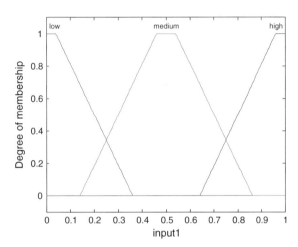

are Mamdani type and have 3 trapezoidal membership functions in each variable (inputs and output). Examples of each variable are shown in Figs. 7 and 8. The result for each case is shown in column 6 and column 7 in Table 3.

4.3 Optimized Fuzzy Integration

For each case previously established, 30 evolutions were performed. In this case, the maximum number of membership functions for each variable was equal to 5. The total number of possible rules was 625, but as described above, depending on the number of membership functions in each variable is the number of rules used. As previously mentioned, the proposed HGAs perform the optimization of parameters of a fuzzy inference system; the number of membership functions in the output is fixed (5 membership functions). In this case in Table 4, the obtained results are

Fig. 8 Example variable of
type-2 fuzzy integrator

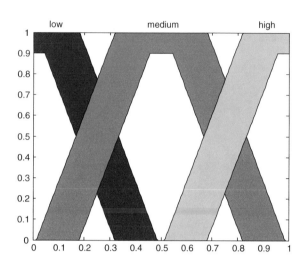

Table 3 Non-optimized results

Case	Face	Iris	Ear	Voice	Non-optimized fuzzy integrator	
					Type-1	Type-2
1	FT4 45.78 %	IT5 94.37 %	ET5 97.40 %	VT2 91.77 %	92.20 % 0.0779	94.80 % 0.0519
2	FT1 87.01 %	IT4 82.57 %	ET2 77.92 %	VT1 87.79 %	86.79 % 0.1320	85.71 % 0.1429
3	FT2 85.71 %	IT2 81.81 %	ET4 79.22 %	VT5 91.23 %	84.93 % 0.1506	84.29 % 0.1571
4	FT6 37.01 %	IT7 63.20 %	ET6 57.14 %	VT10 86.36 %	61.58 % 0.3842	64.82 % 0.3517
5	FT4 45.78 %	IT7 63.20 %	ET9 82.47 %	VT4 91.88 %	66.88 % 0.3312	70.45 % 0.2955

Table 4 The results obtained
with type-1 fuzzy logic

Case	Type-1 fuzzy logic		
	Number of evolutions	Best	Average
1	15	100 % 0	99.38 % 0.0062
2	12	93.72 % 0.0627	91.22 % 0.0877
3	9	95.97 % 0.0402	95.44 % 0.0455
4	10	87.12 % 0.1287	85.22 % 0.1478
5	9	95.78 % 0.0422	93.90 % 0.0610

Table 5 The results obtained with type-2 fuzzy logic

Case	Type-2 fuzzy logic		
	Number of evolutions	Best	Average
1	15	100 % 0	99.35 % 0.0064
2	18	94.70 % 0.0503	91.62 % 0.0838
3	21	97.14 % 0.0285	94.12 % 0.0588
4	20	89.80 % 0.1017	87.24 % 0.1275
5	21	96.42 % 0.0357	94.21 % 0.0579

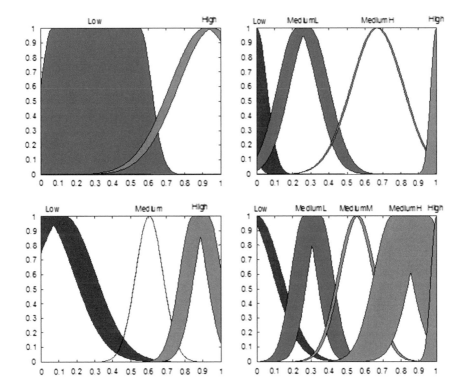

Fig. 9 Best fuzzy integrator for case #2

shown (bests and average) when the HGA considered better Type-1 FL than Type-2 FL, and in Table 5 when the HGA considered better Type-2 FL than Type-1 FL.

In Fig. 9, the best fuzzy integrator for the case #2 is shown; it is a fuzzy integrator of Sugeno type and with Gaussian Membership Functions, and in Fig. 10, the convergence of that evolution is shown.

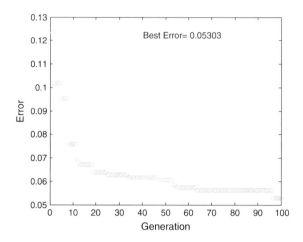

Fig. 10 Convergence of evolution #19 (case #2)

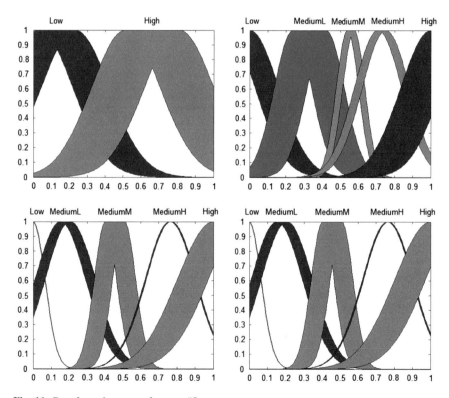

Fig. 11 Best fuzzy integrator for case #5

Fig. 12 Convergence of evolution #17 (case #5)

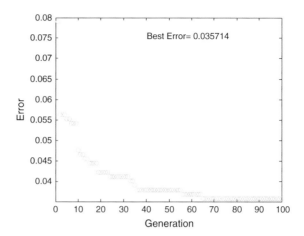

In Fig. 11, the best fuzzy integrator for the case #5 is shown; it is a fuzzy integrator of Sugeno type and with Gaussian Membership Functions, and in Fig. 12, the convergence of that evolution is shown.

5   Conclusions

In this paper, a new HGA was presented; the main idea of this HGA is to perform the optimization of fuzzy systems. Some parameters were optimized such as type of FL, type of system, number of membership functions in each inputs, type of membership functions in each variable, their parameters, and the consequences of the fuzzy rules. The optimization provides better results than when the fuzzy integrators are not optimized. When the HGA was used, in almost all the cases when type-2 was chosen by the HGA, better results were obtained. In this work, the objective function was the minimization of the error of recognition. In the future, the optimization of the number of membership function in the output will be implemented to obtain better results.

References

1. Abiyev, R., Altunkaya, K.: Personal Iris Recognition Using Neural Network. Near East University, Department of Computer Engineering, Lefkosa, North Cyprus (2008)
2. Moreno, B., Sanchez, A., Velez, J.F.: On the use of outer ear images for personal identification in security applications. In: IEEE 33rd Annual International Carnahan Conference on Security Technology, pp. 469–476 (1999)
3. Zhang, Z., Zhang, C.: An Agent-Based Hybrid Intelligent System for Financial Investment Planning. PRICAI, pp. 355–364 (2002)

4. Melin, P., Mendoza, O., Castillo, O.: An improved method for edge detection based on interval type-2 fuzzy logic. Expert Syst. Appl. **37**(12), 8527–8535 (2010)
5. Melin, P., Mendoza, O., Castillo, O.: Face recognition with an improved interval type-2 fuzzy logic sugeno integral and modular neural networks. IEEE Trans. Syst. Man Cybern. Part A **41** (5), 1001–1012 (2011)
6. Melin, P., Sánchez, D., Castillo, O.: Genetic optimization of modular neural networks with fuzzy response integration for human recognition. Inf. Sci. **197**, 1–19 (2012)
7. Hidalgo, D., Castillo, O., Melin, P.: Type-1 and type-2 fuzzy inference systems as integration methods in modular neural networks for multimodal biometry and its optimization with genetic algorithms. Inf. Sci. **179**(13), 2123–2145 (2009)
8. Muñoz, R., Castillo, O., Melin, P.: Face, fingerprint and voice recognition with modular neural networks and fuzzy integration. In: Bio-inspired Hybrid Intelligent Systems for Image Analysis and Pattern Recognition, pp. 69–79 (2009)
9. Azamm, F.: Biologically inspired modular neural networks, Ph.D. thesis, Virginia Polytechnic Institute and State University, Blacksburg, Virginia (2000)
10. Khan, A., Bandopadhyaya, T., Sharma, S.: Classification of stocks using self organizing map. Int. J. Soft Comput. Appl. **4**, 19–24 (2009)
11. Santos, J.M., Alexandre, L.A., Marques, de Sá J.: Modular neural network task decomposition via entropic clustering. ISDA (1), 62–67 (2006)
12. Auda, G., Kamel, M.S.: Modular neural networks a survey. Int. J. Neural Syst. **9**(2), 129–151 (1999)
13. Vázquez, J.C., Lopez, M., Melin, P.: Real time face identification using a neural network approach. In: Soft Computing for Recognition Based on Biometrics, pp. 155–169 (2010)
14. Zadeh, L.A.: Fuzzy Sets. J. Inf. Control **8**, 338–353 (1965)
15. Melin, P., Castillo, O.: Hybrid Intelligent Systems for Pattern Recognition Using Soft Computing: An Evolutionary Approach for Neural Networks and Fuzzy Systems, 1st edn, pp. 119–122. Springer, Berlin (2005)
16. Wang, W., Bridges, S.: Genetic algorithm optimization of membership functions for mining fuzzy association rules. Department of Computer Science Mississippi State University, 2 Mar 2000
17. Raikova, R.T., Aladjov, HTs: Hierarchical genetic algorithm versus static optimization investigation of elbow flexion and extension movements. J. Biomech. **35**, 1123–1135 (2002)
18. Haupt, R., Haupt, S.: Practical Genetic Algorithms, 2 edn, pp. 42–43. Wiley-Interscience, New York (2004)
19. Mitchell, M.: An Introduction to Genetic Algorithms, 3rd edn. A Bradford Book (1998)
20. Coley, A.: An Introduction to Genetic Algorithms for Scientists and Engineers, Wspc, Har/Dskt edn (1999)
21. Huang, J., Wechsler, H.: Eye location using genetic algorithm. In: Second International Conferenceon Audio and Video-Based Biometric Person Authentication, pp. 130–135 (1999)
22. Nawa, N., Takeshi, F., Hashiyama, T., Uchikawa, Y.: A study on the discovery of relevant fuzzy rules using pseudobacterial genetic algorithm. IEEE Trans. Ind. Electron. **46**(6), 1080–1089 (1999)
23. Tang, K.S., Man, K.F., Kwong, S., Liu, Z.F.: Minimal fuzzy memberships and rule using hierarchical genetic algorithms. IEEE Trans. Ind. Electron. **45**(1), 162–169 (1998)
24. Wang, C., Soh, Y.C., Wang, H., Wang, H.: A hierarchical genetic algorithm for path planning in a static environment with obstacles. In: Canadian Conference on Electrical and Computer Engineering IEEE CCECE 2002, vol. 3, pp. 1652–1657 (2002)
25. Worapradya, K., Pratishthananda, S.: Fuzzy supervisory PI controller using hierarchical genetic algorithms. In: Control Conference, 2004. 5th Asian, vol. 3, pp. 1523–1528 (2004)
26. Database of Face. Institute of Automation of Chinese Academy of Sciences (CASIA). Found on the web page: http://biometrics.idealtest.org/dbDetailForUser.do?id=9 (Accessed 11 Nov 2012)

27. Database of Human Iris. Institute of Automation of Chinese Academy of Sciences (CASIA). Found on the web page: http://www.cbsr.ia.ac.cn/english/IrisDatabase.asp (Accessed 21 Sep 2009)
28. Database Ear Recognition Laboratory from the University of Science and Technology Beijing (USTB). Found on the web page: http://www.ustb.edu.cn/resb/en/index.htm asp (Accessed 21 Sep 2009)

Optimization by Cuckoo Search of Interval Type-2 Fuzzy Logic Systems for Edge Detection

C.I. Gonzalez, Juan R. Castro, Olivia Mendoza, Patricia Melin and Oscar Castillo

Abstract This paper presents the optimization of the antecedent parameters for a system of edge detection based on Sobel technique combined with interval type-2 fuzzy logic. For the optimization of the fuzzy inference system, the cuckoo search (CS) algorithm is applied, the idea is to find the design parameters of an IT2-FLS and achieve better results in applications of edge detection for digital images.

1 Introduction

One of the main problems that we have to consider with the fuzzy inference system is how to design a model, such as the number of inputs and outputs, number of membership functions, how to parameterize these membership functions and the number of ideal fuzzy rules to get a good result. Most of the investigations present their design of fuzzy inference system obtained under experimentation, which reaches an acceptable result after a series of test. On the other hand, some investigations are conducted using bio-inspired algorithms to find optimal solutions for fuzzy logic systems design [1, 2]. There are different bio-inspired algorithms used to solve optimization problems such as genetic algorithms (GAs) [3], ant colony optimization (ACO) [4], bee algorithms (BA), particle swarm optimization

C.I. Gonzalez (✉) · J.R. Castro · O. Mendoza
School of Engineering, University of Baja California, Tijuana, Mexico
e-mail: Claudia.gonzalez18@uabc.edu.mx

J.R. Castro
e-mail: jrcastror@uabc.edu.mx

O. Mendoza
e-mail: omendoza@uabc.edu.mx

P. Melin · O. Castillo
Division of Graduate Studies, Tijuana Institute of Technology, Tijuana, Mexico
e-mail: pmelin@tectijuana.mx

O. Castillo
e-mail: ocastillo@tectijuana.mx

© Springer International Publishing Switzerland 2016
L.A. Zadeh et al. (eds.), *Recent Developments and New Direction in Soft-Computing Foundations and Applications*, Studies in Fuzziness and Soft Computing 342, DOI 10.1007/978-3-319-32229-2_11

(PSO) [5–7], the firefly algorithms (FAs) [8, 9], bat-inspired algorithms (BATs) [10], and cuckoo search (CS) algorithms [11–14]. The main contribution of this paper is to optimize a fuzzy logic system for edge detection based on Sobel technique and interval type-2 fuzzy logic system (IT2-FLS), applying the CS algorithm [11–14]. The reason for using the CS algorithm is the fact that there are fewer parameters to be fine-tuned in CS than in other methods such as PSO and GA algorithms [3, 5].

In Mendoza and Melin [15], an edge detection method based on Sobel technique combined with interval IT2-FLS is proposed; the parameter of design (antecedents, consequents and rules) of the IT2-FLS were performance under experimentation. This paper is the principal reference for our proposal; the idea is to find the design of antecedent parameters for an IT2-FLS implementing CS algorithm and achieve better results than the non-optimized IT2-FLS in applications of edge detection for digital images

In order to evaluate objectively the performance of both edge detectors, for the simulation results we used the same synthetic images, these images are presented in Sect. 6. The quality of edge detection is evaluated with the figure of merit (FOM) of Pratt, described in Sect. 4.

The rest of the paper is organized as follows. The basic concepts of the CS bio-inspired algorithm are defined in Sect. 2. In Sect. 3, the Sobel operator is described. The model proposed to achieve the optimization of IT2-FSL by CS is presented in Sect. 5. In Sect. 6, the simulation results are shown, and finally, conclusions in Sect. 7.

2 Overview of Cuckoo Search

2.1 Cuckoo Behavior

The cuckoo birds are a fascinating species and of great interest for its study, because they have an aggressive reproduction strategy. Some parasitic cuckoos lay their eggs in communal nests, so they can be breeding by other species, but some host birds can engage direct conflict with the intruding cuckoos. If host birds discover the eggs are not their own, they will either throw these alien eggs away or simply abandon the nest and build a new nest; therefore, this reduces the probability of their reproductivity.

Some female cuckoo species such as the Tapera are specialists in mimicking the color and pattern of the eggs of other species; this strategy helps to reduce the probability that their eggs can be abandoned and thus increases their reproductivity [12–14, 16].

Another strategy of the parasitic cuckoos is that they choose a nest where the host nests just laid their own eggs. Most of the time, the cuckoo eggs are hatched earlier than the host bird eggs. The first instinct of the cuckoo chicks is to evict the host bird's eggs propelling them blindly out of the nest, which increases the possibility of being fed by the bird host [17–19].

In Yang and Deb [13], cuckoo breeding behavior was the inspiration to develop bio-inspired algorithm described in Sect. 2.2.

2.2 Cuckoo Search (CS) Algorithm

CS idealized such breeding behavior and thus can be applied for various optimization problems. It seems that it can outperform other metaheuristic algorithms in some applications [13].

For simplicity, CS uses the following three idealized rules:

1. Each cuckoo lays one egg at a time and dumps its egg in a randomly chosen nest;
2. The best nests with high-quality eggs will carry over to the next generations;
3. The number of available host nests is fixed, and the egg laid by a cuckoo is discovered by the host bird with a probability $p_a \in [0, 1]$. In this case, the host bird can either throw the egg away or abandon the nest and build a completely new nest.

Based on these three rules, the basic steps of the CS can be summarized as the pseudocode shown in Fig. 1 [13].

When generating new solutions $x(t+1)$ for, say, a cuckoo i, a Lévy flight is performed by (1)

$$x_i^{(t+1)} = x_i^{(t)} + \alpha \oplus \text{Lévy}(\lambda) \qquad (1)$$

where $\alpha > 0$ is the step size which should be related to the scales of the problem of interests. In most cases, we can use $\alpha = 1$ [12, 20–22].

Fig. 1 Pseudocode of the cuckoo search (CS) via Lévy flight

Objective function: f(x), x= (x₁,..., x_d):
Generate initial population of n host nests x_i (i = 1,2,...,n)
While *(t < MaxGeneration) or (stop criterion)*
  Get a cuckoo randomly by performing Lévy flights;
    Evaluate its quality/fitness F_i
  Choose a nest among n (say, j) randomly;
if (F_i>F_j),
      Replace j by the new solution;
end if
  A fraction (Pa) of the worse nests
are abandoned and new ones are built;
  Keep the best solutions/nests;
  Rank the solutions/nests and find the current best;
end while
Postprocess results and visualization;
End

Some advantage of using CS algorithm is that this has few parameters to be fine-tuned, basically these parameters are the population size or number of nest (n), iterations number (t), and the parameter p_a. On the other hand, some research also indicate that the convergence rate is insensitive to the parameter p_a; therefore, we do not have to fine-tuned these parameters for a specific problem [13, 18].

3   Sobel Operator

A number of edge detectors based on a single derivative have been developed by various researchers. Among them most important operators are the Robert operator, Sobel Operator, Prewitt operator, Canny operator and Krisch operator [23–26]. In each of these operator-based edge detection strategies, we compute the gradient magnitude. If the magnitude of the gradient is higher than a threshold, then we detected the presence of an edge.

The classic Sobel operator is a 3×3 neighborhood-based gradient operator. The convolution masks for the Sobel operator on a gray-scale digital image are defined by Sobel_x (2) and Sobel_y (3).

$$\text{Sobel}_x = \begin{bmatrix} -1 & -2 & -1 \\ 0 & 0 & 0 \\ 1 & 2 & 1 \end{bmatrix} \tag{2}$$

$$\text{Sobel}_y = \begin{bmatrix} -1 & 0 & 1 \\ -2 & 0 & 2 \\ -1 & 0 & 1 \end{bmatrix} \tag{3}$$

The two masks are separately applied on the input image to yield two gradient components g_x (4) and g_y (5) in the horizontal and vertical orientations, respectively; the positions of the input image (f) are shown in Fig. 2, where x-axis represents the horizontal positions and y-axis vertical positions.

Fig. 2 Positions of the input image (f)

(x-1, y-1)	(x-1, y)	(x-1, y+1)
(x, y-1)	(x, y)	(x, y+1)
(x+1, y-1)	(x+1, y)	(x+1, y+1)

$$g_x = [f(x+1, y-1) + 2f(x+1, y) + f(x+1, y+1)]$$
$$- [f(x-1, y-1) + 2f(x-1, y) + f(x-1, y+1)]$$
(4)

$$g_y = [f(x-1, y+1) + 2f(x, y+1) + f(x+1, y+1)]$$
$$- [f(x-1, y-1) + 2f(x, y-1) + f(x+1, j-1)]$$
(5)

The gradient magnitude is obtained with (6).

$$G[f(x, y)] = \sqrt{g_x^2 + g_y^2}$$
(6)

where g_x (4) is the result of the convolution of the input image (f) with the filter Sobel$_x$ (2), and g_y (5) is the result of the convolution of the input image (f) with the filter Sobel$_y$ (3) [15, 23, 27].

The result of an edge image based on the Sobel operator (6) is shown in Fig. 3.

Fig. 3 Edge detector based on the Sobel operator

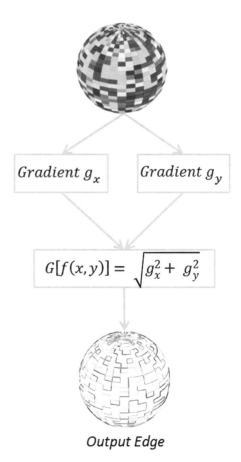

Fig. 4 a Sphere image.
b Reference image or ideal
edge

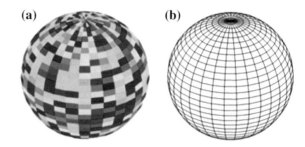

4 Metrics

There are different types of methods to evaluate the detected edge of an image that use different parameters for assessing the abrupt change of color in the pixels. One of the most frequently used techniques is the FOM of Pratt [28, 29]. It represents the deviation of the actual (calculated) edge point from the ideal edge and it is defined in (7).

$$\text{FOM} = \frac{1}{\max(I_I, I_A)} \sum_{i=1}^{I_A} \frac{1}{1 + \propto d_i^2} \tag{7}$$

where I_A is the actual number of detected edge points, I_I is the number of edge points on the ideal edge, $d(i)$ is the distance between the edge of the current pixel and its correct position in the reference image and α is scaling constant (usually 1/9) [30]; i.e., the reference image (I_I) or ideal edge of Fig. 4a is Fig. 4b. If the result of the FOM (7) is 1 or very close to 1, it means that the detected edge I_A is the same or very similar to the ideal edge (I_I). Otherwise, the closer we are to 0, then there is a high difference between the edge detected (I_A) and ideal edge (I_I).

5 Cuckoo Search for Parameter Optimization of the Interval Type-2 Fuzzy Logic System

In this section, the methodology for the optimization of an interval IT2-FLS [30, 31] for edge detection is described. The idea is to find the parameters of the membership functions of the antecedents. For the IT2-FLS, two inputs are required which are the gradients with respect to x-axis and y-axis, calculated with (4) and (5), for this case study we call DH and DV, respectively. For all the fuzzy variables, inputs and output, the membership functions are Gaussian membership functions with uncertain mean. In the input membership functions, we used three linguistic variables: LOW, MIDDLE, HIGH and for the output (EDGES) it was represented by the linguistic variables: BACKGROUND and EDGES. In Fig. 5, the general structure of the system to be optimized is shown.

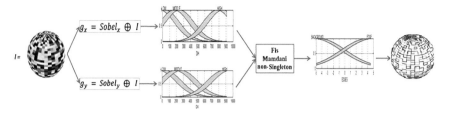

Fig. 5 Interval type-2 fuzzy inference system for edge detection

5.1 Optimization of Antecedent Parameters by CS

The optimization of antecedent parameters was based on the IT2-FLS, i.e., shown in Fig. 5.

The steps for the development of this optimization problem applying the CS algorithm are described as follows:

1. *Objective function*: $f(x), x = (x_1, \ldots, x_d)$: the objective function is obtained with (8, 9);

$$\text{FOM} = \frac{1}{\max(I_I, I_A)} \sum_{i=1}^{I_A} \frac{1}{1 + \propto d_i^2} \tag{8}$$

$$\text{Obj} = 1 - \text{FOM} \tag{9}$$

where the FOM represents the metric that is obtained after evaluating the detected edge in the IT2-FLS.

2. *Generate initial population of n host* nests $x_i (i = 1, 2, \ldots, n)$: the number of solutions of each nest is defined by the number of parameters to optimize; for this case study, we have two inputs with three Gaussian membership functions and to generate each membership function three parameters are needed (σ, m_1, m_2); therefore, we have 18 solutions that represent the parameters of input membership functions. In Table 1, the parameters to optimize are presented.

3. **While** *(t < MaxGeneration) or (stop criterion)*
   Get a cuckoo randomly by performing Lévy flights;
   Evaluate its quality/fitness **F_i**: *the Objective function is obtained with* (8, 9);
   Choose a nest among n (say, j) randomly;
   if ($F_i > F_j$),

Table 1 Parameters to optimize using cuckoo search

Input 1									Input 2								
LOW			MIDDLE			HIGH			LOW			MIDDLE			HIGH		
σ	m_1	m_2	σ	m_1	m_2	σ	m_1	m_2	σ	m_1	m_2	σ	m_1	m_2	σ	m_1	m_2

Replace j by the new solution;

end if
*A fraction (**Pa**) of the worse nests are abandoned and new ones are built;*
Keep the best solutions/nests;
Rank the solutions/nests and find the current best;

end while
4. *Postprocess results and visualization;*

End

6   Simulations Results

In order to evaluate objectively the performance of the proposed edge detector, for the simulation results we perform all the experiments using synthetic images; these synthetic images were built with the plot of three mathematical functions as the original image and as ground truth reference for ideal edges, in Table 2 synthetic images and reference images are shown. The reference image is necessary to evaluate the quality of the edge detection defined by (7); on the other hand, this is important to evaluate the objective function of the CS algorithm used in this paper.

Table 2 Synthetic images for simulation results

Name of figure	Synthetic image	Reference image
Sphere		
Peaks		
Doughnut		

Table 3 Simulation results using Sobel operator	Synthetic image	FOM	Error
	Sphere	0.7494	0.2506
	Peaks	0.7643	0.2357
	Doughnut	0.7876	0.2124

6.1 Simulation Results with Sobel Operator Edge Detector

In the first test, the edge detector was performance with the traditional Sobel operator, without using a fuzzy logic system, as was presented in Fig. 3. The FOM was obtained with (7); the ideal FOM for any image is 1 or an ERROR of 0 that means that all edges have been detected correctly. The results for the synthetic images of Table 2, as shown in Table 3, were an ERROR of 0.2506 was obtained with the image sphere, 0.2357 for a peaks image and 0.2124 for the doughnut image; remember that the ideal ERROR is 0; therefore, this experiment is a little far from the ideal ERROR or ideal edge.

6.2 Simulation Results with Edge Detection Based on Sobel Technique and IT2-FLS

In the second test, the edge detection based on Sobel technique and interval IT2-FLS was applied. For the IT2-FLS, the obtained gradients with (4) and (5) are used as inputs and the fuzzy system was built using the membership functions presented in Fig. 5; in this test the antecedents, consequents, and fuzzy rules were obtained based on the experimentation; where we consider the work presented in Mendoza and Melin [15]. The fuzzy rules used for this test are described in Table 4. In Table 5, we can find the values of FOM obtained for the sphere, peaks, and doughnut synthetic images, and their ERROR, respectively.

Table 4 Fuzzy rules	
	1. If (DH is LOW) and (DV is LOW), then (E is BACKGROUND)
	2. If (DH is MIDDLE) or (DV is MIDDLE), then (E is EDGE)
	3. If (DH is HIGH) or (DV is HIGH), then (E is EDGE)

Table 5 Simulation results using interval type-2 fuzzy systems	Synthetic image	FOM	Error
	Sphere	0.9599	0.0401
	Peaks	0.9645	0.0355
	Doughnut	0.9535	0.0465

Table 6 Antecedents optimization using cuckoo search

Interval type-2 fuzzy logic system		
Synthetic image	FOM	Error
Sphere	0.9619	0.0381
Peaks	0.9663	0.0337
Doughnut	0.9667	0.0333

6.3 Simulation Results for Optimizations of an Interval Type-2 Fuzzy Logic System by CS

In another test, the parameters of the antecedents of the IT2-FLS were optimized by CS algorithm, the simulation results of these are shown in Table 6. For these simulations, we used the same parameters of design: consequents and fuzzy rules presented in Sect. 6.2. For the development of this test, ten experiments with the CS algorithm were executed, for each image presented in Table 2.

In order to achieved better performance in the experiments, the principal parameters of CS algorithm were varied, such as the number of nests (Nest), probability of nest loss (Pa), and the number of iterations (*I*). For the number of nests, we used values between 15 and 30, but the best results were obtained with 25 nests. The parameter Pa varied from 20 to 35 and the iterations used values between 5 and 100 achieve good results with 10 iterations. In summary, the values that we proposed for this case of study are Nest = 25, Pa = 30, and $I = 10$.

In Table 6, the average of the ten simulations obtained for each image of Table 2 are presented, we show the values of FOM and ERROR obtained for the three images.

Based on test performance in Sects. 6.1, 6.2, and 6.3, a comparative analysis was made. In Table 7, the average of the FOM and ERROR of the sphere, peaks, and doughnut images obtained in the results of Tables 4, 5, and 6 is presented.

In the results of Table 7, we can note that the FOM and ERROR values were better using the optimization of IT2-FLS by CS, with a FOM of 0.9650 and an ERROR equal to 0.0350, and this improved the results achieved by the non-optimized IT2-FLS whose FOM was 0.9593 and ERROR 0.0407; in contrast, the results obtained by the traditional Sobel edge detector were lower than both fuzzy detectors.

Table 7 FOM and error of Sobel operator, non-optimized Sobel + IT2-FLS, and Sobel + IT2-FLS optimized using CS

Edge detector	Algorithm	FOM	Error
Sobel		0.7671	0.2329
Sobel + IT2-FLS		0.9593	0.0407
Sobel + IT2-FLS	CS	**0.9650**	**0.0350**

Table 8 Edge detection by Sobel operator, non-optimized Sobel + IT2-FLS, and Sobel + IT2-FLS optimized using CS

Sobel operator	Sobel + IT2-FLS	Sobel + IT2-FLS (CS)

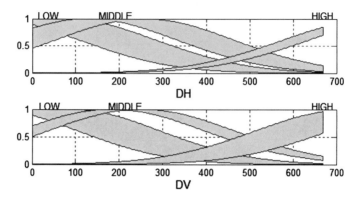

Fig. 6 IT2-FLS membership functions for the sphere image

The visual edge detection for the Sobel, Sobel + IT2-FLS, and optimized Sobel + IT2-FLS are shown in Table 8; graphically, we cannot see the difference in the detected edges; for that reason, the metric (7) described in Sect. 4 was applied.

An example of the input membership functions DH and DV of the IT2-FLS after the optimization by CS is shown in Fig. 6. This membership functions corresponding to the sphere image.

7 Conclusions

In summary, in this work several simulations results were presented, the objective was to obtain the design parameters of the antecedents of an interval IT2-FLS. For the optimization, the bio-inspired algorithms CS were applied.

In the results of Table 7 when the CS algorithm is applied to optimize the design parameters of the antecedents of an IT2-FLS, better results are obtained, in contrast to the non-optimized IT2-FLS. On the other hand, the results obtained with the non-optimized IT2-FLS were better than the traditional Sobel operator.

Acknowledgments We thank the MyDCI program of UABC University, the Division of Graduate Studies and Research of Tijuana Institute of Technology and the financial support provided by CONACYT Contract Grant Number: 44524.

References

1. Arora, J.: Introduction to Optimum Design. McGraw-Hill, New York (1989)
2. Talbi, E.G.: Metaheuristics: From Design to Implementation. Wiley, New York (2009)
3. Man, K.F., Tang, K.S., Kwong, S.: Genetic Algorithms: Concepts and Designs. Springer, Berlin (1999)

4. Dorigo, M., Birattari, M., Stutzle, T.: Ant colony optimization. Comput. Intell. **1**(4), 28–39 (2006)
5. Kennedy, J., Eberhart, R.: Particle swarm optimization. IEEE Int. Conf. Neural Networks **4**, 1942–1948 (1995)
6. Clerc, M., Kennedy, J.: The particle swarm—explosion, stability, and convergence in a multidimensional complex space. Evol. Comput. **6**(1), 58–73 (2002)
7. James, K., Eberhart, R.C.: Swarm Intelligence. Kaufmann, San Francisco (2001)
8. Yang, X.S.: Firefly algorithms for multimodal optimization. In: Stochastic Algorithms: Foundations and Applications. Lecture Notes Computer Science, vol. 5792, pp. 169–178 (2009)
9. Lukasik, S., Zak, S.: Firefly algorithm for continuos constrained optimization tasks. Lect. Notes Artif. Intell. **5796**, 97–106 (2007)
10. Yang, X.: A new metaheuristic bat-inspired algorithm. Stud. Comput. Intell. **284**, 65–74 (2010)
11. Zhou, Y., Zheng, H.: A novel complex valued cuckoo search algorithm. Sci. World J. **2013**(1), 597–803 (2013)
12. Yang, X., Press, L.: Nature-Inspired Metaheuristic Algorithms, 2nd edn. (2010)
13. Yang, X., Deb, S.: Cuckoo search via Lévy flights. In: World Congress on Nature and Biologically Inspired Computing, pp. 210–214 (2009)
14. Yang, X.-S., Deb, S., Karamanoglu, M., He, X.: Cuckoo search for business optimization applications. In: 2012 National Conference on Computing and Communication Systems, pp. 1–5 (2012)
15. Mendoza, O., Melin, P.: Quantitative evaluation of fuzzy edge detectors applied to neural networks for image recognition. Advances in research and developments in digital systems, pp. 324–335. In: Stochastic Algorithms: Foundations and Applications. Lecture Notes Computer Science, vol. 5792, pp. 169–178 (2011)
16. Payne, R.B., Sorenson, M.D., Klitz, K.: The cuckoos. Oxford University Press, Oxford (2005)
17. Shlesinger, M.F.: Search research. Nature **443**, 281–282 (2006)
18. Yang, X., Deb, S.: Engineering optimisation by cuckoo search. Math. Model. Numer. Optim. **1**(4), 330–343 (2010)
19. Gandomi, A.H., Yang, X.-S., Alavi, A.H.: Cuckoo search algorithm: a metaheuristic approach to solve structural optimization problems. Eng. Comput. **29**(1), 17–35 (2011)
20. Brown, C., Liebovitch, L.S., Glendon, R.: Lévy flights in Dobe Ju/'hoansi foraging patterns. Hum. Ecol. **35**, 129–138 (2007)
21. Pavlyukevich, I., Flights, L.: Non-local search and simulated annealing. Comput. Phys. **226**, 1830–1844 (2007)
22. Shlesinger, M.F., Zaslavsky, G.M., Frisch, U.: Lévy flights and related topics in physics. Springer, Berlin (1995)
23. Sobel, I.: Camera models and perception. Ph.D. thesis, Stanford University, Stanford, CA (1970)
24. Canny, J.: A computational approach to edge detection. IEEE Trans. Pattern Anal. Mach. Intell. **8**(2), 679–698 (1986)
25. Prewitt, J.M.S.: Object enhancement and extraction. In: Lipkin, B.S., Rosenfeld, A. (eds.) Picture analysis and psychopictorics, pp. 75–149. Academic Press, New York (1970)
26. Kirsch, R.: Computer determination of the constituent structure of biological images. Comput. Biomed. Res. **4**, 315–328 (1971)
27. Mendoza, O., Melin, P., Licea, G.: A hybrid approach for image recognition combining type-2 fuzzy logic, modular neural networks and the Sugeno integral. Inf. Sci. (Ny) **179**(13), 2078–2101 (2009)
28. Perez-Ornelas, F., Mendoza, O., Melin, P., Castro, J.R.: Interval type-2 fuzzy logic for image edge detection quality evaluation. In: 2012 Annual Meeting of the North American Fuzzy Information Processing Society (NAFIPS), no. 1, pp. 1–6 (2012)

29. Abdou, I., Pratt, W.: Quantitative design and evaluation of enhancement/thresholding edge detectors. Proc. IEEE **67**(5), 753–763 (1979)
30. Castillo, O., Melin, P.: Type-2 fuzzy logic theory and applications. Springer, Berlin (2008)
31. Castro, J.R., Castillo, O., Melin, P.: An Interval Type-2 Fuzzy Logic Toolbox for Control Applications, pp. 1–6 (2007)

Part V
Classification and Clustering

Comparing the Properties of Meta-heuristic Optimization Techniques with Various Parameters on a Fuzzy Rule-Based Classifier

A. Tormási and L.T. Kóczy

Abstract In this paper, the results of meta-heuristic optimization techniques with various parameter settings are presented. A formerly published Fuzzy-Based Recognizer (FUBAR): A fuzzy rule-based classification algorithm was used to analyze and evaluate the behavior of the used meta-heuristic optimization algorithms for rule-base optimization. Besides the reached accuracy, the execution time, the CPU load of the algorithms, and the effects of the shapes of the fuzzy membership functions in the initial rule-base are also investigated.

1 Introduction

The classification problems and related areas are well researched, with well-developed methodology as much in theory as in applications; however, some special problems cannot be solved with traditional methods or their evaluation requires unacceptable amount of resources. In such situations, computational intelligence techniques such as artificial neural networks [1], fuzzy systems [2–7], and meta-heuristic algorithms [8–12] may provide an alternative solution or remarkable improvement with acceptable compromises and better resource consumption [13–20].

Gesture or single-stroke ("unistroke") and multi-stroke character and handwriting recognition are a special and very complex subarea of classification problems, in which the input—based on imprecise data with high amount of noise—can be grouped into a finite number of classes ("alphabet").

A. Tormási (✉) · L.T. Kóczy
Department of Automation, Széchenyi István University, Győr, Hungary
e-mail: tormasi@sze.hu

L.T. Kóczy
e-mail: koczy@sze.hu; koczy@tmit.bme.hu

L.T. Kóczy
Department of Telecommunication and Mediainformatics, Budapest University
of Technology and Economics, Budapest, Hungary

© Springer International Publishing Switzerland 2016
L.A. Zadeh et al. (eds.), *Recent Developments and New Direction
in Soft-Computing Foundations and Applications*, Studies in Fuzziness
and Soft Computing 342, DOI 10.1007/978-3-319-32229-2_12

157

These kinds of problems cannot be solved with closed mathematical formulas; the most general methods are usually resource consuming as a result of robustness and complex geometrical transformations such as trapezoid correction and rotation. The ideal recognition algorithm should work on portable devices with an acceptable recognition rate [21]. Most of the commercial products could not satisfy the users' requirements in accuracy and response time.

One of the first successful recognizers was the Apple Newton's handwriting recognizer [22], which uses an artificial neural network; the other great system was the Palm's Graffiti [23], and it uses a single-stroke alphabet, which reduces the complexity of the problem and gives the possibility to deal easily with the segmentation with their devices input interface. The solution requires from the user to learn the simplified symbol set. The method could reach 91 % accuracy according to the results in [24]. The Graffiti was a great recognition system in its time with acceptable compromises and special advantages. The second version of the Palm's recognizer, the Graffiti 2, which was a slightly improved version of the original system supporting multi-stroke letters, could reach only 86.03 % accuracy as presented in [25].

Some new recognition methods such as $1 [26], $N [27], and $P [28] reached impressive results in accuracy and resource requirements as well—compared to other methods found in literature.

Fuzzy-Based Recognizer (FUBAR) is a family of single-stroke [19, 29–34] and multi-stroke [20] recognition algorithms. The method uses Takagi–Sugeno inference method [4] to recognize the symbols and stores the knowledge in fuzzy rule-bases. A better accuracy requires determining a quasi-optimal initial rule-base for the system.

The results of meta-heuristic optimization techniques applied on the rule-base of multi-stroke FUBAR showed that these methods can significantly increase the recognition rate of the system [35]. These results motivated a deeper investigation of the properties of these meta-heuristic methods presented in this paper.

The basic concepts of the used meta-heuristic algorithms are summarized after the introduction, in Sect. 2. A brief conceptual summary of the FUBAR algorithms is in Sect. 3. The details of the test environment and the settings are in Sect. 4. The results are presented in Sect. 5; conclusions and future directions are in Sect. 6.

2 Meta-heuristic Optimization Techniques Used

The concept of each meta-heuristic optimization algorithm included in the experiment is summarized in the following subsections. The knowledge in stored in the membership functions of the rule antecedents; the accuracy of the system could be increased with the slight modification of these membership functions. The algorithms were modified to use multiple populations as a result of the characteristics of the rule-base optimization problem.

2.1 Bacterial Evolutionary Algorithm

The bacterial evolutionary algorithm (BEA) [9] is inspired by the evolutional processes of bacteria. Each bacterium in the population represents a solution in the problem space.

The first step of the algorithm is the bacterial mutation. Each bacterium is selected individually and cloned a maximal number of times. Each randomly selected allele of the clones is modified randomly, and the modified allele of the alternative bacterium (or the original one's) with the best result is copied to all other clones; this step is repeated until all the alleles were not selected.

The second step of the algorithm is the gene transfer (infection), in which the population is sorted by the goodness of the bacteria and divided into two subsets; the set of good and the set of bad bacteria. A randomly selected allele of a bacterium selected randomly from the group of good bacteria is copied to a randomly selected bad bacterium. This step is repeated until the algorithm reaches the maximum number of infections.

The steps of the algorithm are repeated until it reaches the maximum number of generations or satisfies other termination conditions (like 100 % result).

2.2 Imperialist Competitive Algorithm

The imperialist competitive algorithm (ICA) [10] uses the analogy of politics and strategy instead of biological systems. The countries are representing the solutions.

The countries can be categorized into the group of conquerors and the group of colonies. The conquerors are the ones with higher strength (aggregated fitness of the empires and colonies) and the colonies are the ones with low results. The strength of the conqueror countries could be increased by the assimilation of colonies.

In the revolution step, the colonies may rise up against the other countries.

2.3 Particle Swarm Optimization Algorithm

The particle swarm optimization (PSO) [11] uses the simplified model of the dynamics of movements of various animal swarms (or particles).

The solutions in the search domain are represented by the particles; each particle has a position and a speed vector. The evolution of the population does not use evolutionary operators unlike in genetic algorithms.

The orientation and the speed of each particle are influenced by all other particles. An individual particle moves toward the particle with the best local or global solution and influenced by its personal best position.

2.4 Big Bang Big Crunch Algorithm

The big bang big crunch algorithm (BBBC) [12] models the concept of the big bang big crunch theory from physics. The population is determined randomly. The solutions in the space are called points; the fitness functions are calculated individually. The position of the "black hole" represents the center of gravity for the population; it is determined by the coordinates of points weighted with their fitness.

The position of the best point in the population and the position of the population's "black hole" are used to determine the center of the search domain in the next generation. These steps are repeated until the maximum number of generations (big bang big crunch cycles) or other termination condition is not satisfied.

This algorithm has special properties compared to the previously described methods; the dimensions of the search domain are handled separately, which means the execution time increases linearly by the increase of the dimensions and narrows the search domain remarkably fast. It is also important to highlight a great disadvantage of the algorithm: It easily converges to local optimum solutions.

3 Properties of the Fuzzy Rule-Based Classifier Used

The system used during the evaluation is a multi-stroke FUBAR algorithm [20]. The algorithm is able to classify 26 different letters. The concept is detailed and properties are investigated of various FUBAR algorithms were investigated in formerly published works [19, 29–35].

It uses a list of x and y coordinates in chronological order to describe input strokes. The features extracted from the input stroke are the width/height ratio of the stroke and the average number of fuzzified points in the rows and columns of a fuzzy grid [29] (a grid in which the boundaries of columns and rows are defined by a fuzzy set) drawn around the stroke.

The knowledge is stored in the rule-base of the system. The properties (features) of the input strokes are described with fuzzy sets in the antecedents of the fuzzy rules; the outputs of the rules are representing the degree of matching between the input stroke and the one stored in the rule. The FUBAR method uses a Takagi–Sugeno method for the inference. The algorithm returns with the letter assigned to the rule with the highest matching value.

4 Test Environment and Parameter Settings

The parameters describing the candidate solutions (genes in the case of BEA) are the rule antecedents with the break points of the trapezoidal membership functions. The fitness functions of the algorithms—representing the goodness of a solution—were to reach the maximal average recognition rate for 26 different symbols with 60 (training) samples per symbol.

The methods were used to determine only the initial rule-base, which means that it does not contain any details about the user-specific writing styles.

The size of the population was 10 in each algorithm, and the maximum number of cycles (generations) was 50 to 100—each time increased by 10. The evaluation of the fitness values of solutions in a given population was processed on 24 threads in parallel.

Three different kind of initial rule-bases were used for the tests: (1) In the first case, each membership function starts as a triangle, with the height of 1 and the triangular membership functions could be modified to trapezoidal ones during the optimization, (2) in the second kind, the membership functions had a value of 1 in a single point and 0 in all other, which could be also extended to trapezoid membership functions, and (3) in the third case, none of the points of the trapezoidal membership functions was given.

Two kinds of results are presented for each algorithm. The first one is based on the 60 (training) samples per letter, which is used to determine the (quasi)optimal representation of the fuzzy sets describing letter features. These results are able to show how many generations were required to reach the given accuracy. It was shown in [35] that it is possible to reach 100 % recognition rate for the same training sets; by knowing these values, it is also possible to determine how successful the training was (higher results are the better). The second kinds of results are based on 120 (validation) samples per letters (provided by test subjects) evaluated with the FUBAR algorithm using the optimized initial rule-base. The validation test shows the general (not user-specific) accuracy of the system, since the samples were collected from distinct users. These results can be compared to the results for the validation samples; it shows that how general is the knowledge extracted from the training (or validation) samples. Both types of results should be considered to determine the behavior of the used optimization algorithms for the given problem.

However, the results shown below are not to represent the capabilities of the FUBAR systems (it was not the aim of the research); it shows how well is the meta-heuristic techniques with certain parameter settings to determine the ideal rule-base.

4.1 Hardware and Software Environment

The experiments were executed on a PC assembled with eight-core CPU (3.6 GHz), 4 GB RAM. The operating system was a Microsoft Windows 8 (64 bit) and the software was implemented in Java programming language and used with Oracle Java 7.0.170 runtime environment.

4.2   Parameter Settings of Bacterial Evolutionary Algorithm

The maximum number of clones per bacteria and the maximum number of infections in a generation were set to 10.

Each allele of the clones should be modified according to the original algorithm, but as a result of its execution time it was limited to 10 alleles. Each allele could be modified once per cycle and the modifications with the same result as before the modification were not accepted.

4.3   Parameter Settings of Imperialist Competitive Algorithm

The number of imperialists was set to 9 and the number of revolutions was set to 10 in the ICA. The assimilation factor of the algorithm was set to 0.5.

4.4   Parameter Settings of Particle Swarm Algorithm

The maximum velocity of the particles was set to 0.6; the learning constants C_1 was set to 4.0, while C_2 was 3.0.

4.5   Parameter Settings of Big Bang Big Crunch Algorithm

The BBBC algorithm's beta parameter was set to 0.5 and the iterations was set to 10 with linear population.

5   Results

The goal of the investigation was not to determine an optimal rule-base for the FUBAR recognizer, but to reach the best possible solution with meta-heuristic optimization algorithms for the given parameters and compare the results. However that was shown, it is possible to find a rule-base, which is able to reach 100 % accuracy for the training samples [35]. The investigated properties are the reached accuracies for the training samples—60 samples for each of the 26 symbols—,the validation samples—120 samples per symbol—and the execution time.

Each algorithm was executed ten times with the same parameter settings. The best and average results of these test runs are detailed below in this section. The average accuracy was calculated according to the number of mistakes for the training and the validation samples (the lower is the better).

The accuracy of the system for each type of the initial rule-bases was 0 % in each case.

5.1 Execution Time and CPU Load of the Algorithms

The execution time of the algorithms is represented in the number of generations evaluated in a second. The results of the investigated meta-heuristic techniques for the previously described parameters in each case were alike. The results are the followings:

1. *BEA*: 0.0667 generation/second
2. *ICA*: 0.667 generation/second
3. *PSO*: 1 generation/second
4. *BBBC*: 4 generation/second

The CPU load of the algorithms was between 87 and 99 % in all cases; the results include the processor time required by the GUI of the application.

5.2 Results from Triangular Membership Functions

The results of the optimization algorithms using triangular membership functions in the antecedents of the initial rule-base are presented in this section. The best results of various meta-heuristic algorithms for the 60 training samples are shown in Table 1.

The results showed that the ICA algorithm reached the best recognition rate for the training samples in all cases. The second best results slightly below the best ones were

Table 1 Best results of meta-heuristic algorithms for the training set from triangular membership functions

Number of generations	Accuracy of meta-heuristic optimization techniques			
	BEA[a] (%)	ICA[b] (%)	PSO[c] (%)	BBBC[d] (%)
50	*3.85*	83.01	53.33	82.56
60	*3.85*	80.58	56.09	76.09
70	*3.85*	83.46	*61.54*	80.32
80	*3.85*	86.35	60.9	83.65
90	*3.85*	84.62	*61.54*	79.23
100	*3.85*	**99.29**	59.29	*84.55*

Italics Best result(s) reached by the given algorithm
Bold Best overall (considering all algorithms with all various number of generations) result
Underline Best result(s) reached in the given number of generations
[a]Bacterial evolutionary algorithm
[b]Imperialist competitive algorithm
[c]Particle swarm optimization
[d]Big bang big crunch algorithm

Table 2 Best results of meta-heuristic algorithms for the validation set from triangular membership functions

Number of generations	Meta-heuristic optimization technique			
	BEA[a] (%)	ICA[b] (%)	PSO[c] (%)	BBBC[d] (%)
50	*3.85*	79.49	51.31	79.58
60	3.85	78.4	54.97	70.74
70	3.85	78.81	*60*	76.54
80	3.85	81.96	57.88	*80.9*
90	3.85	82.15	59.17	76.09
100	3.85	***96.03***	57.47	79.36

Italics Best result(s) reached by the given algorithm
Bold Best overall (considering all algorithms with all various number of generations) result
Underline Best result(s) reached in the given number of generations
[a]Bacterial evolutionary algorithm
[b]Imperialist competitive algorithm
[c]Particle swarm optimization
[d]Big bang big crunch algorithm

achieved with the BBBC, while the PSO performed significantly worse. The BEA algorithm could not perform considerable results. The best result was achieved in 1 generation of 50 with BEA, in 92 generations of 100 with ICA, in 50 generations of 70 (and 84 of 100) with PSO, and in 99 generations of 100 with BBBC.

The best results of various meta-heuristic algorithms for the 120 validation sample set are shown in Table 2.

The best average recognition rate was achieved with the ICA algorithm for the validation set.

5.3 Results from Single Discrete Point Membership Functions

Discrete membership functions, which have a single point with the membership value 1 (and 0 in others) in the antecedents, were used as initial rule-base in this experiment. The results of the optimization algorithms are detailed in this subsection. The best results of various meta-heuristic algorithms for the 60 training samples are shown in Table 3.

The results showed that the BBBC algorithm reached the best recognition rate (19.23 %) for the training samples. In this particular case, the BEA and ICA algorithms were not able to reach any increase in the recognition rate starting from the initial rule-base. The best result was achieved in 45 generations of 50 (51 of 60 and 58 of 70) with PSO and in 5 generations of 50 (38 of 60 and 4 of 80) with BBBC.

The best results of various meta-heuristic algorithms for the 120 validation sample set are shown in Table 4.

The best average recognition rate for the validation set (19.13 %) was achieved with the BBBC algorithm.

Table 3 Best results of meta-heuristic algorithms for the training set from singleton membership functions

Number of generations	Meta-heuristic optimization technique			
	BEA[a] (%)	ICA[b] (%)	PSO[c] (%)	BBBC[d] (%)
50	0	0	*11.54*	___**19.23**___
60	0	0	11.54	___**19.23**___
70	0	0	11.54	15.38
80	0	0	7.69	___**19.23**___
90	0	0	3.85	15.38
100	0	0	3.85	___15.38___

Italics Best result(s) reached by the given algorithm
Bold Best overall (considering all algorithms with all various number of generations) result
Underline Best result(s) reached in the given number of generations
[a]Bacterial evolutionary algorithm
[b]Imperialist competitive algorithm
[c]Particle swarm optimization
[d]Big bang big crunch algorithm

Table 4 Best results of meta-heuristic algorithms for the validation set from singleton membership functions

Number of generations	Meta-heuristic optimization technique			
	BEA[a] (%)	ICA[b] (%)	PSO[c] (%)	BBBC[d] (%)
50	0	0	11.51	___**19.13**___
60	0	0	*11.54*	___18.85___
70	0	0	11.47	___15.35___
80	0	0	7.69	___18.75___
90	0	0	3.85	___15.29___
100	0	0	3.85	___15.1___

Italics Best result(s) reached by the given algorithm
Bold Best overall (considering all algorithms with all various number of generations) result
Underline Best result(s) reached in the given number of generations
[a]Bacterial evolutionary algorithm
[b]Imperialist competitive algorithm
[c]Particle swarm optimization
[d]Big bang big crunch algorithm

5.4 Results Without Initial Membership Functions

None of the break points in the antecedents' membership functions was predetermined for this set of tests. Each solution of the algorithms had to determine the trapezoidal membership functions from scratch. The highest average recognition rates of various meta-heuristic algorithms for the 60 training samples are shown in Table 5.

The results showed that the ICA algorithm reached the best recognition rate for the training samples; in some cases, the BBBC algorithm performed the same or

Table 5 Best results of meta-heuristic algorithms for the training set from undefined membership functions

Number of generations	Meta-heuristic optimization technique			
	BEA[a] (%)	ICA[b] (%)	PSO[c] (%)	BBBC[d] (%)
50	3.85	72.24 %	57.05 %	77.24 %
60	3.85	76.73 %	51.99 %	80.51 %
70	3.85	78.53 %	56.92 %	79.49 %
80	3.85	76.22 %	49.94 %	76.03 %
90	57.69	**83.91 %**	53.78 %	82.37 %
100	*61.47*	80.71 %	*65.38 %*	76.47 %

Italics Best result(s) reached by the given algorithm
Bold Best overall (considering all algorithms with all various number of generations) result
Underline Best result(s) reached in the given number of generations
[a]Bacterial evolutionary algorithm
[b]Imperialist competitive algorithm
[c]Particle swarm optimization
[d]Big bang big crunch algorithm

better results. The best result was achieved in 44 generations of 100 with BEA, in 90 generations of 90 with ICA, in 69 generations of 100 with PSO, and in 88 generations of 90 with BBBC.

The results of various meta-heuristic algorithms with the greatest accuracy for the 120 validation sample sets are shown in Table 6.

The best average recognition rate was achieved with the ICA algorithm for the validation set. Similarly to the results for the training set, the BBBC algorithm was able to reach better results in some cases.

Table 6 Best results of meta-heuristic algorithms for the validation set from undefined membership functions

Number of generations	Meta-heuristic optimization technique			
	BEA[a] (%)	ICA[b] (%)	PSO[c] (%)	BBBC[d] (%)
50	3.85	69.97	55.8	72.53
60	3.85	73.43	51.86	77.21
70	3.85	76.12	54.36	75.99
80	3.85	72.24	49.17	74.81
90	56.89	*81.79*	51.7	*78.56*
100	*60.58*	76.83	*63.97*	73.85

Italics Best result(s) reached by the given algorithm
Bold Best overall (considering all algorithms with all various number of generations) result
Underline Best result(s) reached in the given number of generations
[a]Bacterial evolutionary algorithm
[b]Imperialist competitive algorithm
[c]Particle swarm optimization
[d]Big bang big crunch algorithm

6 Conclusion and Future Work

It is important to highlight that this research was not about reaching the highest average recognition rate with the FUBAR system by optimizing the rule-base parameters. The aim of the work was to investigate the properties of the presented optimization methods for the given (model identification) problem with certain parameters.

In [20, 29, 30, 35], it was shown that the FUBAR algorithms could reach over 97 % average recognition rate for 26 multi-stroke and over 99 % accuracy for 26 single-stroke letters. The user acceptance threshold for handwritten recognition was determined in 97 % [36]; this means that the users consider the system unusable (or uncomfortable) if it operates with lower accuracy.

The investigation of the behavior of meta-heuristic methods in rule-base identification is important for the FUBAR methods and other complex fuzzy systems. If the initial rule-base of a FUBAR method reaches a better accuracy than with another rule-base, then it provides better user experience also. The presented results may give a better perspective on the algorithms used in rule-base identification and it also can be used as a learning phase for adapting to user-specified writing styles to increase the accuracy further.

For the big amount of data to process in systems such as HandSpy [37], the used algorithms should reach high average recognition rate, while keeping the computational cost as low as possible. The functionalities of HandSpy could be extended by combining it with FUBAR methods and meta-heuristic optimization.

The used meta-heuristic techniques were applied to optimize the initial rule-base —determined statistically from the training set and modified randomly—of the multi-stroke FUBAR [35]. The membership functions (fuzzy sets) in the antecedent part of the rules are changed (slightly modified or tuned) during the optimization. These changes do not have any effect on the computational cost or time of the recognizer; it modifies only the description of the template alphabet stored as fuzzy rules; however, it could be modified to apply changes to the rules based on the samples originated from the user, which way the knowledge stored in the system would be optimized to match the given person's writing style.

The results showed that the various meta-heuristic optimization techniques have significantly different results and execution time for the presented problem. In some cases, the algorithms could not reach as good results with a higher number of generations as with a lower one; this reflects the nondeterministic behavior of the algorithms and their properties causing a convergence to local optimum solutions.

The BBBC algorithm had the fastest processing of generations (4 generation/s) and BEA had the highest processing time (0.067 generation/s). The best accuracy was achieved with the ICA algorithm starting from triangular membership functions in 99 generations. The second best algorithm was the BBBC, and it reached the highest increase in average recognition rate considering the processing time.

The presented results are interesting compared to the ones included in [35], where the initial rule-bases were predetermined statistically from the training set before the optimization process. However, further analyses are required for a more detailed view of the algorithms.

The work will be continued with extending the experiments to various population size and various algorithm-dependent parameter values.

After the detailed investigation of the algorithms and their possible extensions, some improvements might be presented; the first idea is to analyze the behavior of the BBBC algorithm with a new operation in the algorithm, which may prevent it from converging to local optimum. The basic idea is to add an entity ("anomaly" in physics analogy), which randomly modifies and has an effect on the center of gravity for the given cycle.

Acknowledgments This paper is partially supported by the TÁMOP-4.2.2.A-11/1/KONV-2012-0012 and Hungarian Scientific Research Fund (OTKA) K105529, K108405.

References

1. Jang, J.-S.R.: Rule extraction using generalized neural networks. In: Proceedings of the 4th IFSA World Congress, pp. 82–86 (1991)
2. Zadeh, L.A.: Fuzzy sets. Inf. Control **8**, 338–353 (1965)
3. Mamdani, E.H., Assilian, S.: An experiment in linguistic synthesis with a fuzzy logic controller. Int. J. Man Mach. Stud. **7**, 1–13 (1975)
4. Takagi, T., Sugeno, M.: Fuzzy identification of systems and its applications to modeling and control. IEEE Trans. Syst. Man Cybern. **SMC-15**, 116–132 (1985)
5. Ishibuchi, H., Nakashima, T.: Effect of rule weights in fuzzy rule-based classification systems. IEEE Trans. Fuzzy Syst. **9**(4), 506–515 (2001)
6. Sugeno, M., Griffin, F.M., Bastian, A.: Fuzzy hierarchical control of an unmanned helicopter. In: Proceedings of IFSA '93, Seoul, pp. 1262–1265 (1993)
7. Kóczy, L.T., Hirota, K.: Approximate inference in hierarchical structured rule bases. In: Proceedings of IFSA '93, Seoul, pp. 1262–1265 (1993)
8. Holland, J.H.: Adaption in Natural and Artificial Systems. The MIT Press, Cambridge (1992)
9. Nawa, N.E., Furuhashi, T.: Fuzzy system parameters discovery by bacterial evolutionary algorithm. IEEE Trans. Fuzzy Syst. **7**(5), 608–616 (1999)
10. Atashpaz-Gargari, E., Lucas, C.: Imperialist competitive algorithm: an algorithm for optimization inspired by imperialistic competition. In: Proceedings of 2007 IEEE Congress on Evolutionary Computation. 7., Singapore, pp. 4661–4666 (2007)
11. Eberhart, R., Kennedy, J.: A new optimizer using particle swarm theory. In: Proceedings of the IEEE International Conference on Neural Networks IV, IEEE Press, Piscataway, NJ, pp. 1942–1948 (1995)
12. Erol Osman, K., Eksin, I.: New optimization method: big bang-big crunch. Adv Eng Softw **37**:106–111 (2006)
13. Kowalski, P.A., Kulczycki, P.: Data sample reduction for classification of interval information using neural network sensitivity analysis. Lecture Notes in Artificial Intelligence, vol. 6304, Springer, Berlin, pp. 271–272 (2010)
14. van den Berg, J., Kaymak, U., van den Bergh, W.M.: Fuzzy classification using probability-based rule weighting. Proceedings of 11th IEEE International Conference on Fuzzy Systems, Hawaii (2002)
15. Ishibuchi, H., Yamamoto, T.: Rule weight specification in fuzzy rule-based classification systems. IEEE Trans. Fuzzy Syst. **13**(4), 428–435 (2005)

16. Kulczycki, P., Kowalski, P.A.: Bayes classification of imprecise information of interval type. Control Cybern **40**(1), 101–123 (2011)
17. Lilik, F., Botzheim, J.: Fuzzy based prequalification methods for EoSHDSL technology. Acta Technica Jurinensis **4**(1), 135–144 (2011)
18. Šarčević, P.: Vehicle classification using neural networks with a single magnetic detector. Issues Challenges Intel Syst Comput Intel SCI **530**, 103–115 (2014)
19. Tormási, A., Botzheim, J.: Single-stroke character recognition with fuzzy method. In: Balas, V.E. et al. (eds.) New Concepts and Applications in Soft Computing SCI, vol. 417, pp. 27–46 (2013)
20. Tormási, A., Kóczy, L.T.: Fuzzy-based multi-stroke character recognizer. Preprints of the Federated Conference on Computer Science and Information Systems, Kraków, pp. 675–678 (2013)
21. LaLomia, M.J.: User acceptance of handwritten recognition accuracy. In: Companion Proceedings of CHI '94. New York, p. 107 (1994)
22. Yaeger, L.S., Webb, B.J., Lyon, R.F.: Combining neural networks and context-driven search for online, printed handwriting recognition in the Newton. AI Magazine **19**(1), 73–90 (1998)
23. Butter, A., Pogue, D.: Piloting Palm: The inside story of Palm, Handspring, and the Birth of the Billion-Dollar Handheld Industry. Wiley, New York (2002)
24. Fleetwood, M.D. et al.: An evaluation of text-entry in palm OS—graffiti and the virtual keyboard. In: Proceedingd of HFES '02, Santa Monica, CA, pp. 617–621 (2002)
25. Költringer, T., Grechenig, T.: Comparing the immediate usability of graffiti 2 and virtual keyboard. In: Proceedings of CHI EA '04, New York, pp. 1175–1178 (2004)
26. Wobbrock, J.O., Wilson, A.D., Li, Y.: Gestures without libraries, toolkits or training: A $1 recognizer for user interface prototypes. In: Proceedings of UIST '07, ACM Press, New York, pp. 159–168 (2007)
27. Anthony, L., Wobbrock, J.O.: A Lightweight multistroke recognizer for user interface prototypes. In: Proceedings of GI'10, Ottawa, pp. 245–253 (2010)
28. Vatavu, R.-D., Anthony, L., Wobbrock, J.O.: Gestures as point clouds: a $P recognizer for user interface prototypes. In: Proceedings of ACM CMI, Santa Monica, pp. 273–280 (2012)
29. Tormási, A., Kóczy, L.T.: Comparing the efficiency of a fuzzy single-stroke character recognizer with various parameter values. In: Greco, S. et al. (eds.) Proceedings of IPMU 2012, Part I. CCIS, vol. 297, pp. 260–269 (2012)
30. Tormasi, A., Kóczy, L.T.: Efficiency and accuracy analysis of a fuzzy single-stroke character recognizer with various rectangle fuzzy grids. In: Proceedings of CSCS '12, Szeged, pp. 54–55 (2012)
31. Tormási, A., Kóczy, L.T.: Improving the efficiency of a fuzzy-based single-stroke character recognizer with hierarchical rule-base. Proceedings of 13th IEEE International Symposium on Computational Intelligence and Informatics, Óbuda, pp. 421–426 (2012)
32. Tormási, A., Kóczy, L.T.: Improving the accuracy of a fuzzy-based single-stroke character recognizer by antecedent weighting. In: Proceedings of 2nd World Conference on Soft Computing, Baku, pp. 172–178 (2012)
33. Tormási, A., Kóczy, L.T.: Dynamic fuzzy rule weight optimization for a fuzzy based single-stroke character recognizer. In: Proceedings of IEEE 17th International Conference on Intelligent Engineering Systems, INES 2013, Costa Rica, pp. 119–124 (2013)
34. Tormási, A., Kóczy, L.T.: Improved fuzzy-based single-stroke character recognizer. In: Proceedings of IFSA 2013 World Congress, Edmonton, pp. 430–435 (2013)
35. Tormási, A., Kóczy, L.T.: Identification of the initial rule-base of a multi-stroke fuzzy-based character recognition method with meta-heuristic techniques. Technical transactions— automatic control, Poland (in press)
36. LaLomia, M.J.: User acceptance of handwritten recognition accuracy. In: Companion Proceedings of CHI '94, New York, p. 107 (1994)
37. Monteiro, C., Leal, J.P.: Managing experiments on cognitive processes in writing with HandSpy. Comput. Sci. Inf. Syst. **10**(4), pp. 1747–1773 (2013) (Novi Sad)

A Neural Network with a Learning Vector Quantization Algorithm for Multiclass Classification Using a Modular Approach

Jonathan Amezcua, Patricia Melin and Oscar Castillo

Abstract This work describes a learning vector quantization (LVQ) method for unsupervised neural networks for classification tasks. We work with a modular architecture of this method, so we can classify three classes per module. We also work with three different databases, the arrhythmia database from MIT-BIH, which contains 15 different classes, a character database from UCI with 26 different classes, and finally a vehicle silhouettes database also from UCI with 4 different classes.

Keywords LVQ · Clustering · Neural networks · Classification

1 Introduction

Classification tasks consist of assigning objects to only one of many predefined categories, and it is a general problem that encompasses many diverse applications. Spam detection in e-mail messages, categorization of cells as benign or malignant, and classification of galaxies based on their shape are some examples of these tasks [1].

The input data for a classification task are a collection of records. Each record is characterized by a tuple (x, y) where x is a set of attributes and y is a special attribute designated as the class.

Some classification models are very useful for different purposes; the most used are the descriptive model and predictive model. There are different techniques to develop these classification models, and each one uses its own learning algorithm to identify a model which best fits the relationship between the set of attributes and the

J. Amezcua (✉) · P. Melin · O. Castillo
Division of Graduate Studies, Tijuana Institute of Technology, Tijuana, Mexico
e-mail: jonathan.aguiluz@yahoo.com

P. Melin
e-mail: pmelin@tectijuana.edu.mx

© Springer International Publishing Switzerland 2016
L.A. Zadeh et al. (eds.), *Recent Developments and New Direction in Soft-Computing Foundations and Applications*, Studies in Fuzziness and Soft Computing 342, DOI 10.1007/978-3-319-32229-2_13

class to which the input data belong. The model generated by the learning algorithm must both fit the input data and also correctly predict the class of records that do not belong to the set of input records [1].

In this work, we used unsupervised neural networks with a learning vector quantization (LVQ) algorithm, as a classification technique using a modular approach for classification of the MIT-BIH arrhythmia dataset [2] with 15 classes and a letter recognition dataset from UCI [3] with 26 classes, and we also worked with 4 classes of the UCI vehicle silhouettes dataset [4] with a monolithic neural network with the LVQ algorithm.

2 Neural Networks

An artificial neural network (ANN) is a system composed of many simple processing elements connected in parallel, whose function is determined by the network structure, the force on the connections, and the processing performed by elements in the nodes.

Each processing element in the network (neuron) is represented as a node. The connections provide a hierarchical structure that, trying to emulate the physiology of the brain, seeks new processing models to solve specific problems of the real world [5]. What is important in neural network development techniques is its useful behavior to learn, recognize, and enforce relationships between objects and object frames of the real world. In this sense, ANNs are used as tools that can be used to solve difficult problems [2].

Figure 1 shows an artificial neuron. Input values are combined to form an individual global entry. This is achieved by an input function, which is computed from the input vector. The input function is defined as follows:

$$\text{input}_i = (\text{in}_{i1} \cdot \mathbf{w}_{i1}) * (\text{in}_{i2} \cdot \mathbf{w}_{i2}) * \cdots * (\text{in}_{in} \cdot \mathbf{w}_{in}) \tag{1}$$

where $*$ represents an appropriate operator (max, sum, product, etc.), n is the number of inputs to neuron $\mathbf{N}_i$, and $\mathbf{w}_i$ represents the weight.

Fig. 1 Artificial neuron

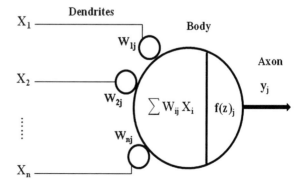

The input values are multiplied by the weights previously entered to the neuron. Therefore, weights that generally are not restricted change the influence measure over the input values. That is, they allow a large input value to have only a small influence, if these are small enough.

Another important function in a processing unit is the activation function. Biologically, a neuron can be active or inactive; that is, it has an activation state. Artificial neurons also have different activation states, some of which only two states, as well as biological; however, others may take any value within a given set.

The activation function computes the state of activity of the neuron, transforming the global entry (minus a threshold, Θi) into an activation state, whose range is typically between 0 and 1 or between -1 and 1. This is because a neuron can be totally inactive (0 or -1) or active as well (1).

The last element needed by an artificial neuron is an output function. The resulting value of this function is the output of the ith neuron (out_i); thus, the output function determines which value is transferred to the associated neurons. If the activation function is below a certain threshold, no output is passed to the subsequent neuron. Any value is not allowed as an input to a neuron; therefore, the output values are within the range [0, 1] or [-1, 1]. They can also be binary {0, 1} or {-1, 1}.

Neural networks, with their remarkable ability to derive meaning from complicated or imprecise data, can be used to extract patterns and detect trends that are too complex to be noticed by humans or other computer techniques. A trained neural network can be thought of as an expert in the category of information that has been analyzed [6, 7].

Neural networks are mainly used in four ways: models of biological nervous systems, artificial intelligence, real-time adapter processes, or to implement simple hardware control applications, such as robotics, data analysis, and pattern recognition [6, 8, 9].

2.1 Modular Neural Networks

An artificial neural network is modular if the computation performed by the network can be decomposed into two or more modules [10]; this means that each network is transformed into a single module that can be combined with other modules, which are integrated together by an integrating unit. Modular neural networks have also a biological background: Natural neural systems are composed of a hierarchy of networks built of elements specialized for different tasks. In general, combined networks are more powerful than flat unstructured ones [11]. Figure 2 shows a classic architecture for a modular neural network.

Fig. 2 Architecture for modular neural networks

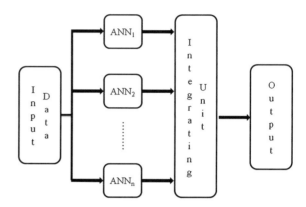

Some features of modular neural networks are as follows:

Robustness: A modular approach of an artificial neural network can have more strength and adds fault tolerance capabilities. Damage to one part of the system can result in a loss of some of the capabilities of the system, but overall, the system can still function partially [12].

Complexity reduction: The complexity of a monolithic neural network dramatically increases when the dimensionality of the data increases as well. In contrast, the modular neural networks can avoid the complexity issue because specialized modules can learn smaller tasks, although the overall task is more complex and difficult [13, 14].

Learning: With modular neural networks, the modules can be trained in an individual way, and then, they can be integrated by an integrating unit [10].

Scalability: This is an important feature of modular neural networks. Modular neural networks present an architecture for the addition of modules which can store any new information to train without having to train all modules [10].

Computational efficiency: If the system can be divided into independent subtasks, and possibly in parallel, then the computational effort is further reduced [13]. A modular neural network can learn a set of assignments faster than a monolithic neural network. A modular neural network can learn a set of assignments faster than a monolithic neural network, and this is because each module in the modular neural network has to learn smaller sets of information [15, 16].

2.2 Learning Vector Quantization

LVQ is an adaptive data classification method based on training data with a desired class of information. However, a supervised training method, LVQ, also applies unsupervised data clustering techniques to preprocess the dataset and obtain cluster centers [13].

Fig. 3 LVQ representation

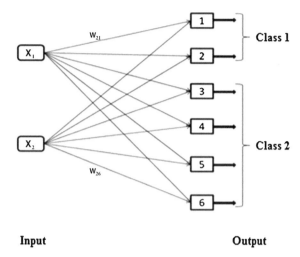

Input   Output

The LVQ learning algorithm involves two steps. In the first step, an unsupervised learning data clustering method is used to locate the several cluster centers without using class information. Once the clusters are obtained, their classes must be labeled before moving to the second step of supervised learning. Such labeling is achieved by the so-called *voting method*. The clustering process of LVQ is based on the general assumption that similar input patterns generally belong to the same class [13].

In the second step, class information is used to fine-tune the cluster centers to minimize the number of misclassified cases. First, a weight vector (or cluster center) **w** that is closest to the input vector **x** must be found. If **x** and **w** belong to the same class, **w** moves toward **x**; otherwise, **w** moves away from the input vector **x** [13].

Figure 3 shows a representation for LVQ, where the input dimension is 2 and the input space is divided into six clusters. The first two clusters belong to class 1, while the other clusters belong to class 2.

After learning, the LVQ network method classifies an input vector by assigning it to the same class as the output unit that has the weight vector (cluster center) closest to the input vector [13].

3 Proposed Models

In this section, we describe the databases that have been considered and also the LVQ network architectures for each dataset.

3.1 Case Study

As a first stage of this work, we considered two classes from the arrhythmia dataset to work with an LVQ network and observe how its works with the distinct parameters of the method. The cardiac cycles 115 and 109 were taken to work with a monolithic LVQ network. Each cycle contains 100 records of which 70 were taken for training and the other 30 records for testing.

For this case, 15 experiments were performed; good results were achieved with a learning rate of 0.1, which is the default learning rate of the method. In experiments with more accurate learning rate, the method was slower and the results were not significantly improved.

With the observed behavior of the parameters of this set of experiments, we started working with the vehicle silhouettes dataset.

3.2 Vehicle Silhouettes Dataset

This dataset is from UCI [4], consisting of 4 classes with 18 attributes for each class. This dataset has the purpose of classifying a vehicle based on data extracted from the silhouette of the vehicle. Main idea here is to use a monolithic LVQ network to work with the 4 classes of the dataset.

For this dataset, we used 177 records, in which 70 % of the records (124) was used to train the LVQ network and the 30 % (53) was for testing. Figure 4 shows the architecture we used for this dataset.

Fig. 4 LVQ network architecture for the vehicle dataset

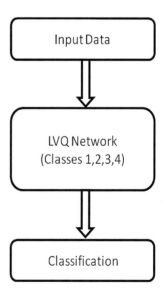

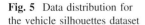

Fig. 5 Data distribution for the vehicle silhouettes dataset

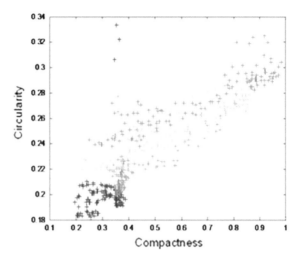

In Fig. 5, we can see how the data for the vehicle silhouettes dataset are distributed. This algorithm was difficult for the LVQ network to achieve good results, and this is because, as is shown in Fig. 5, some data are widely dispersed, even those belonging to the same class.

3.3 Letter Recognition Dataset

This dataset is also from UCI [3], consisting of 26 classes. The objective is to identify each of a large number of black-and-white rectangular pixel displays as one of the 26 capital letters in the English alphabet. Character images were based on 20 different fonts, and each letter within these 20 fonts was randomly distorted to produce a file of 20,000 unique stimuli. Each stimulus was converted into 16 primitive numerical attributes which were then scaled to fit into a range of integer values from 0 through 15.

We take 700 records from each class to perform the experiments. Four ninety (70 %) records of each class was taken to train the LVQ network and 210 records for testing. Figure 6 shows the architecture used for these experiments, with 5 modules, with each of the first 4 modules containing 5 classes and the fifth module 6 classes. In the integrating unit, we used the winner-take-all method to obtain a classification result [17].

The combination of classes in each module for this architecture was performed randomly. Figure 7 shows how is the data distribution for this letter recognition dataset.

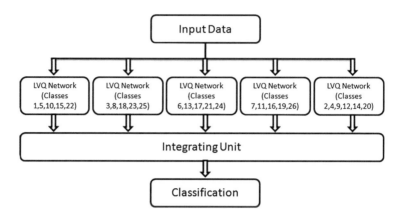

Fig. 6 LVQ network architecture for the letter recognition dataset

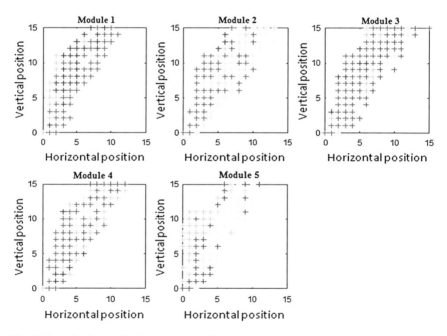

Fig. 7 Data distribution for the letter recognition dataset

3.4 Arrhythmia Dataset

This dataset was obtained from MIT-BIH. It contains 48 half-hour excerpts of two-channel ambulatory ECG recordings, obtained from 47 subjects studied by the BIH Arrhythmia Laboratory. Twenty-three recordings were chosen at random from a set of 4000 · 24-h ambulatory ECG recordings collected from a mixed population of inpatients (about 60 %) and outpatients (about 40 %) at Boston's Beth Israel Hospital; the remaining 25 recordings were selected from the same set to include less common but clinically significant arrhythmias that would not be well represented in a small random sample [2, 18, 19].

For this dataset, we have a total of 15 classes and 100 records for each class. For each record, 76 V were taken; we have 70 % of the records of each class for training and 30 % for testing.

For this dataset, a preprocessing step was necessary. The extraction of records was carried out using a tool provided by MIT-BIH called *PhysioNet_ECG_Exporter.m*, this tool graphically shows the representation of arrhythmias and also allows to obtain the range of values with which you want to work, and as we mention above, for this research we extracted 100 records [2].

We developed a LVQ network architecture with 5 modules to work with this dataset. Once again, the combination of classes for each module of this architecture was done randomly. The classes that were considered are as follows:

- Normal
- LBBB
- RBBB
- PVC
- Fusion paced and normal
- Paced
- Nodal
- Fusion ventricular and normal
- Atrial
- Ventricular flutter wave
- Paced maker fusion
- Aberrated APC
- Blocked APC
- Atrial escape
- Fusion PVC

The challenge in working with this database was to properly combine classes in each module, since this process was done manually [20]. Figure 8 shows the LVQ network for the arrhythmia dataset, and we have three classes per module and a winner-take-all method as our integrating unit [17].

Figure 9 shows the data distribution in each module for this architecture of the arrhythmia dataset. As shown, the data in each class are similar; also there is some grade of overlap between classes. This is an important issue of the algorithm LVQ neural networks, because the more similar the data, the more complicated the method.

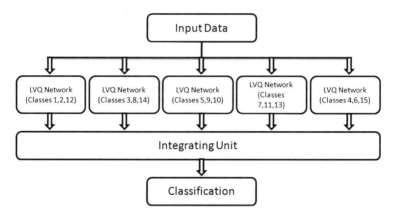

Fig. 8 LVQ network architecture for the arrhythmia dataset

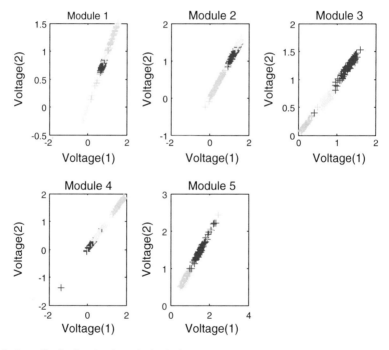

Fig. 9 Data distribution for the arrhythmia dataset

4 Experimental Results

This section shows the results obtained from the experiments described above, and for each of the databases, a total of 15 experiments were performed.

Table 1 Case study results

Experiment	Epochs	Clusters	Time	Class (%)
1	200	10	00:01:31	98.33
2	200	12	00:01:28	98.33
3	200	14	00:01:33	96.66
4	400	14	00:03:02	98.33
5	300	16	00:02:20	98.33
6	440	16	00:03:35	100
7	207	18	00:01:38	100
8	500	18	00:04:07	98.33
9	500	20	00:04:07	98.33
10	124	18	00:00:58	100
11	4	16	00:00:07	100
12	33	14	00:00:20	100
13	35	12	00:00:23	100
14	28	10	00:00:13	100
15	40	8	00:00:27	100

4.1 Case Study Results

Table 1 shows the results obtained from our case study, and 15 experiments were performed.

As can be noted in the above table, the best results were obtained in the last 5 experiments, not only achieving the best classification rate, but also the shortest possible time.

4.2 Vehicle Silhouettes Dataset Results

Below in Table 2 are the results of the vehicle silhouettes dataset. Once again, 15 experiments were performed, having the experiment 4 as the best result.

For this set of experiments, we used a default learning rate of 0.1 for the first 10 experiments and a learning rate of 0.01 for the last five experiments.

4.3 Letter Recognition Dataset Results

In Table 3, we show results for letter recognition dataset, and same as above, 15 experiments were done.

As we mentioned earlier, for this set of experiments, we used an architecture with 5 modules; we also used a learning rate of 0.1 for all modules.

Table 2 Vehicle dataset results

Experiment	Epochs	Clusters	Time	Class (%)
1	1000	20	00:29:59	88.67
2	1000	18	00:29:14	89.62
3	800	18	00:24:21	87.73
4	800	16	00:24:25	91.98
5	700	15	00:21:45	85.84
6	700	14	00:19:59	88.20
7	600	13	00:15:41	83.96
8	500	12	00:14:06	82.07
9	500	15	00:14:13	84.90
10	500	18	00:14:49	86.79
11	500	18	00:14:59	89.15
12	600	16	00:17:07	88.67
13	600	18	00:18:08	90.09
14	800	18	00:24:19	90.56
15	1000	20	00:26:39	91.98
			Average	88.01

Table 3 Letter recognition dataset results

Experiment	Time	Epochs per module	Clusters per module	Class (%)
1	18:54:20	2000	65	99.23
2	18:58:01	2000	65	99.04
3	19:05:17	2000	65	98.66
4	18:53:52	2000	65	98.66
5	18:41:32	2000	65	99.23
6	18:45:02	2000	65	99.04
7	18:51:14	2000	65	98.66
8	18:51:11	2000	65	98.66
9	18:53:52	2000	65	98.95
10	18:37:54	2000	65	98.57
11	18:51:56	2000	65	97.80
12	18:48:37	2000	65	99.14
13	18:48:28	2000	65	99.04
14	22:08:14	2000	65	98.57
15	22:04:01	2000	65	97.80
			Average	98.74

Table 4 Arrhythmia dataset results

Experiment	Time	Epochs	Clusters	Class (%)
1	00:18:53	81	30	98.88
2	00:21:35	108	30	98.88
3	00:22:06	114	30	100
4	00:14:36	58	30	98.88
5	00:13:13	56	30	100
6	00:16:09	67	30	98.88
7	00:20:40	76	30	98.88
8	00:16:30	69	30	98.88
9	00:13:48	57	30	98.88
10	00:19:53	82	30	98.88
11	00:20:00	74	30	98.88
12	00:15:17	79	30	98.88
13	00:15:49	56	30	98.88
14	00:15:50	35	30	100
15	00:20:12	77	30	98.89
			Average	99.10

4.4 Arrhythmia Dataset Results

Table 4 shows the results obtained from the arrhythmia dataset experiments. The best result for this set of experiments was obtained experiment 5, achieving a 100 % of classification and the shortest time. For this set of experiments, we have a learning rate between 0.1, 0.01, and 0.0001.

5 Discussion and Future Research

This work introduced an unsupervised neural network method with a LVQ algorithm for multiclass classification using a modular approach. As we have shown, we were working with a maximum of three classes per module and the method achieved good results. However, datasets are an important aspect to consider when working with this classification method, as getting the best classification results is not certain as you increase the number of classes per module.

As future work, we will experiment with the MIT-BIH arrhythmia dataset and a modular architecture, having not only three classes per module, but also increasing the number of classes per module.

References

1. Pang-Ning, T., Steinbach, M., Kumar, V.: Introduction to Data Mining, pp. 145–148. Pearson Addison Wesley, Boston (2006)
2. MIT-BIH Arrhythmia Database.: PhysioBank, Physiologic Signal Archives for Biomedical Research. Site: http://www.physionet.org/physiobank/database/mitdb/. Last accessed 15 Oct 2012
3. UCI Letter Recognition Database.: Site: http://archive.ics.uci.edu/ml/datasets/Letter+Recognition. Last accessed 15 Oct 2012
4. UCI Statlog Vehicle Silhouettes Database.: Site: http://archive.ics.uci.edu/ml/datasets/Statlog+(Vehicle+Silhouettes). Last accessed 15 Oct 2012
5. Freeman, J.A., Skapura, D.: Neural Networks. Algorithms, Applications and Programming Techniques, p. 306. Addison-Wesley, Boston (1993)
6. Fernández, F., Hervás-Martínez, C., Ruiz, R., Riquelme, J.: Evolutionary generalized radial basis function neural networks for improving prediction accuracy in gene classification using feature selection. Appl. Soft Comput. 1787–1800 (2012)
7. Janeela, T., Raj, V.J.: Fuzzy based genetic neural networks for the classification of murder cases using Trapezoidal and Lagrange Interpolation Membership Functions. Appl. Soft Comput. 743–754 (2013)
8. Zadeh, A.E., Moussavi, E.: Classification of communications signals using an advanced technique. Appl. Soft Comput. 428–435 (2011)
9. Zhigang, Z., Jun, W.: Advances in Neural Network Research and Applications. Springer©, Berlin, pp. 283–371 (2010)
10. Sánchez, D., Melin, P.: Modular neural network with fuzzy integration and its optimization using genetic algorithms for human recognition based on Iris, ear and voice biometrics. In: Soft Computing for Recognition Based on Biometrics Studies in Computational Intelligence, pp. 85–102, 1st edn. Springer, Berlin (2010)
11. Rojas, R.: Neural Networks. A Systematic Introduction, Vol. 16, pp. 413-416. Springer, Berlin (1996)
12. Lee, T.: Structure Level Adaptation for Artificial Neural Networks. Kluwer Academic Publishers, Berlin (1991)
13. Jang, J., Sun, C., Mizutani, E.: Neuro-fuzzy and Soft Computing, pp. 308–310. Prentice Hall, NJ (1997)
14. Terrance, J., Rosenberg, C.R.: Parallel networks that learns to pronounce english text. Complex Syst. 1, 145–168 (1987)
15. Jain, A., Kumar, A.: An evaluation of artificial neural network technique for the determination of infiltration model parameters. Appl. Soft Comput. 272–282 (2006)
16. Rabuñal, J., Dorado, J.: Artificial Neural Networks in Real-Life Applications. IGP©, Hershey (2006)
17. Biehl, M., Ghosh, A., Hammer, B.: Learning vector quantization: the dynamics of winner-takes-all algorithms (Original Research Article). Neurocomputing 69(7–9), 660–670 (2006)
18. Zadeh, A.E., Khazaee, A., Ranaee, V.: Classification of the electrocardiogram signals using supervised classifiers and efficient features. Comput. Methods Programs Biomed. 179–194 (2010)
19. Nejadgholi, I., Hasan, M., Abdolali, F.: Using phase space reconstruction for patient independent heartbeat classification in comparison with some benchmark methods. Comput. Biol. Med. 411–419 (2011)
20. Osowski, S., Siwek, K., Siroic, R.: Neural system for heartbeats recognition using genetically integrated ensemble of classifiers. Comput. Biol. Med. pp. 173–180 (2011)

Interval Type-2 Fuzzy Possibilistic C-Means Clustering Algorithm

E. Rubio, Oscar Castillo and Patricia Melin

Abstract In this paper, we present the extension of the fuzzy possibilistic C-means (FPCM) algorithm using type-2 fuzzy logic techniques, with the goal of improving the performance of this algorithm. We also performed the comparison of this proposed algorithm against the interval type-2 fuzzy C-means (IT2FCM) algorithm to observe whether the proposed approach performs better than this algorithm. The proposed extension was realized considering both of the weight exponents (fuzzy and possibilistic), m and η, as interval fuzzy sets.

Keywords Fuzzy sets · Fuzzy C-means · Possibilistic C-means · Interval type-2 clustering algorithm

1 Introduction

Clustering algorithms are used widely in different areas of research, such as pattern recognition [1], data mining [2], classification [3], image segmentation [4, 5], data analysis, and modeling [6]. Clustering algorithms arise due to need and to find data groups that share similar characteristics in a given data set; currently, there are several fuzzy clustering algorithms such as fuzzy C-means (FCM) [1], possibilistic C-means (PCM) [7], fuzzy possibilistic C-means (FPCM) [8], and possibilistic fuzzy C-means (PFCM) [9] among others, the popularity of the fuzzy clustering algorithms is due to the fact that allow to a datum belongs to different data clusters into a given data set.

However, these methods are not able to handle uncertainty in a given data set during the process of data clustering; due to this problem, the FCM and PCM have been extended using type-2 fuzzy logic techniques [10, 11], and the improvement

E. Rubio (✉) · O. Castillo · P. Melin
Division of Graduate Studies and Research, Tijuana Institute of Technology,
Tijuana, Mexico
e-mail: elid.rubio@hotmail.com

O. Castillo
e-mail: ocastillo@hafsamx.org

© Springer International Publishing Switzerland 2016
L.A. Zadeh et al. (eds.), *Recent Developments and New Direction
in Soft-Computing Foundations and Applications*, Studies in Fuzziness
and Soft Computing 342, DOI 10.1007/978-3-319-32229-2_14

185

of these algorithms was called interval type-2 fuzzy C-means (IT2FCM) [12, 13] and interval type-2 possibilistic C-means (IT2PCM) [14] These extensions have been applied to the creation of membership functions [15, 16], classification [17], and image processing [18, 19].

In this work, we present the extension of the FPCM using type-2 fuzzy logic techniques to provide this method with the capability of handling a higher degree of uncertainty.

2   Overview of Interval Type-2 Fuzzy Sets

Type-2 fuzzy sets are an extension of the type-1 fuzzy sets proposed by Zadeh in 1975; this extension was designed to describe uncertainty effectively in situations where the available information is uncertain. These sets include a secondary membership function to model uncertainty of type-1 fuzzy sets [10, 11].

A type-2 fuzzy set in the universal set X is denoted as $\tilde{A}$ and can be characterized by a type-2 fuzzy membership function $\mu_{\tilde{A}} = (x, u)$ as:

$$\tilde{A} = \int_{x \in X} \mu_{\tilde{A}}(x)/x = \int_{x \in X} \left[\int_{u \in J_x} f_x(u)/u \right]/x, J_x \subseteq [0, 1] \tag{1}$$

where J_x is the primary membership function of x, which is the domain of the secondary membership function $f_x(u)$.

The shaded region is shown in Fig. 1a and it is normally called footprint of uncertainty (FOU). The FOU of $\tilde{A}$ is the union of all primary memberships that are within the lower and upper limits of the interval of membership functions and can be expressed as:

$$FOU(\tilde{A}) = \bigcup_{\forall x \in X} J_x = \{(x, u) | u \in J_x \subseteq [0, 1]\} \tag{2}$$

The lower membership function (LMF) and upper membership function (UMF) are denoted by $\underline{\mu}_{\tilde{A}}(x)$ and $\bar{\mu}_{\tilde{A}}(x)$ and are associated with the lower and upper bound of FOU($\tilde{A}$), respectively; that is, the UMF and LMF of $\tilde{A}$ are two type-1 membership functions that bound the FOU as shown in Fig. 1a. By definition, they can be represented as:

$$\underline{\mu}_{\tilde{A}}(x) = \underline{FOU(\tilde{A})} \forall x \in X \tag{3}$$

$$\bar{\mu}_{\tilde{A}}(x) = \overline{FOU(\tilde{A})} \forall x \in X \tag{4}$$

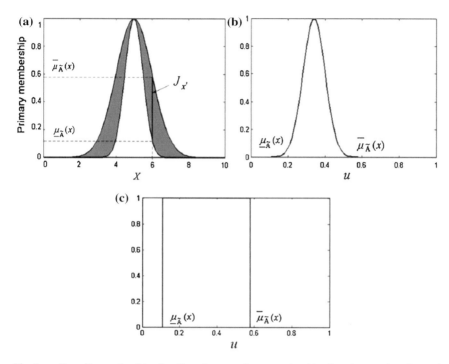

Fig. 1 a Type-2 membership function, **b** secondary membership function, and **c** interval secondary membership function

The secondary membership function is a vertical slice of $\mu_{\tilde{A}} = (x, u)$ as shown in Fig. 1b. Type-2 fuzzy sets are capable of modeling uncertainty, where type-1 fuzzy sets cannot. The computation operations required by type-2 fuzzy systems are considerably and, undesirably, large; this is due to these operations that involve numerous embedded type-2 fuzzy sets which consider all possible combinations of the secondary membership values [10, 11]. However, with the aim of reducing the computational complexity, interval type-2 fuzzy sets were proposed, where the secondary membership functions are interval sets expressed as:

$$\tilde{A} = \int_{x \in X} \left[\int_{u \in J_x} 1/u \right] /x \tag{5}$$

Figure 1c shows the membership function of an interval type-2 fuzzy set. The secondary memberships are all uniformly weighted for each primary membership of x. Therefore, J_x can be expressed as:

$$J_x = \left\{ (x, u) | u \in \left[\underline{\mu}_{\tilde{A}}(x), \bar{\mu}_{\tilde{A}}(x) \right] \right\}$$ (6)

Moreover, FOU($\tilde{A}$) in (2) can be expressed as:

$$\text{FOU}(\tilde{A}) = \bigcup_{\forall x \in X} \left\{ (x, u) | u \in \left[\underline{\mu}_{\tilde{A}}(x), \bar{\mu}_{\tilde{A}}(x) \right] \right\}$$ (7)

As a result of this, the computational complexity using interval type-2 fuzzy sets is reduced only to calculate simple interval arithmetic.

3 Fuzzy Possibilistic C-Means Algorithm

The FPCM algorithm is a combination of the FCM and PCM algorithms proposed by Pal et al. in 1975. This algorithm produces both memberships and possibilities, along with the usual point prototypes or cluster centers for each cluster. This clustering method is also an iterative algorithm which uses the necessary condition to achieve the minimization of the objective function $J_{m,\eta}$ represented by the following equation [8]:

$$J_{m,\eta}(U, T, V, X) = \sum_{i=1}^{c} \sum_{k=1}^{n} \left(\mu_i(x_k)^m + \tau_i(x_k)^\eta \right) \cdot d_{ik}^2$$ (8)

Subject to the constraints $m > 1$, $\eta > 1$, $0 \le \mu_i(x_k)^m$, $\tau_i(x_k)^\eta \le 1$ and

$$\sum_{i=1}^{c} \mu_i(x_k) = 1$$ (9)

$$\sum_{k=1}^{n} \tau_i(x_k) = 1$$ (10)

where the variables in (8) represent the following:

- n is the total number of patterns in a given data set.
- c is the number of clusters, which can be found from 2 to $n - 1$.
- X are data characteristics, where $X = \{x_1, x_2, \ldots, x_n\} \subset R^s$
- V are the centers of the clusters, where $V = \{v_1, v_2, \ldots, v_n\} \subset R^s$
- $U = \mu_i(x_k)$ is a fuzzy partition matrix, which contains the membership degree of each data set x_k to each cluster v_i.
- $T = \tau_i(x_k)$ is a typicality partition matrix, which contains the membership degree of each data set x_k to each cluster v_i.
- d_{ik} is the Euclidean distance between each data x_k of the data set and the centers vi of clusters.

- m is the weighting exponent for fuzzy partition matrix.
- η is the weighting exponent for typicality partition matrix.

The corresponding centers of the clusters and membership degree to each respective data to solve the optimization problem with the constraints in (8) are given by Eqs. (11), (12), and (13), which provide an iterative procedure. The aim is to improve a sequence of fuzzy clusters until no further improvement in $J_{m,\eta}(U,T,V,X)$ can be obtained.

$$
v_i = \frac{\sum\limits_{k=1}^{n} \left(\mu_i(x_k)^m + \tau_i(x_k)^\eta \right) \cdot x_k}{\sum\limits_{k=1}^{n} \left(\mu_i(x_k)^m + \tau_i(x_k)^\eta \right)} \tag{11}
$$

$$
\mu_i(x_k) = \left(\sum_{j=1}^{c} \left(\frac{d_{ik}}{d_{jk}} \right)^{\frac{2}{m-1}} \right)^{-1} \tag{12}
$$

$$
\tau_i(x_k) = \left(\sum_{j=1}^{c} \left(\frac{d_{ik}}{d_{jk}} \right)^{\frac{2}{\eta-1}} \right)^{-1} \tag{13}
$$

Equations (11), (12), and (13) are an iterative optimization procedure. The aim is to improve a sequence of fuzzy clusters until no further improvement in $J_{m,\eta}(U,T,V,X)$ can be made. The FPCM algorithm consists of the following steps:

1. Given a preselected number of clusters c and a chosen value for m, initialize the fuzzy partition matrix and typically the partition matrix with constraint in (9) and (10), respectively.
2. Calculate the center of the fuzzy cluster, v_i for $i = 1, 2,..., c$ using Eq. (11).
3. Use Eq. (12) to update the fuzzy membership $\mu_i(x_k)$.
4. Use Eq. (13) to update the typically membership $\tau_i(x_k)$.
5. If the improvement in $J_{m,\eta}(U,T,V,X)$ is less than a certain threshold (ε), then stop; otherwise, go to step 2.

4   Interval Type-2 Fuzzy Possibilistic C-Means Algorithm

The interval type-2 FPCM algorithm is a proposed extension of the FPCM algorithm using type-2 fuzzy logic techniques. This algorithm produces in the same way that FPCM algorithm both membership and possibility uses weights as exponents m and η for the fuzziness and possibility, respectively, which may be represented by a range rather than a precise value; that is, $m = [m_1, m_2]$, where m_1 and m_2 represent the lower and upper limits of weighting exponents for fuzziness and $\eta = [\eta_1, \eta_2]$, where η_1 and η_2 represent the lower and upper limits of weighting exponents for possibility.

Because the m value is represented by an interval, the fuzzy partition matrix $\mu_i(x_k)$ must be calculated for the interval $[m_1, m_2]$; for this reason, $\mu_i(x_k)$ would be given by belonging interval $[\underline{\mu}_i(x_k), \bar{\mu}_i(x_k)]$ where $\underline{\mu}_i(x_k)$ and $\bar{\mu}_i(x_k)$ represent the lower and upper limit of the belonging interval of datum x_j to a clustering v_i, updating the lower and upper limits of the range of the fuzzy membership matrix can be expressed as:

$$\underline{\mu}_i(x_k) = \min\left\{\left(\sum_{j=1}^{c}\left(\frac{d_{ik}}{d_{jk}}\right)^{\frac{2}{m_1-1}}\right)^{-1}, \left(\sum_{j=1}^{c}\left(\frac{d_{ik}}{d_{jk}}\right)^{\frac{2}{m_2-1}}\right)^{-1}\right\} \tag{14}$$

$$\bar{\mu}_i(x_k) = \max\left\{\left(\sum_{j=1}^{c}\left(\frac{d_{ik}}{d_{jk}}\right)^{\frac{2}{m_1-1}}\right)^{-1}, \left(\sum_{j=1}^{c}\left(\frac{d_{ik}}{d_{jk}}\right)^{\frac{2}{m_2-1}}\right)^{-1}\right\} \tag{15}$$

Because the η value is represented by an interval, fuzzy partition matrix $\tau_i(x_k)$ must be calculated for the interval $[\eta_1, \eta_2]$; for this reason, $\tau_i(x_k)$ would be given by the belonging interval $[\underline{\tau}_i(x_k), \bar{\tau}_i(x_k)]$ where $\underline{\tau}_i(x_k)$ and $\bar{\tau}_i(x_k)$ represent the lower and upper limit of the belonging interval of datum x_j to a clustering v_i, updating the lower and upper limits of the range of the fuzzy membership matrix can be expressed as:

$$\underline{\tau}_i(x_k) = \min\left\{\left(\sum_{j=1}^{c}\left(\frac{d_{ik}}{d_{jk}}\right)^{\frac{2}{\eta_1-1}}\right)^{-1}, \left(\sum_{j=1}^{c}\left(\frac{d_{ik}}{d_{jk}}\right)^{\frac{2}{\eta_2-1}}\right)^{-1}\right\} \tag{16}$$

$$\bar{\tau}_i(x_k) = \max\left\{\left(\sum_{j=1}^{c}\left(\frac{d_{ik}}{d_{jk}}\right)^{\frac{2}{\eta_1-1}}\right)^{-1}, \left(\sum_{j=1}^{c}\left(\frac{d_{ik}}{d_{jk}}\right)^{\frac{2}{\eta_2-1}}\right)^{-1}\right\} \tag{17}$$

Updating the positions of the centroids of clusters should take into account the degree of belonging interval of the fuzzy and possibilistic matrices, resulting in a range of coordinates of the positions of the centroids of the clusters. The procedure for updating cluster prototypes in IT2FPCM requires calculating the centroids for the lower and upper limits of the interval using the fuzzy and possibilistic membership matrices; these centroids for will be given by the following equations:

$$\underline{v}_i = \frac{\sum_{j=1}^{n}\left(\underline{\mu}_i(x_j) + \underline{\tau}_i(x_j)\right)^{m_1} x_j}{\sum_{j=1}^{n}\left(\underline{\mu}_i(x_j) + \underline{\tau}_i(x_j)\right)^{m_1}} \tag{18}$$

$$\bar{v}_i = \frac{\sum\limits_{j=1}^{n} \left(\bar{\mu}_i(x_j) + \bar{\tau}_i(x_k)\right)^{m_1} x_j}{\sum\limits_{j=1}^{n} \left(\bar{\mu}_i(x_j) + \bar{\tau}_i(x_k)\right)^{m_1}} \tag{19}$$

The centroid calculation for the lower and upper limits of the interval results in an interval of coordinates of positions of the centroids of the clusters. Type reduction and defuzzification use the type-2 fuzzy operations. The centroids matrix and the fuzzy partition matrix are obtained by the type reduction as shown in the following equations:

$$v_j = \frac{\underline{v}_j + \bar{v}_j}{2} \tag{20}$$

$$\mu_i(x_j) = \frac{\underline{\mu}_i(x_j) + \bar{\mu}_i(x_j)}{2} \tag{21}$$

Based on all this, the IT2 FCM algorithm consists of the following steps:

1. Establish c, m_1, m_2.
2. Initialize randomly centroids for the lower and upper bounds of the interval.
3. Calculating the update of the fuzzy partition matrices for lower and upper bound of the interval using Eqs. (14) and (15), respectively.
4. Calculating the update of the possibilistic partition matrices for lower and upper bounds of the interval using Eqs. (16) and (17), respectively.
5. Calculate the centroids for the lower and upper fuzzy partition matrix using Eqs. (18) and (19), respectively.
6. Type reduction of the fuzzy partition matrix and centroid, if the problem requires using Eqs. (20) and (21), respectively.
7. Repeat steps 3–5 until $|\tilde{J}_{m,\eta}(t) - \tilde{J}_{m,\eta}(t-1)| < \varepsilon$.

This extension on the FPCM algorithm is intended to realize that this algorithm is capable of handling uncertainty and is less susceptible to noise.

5   Results of the Implementation of the IT2FPCM Algorithm

The IT2FPCM algorithm was tested with the data set of the Iris Flower, which is comprised of 150 instances and 4 attributes (sepal length, sepal width, petal length, petal width) of each sample; this data set contains 3 different classes of flowers (setosa, versicolor, virginica) for each type of flower are 50 instances.

Table 1 shows the mean by class of the Iris flower data set; these data are needed to know how accurate are the defuzzification of centers found by the IT2FPCM algorithm.

Table 2 shows the defuzzification of centers found IT2FPCM algorithm using like parameters m = [1.5, 2.5] and η = [1.5, 2.5].

Table 3 shows the defuzzification of centers found IT2FPCM algorithm using like parameters m = [1.3, 2.3] and η = [1.1, 2.5].

Table 4 shows the defuzzification of centers found by the IT2FCM algorithm using the value of m = [1.5, 2.5].

Table 5 shows the defuzzification of centers found by the IT2FCM algorithm using the value of m = [1.3, 2.3].

Observing the above tables, we do not notice if the IT2FPCM algorithm is better than IT2FCM algorithm for this reason in the Table 6 and we show the calculation of the norm of the difference between the matrix of class centers data set of Iris flower shows and defuzzification of the centers found by the algorithms above mentioned for lower and upper limit:

$$\|CM - CFPCM\| \tag{21}$$

Table 1 Mean by class of the Iris flower data set

Sepal length	Sepal width	Petal length	Petal width
5.006	3.418	1.464	0.244
5.936	2.770	4.260	1.326
6.588	2.974	5.552	2.026

Table 2 Defuzzification of centers found for lower and upper limits using the IT2FPCM algorithm

Sepal length	Sepal width	Petal length	Petal width
5.004	3.403	1.484	0.251
5.893	2.763	4.373	1.404
6.778	3.055	5.649	2.056

Table 3 Defuzzification of centers found for lower and upper limits using the IT2FPCM algorithm

Sepal length	Sepal width	Petal length	Petal width
5.004	3.406	1.481	0.249
5.893	2.760	4.377	1.409
6.794	3.059	5.665	2.060

Table 4 Defuzzification of centers found for lower and upper limits using the IT2FCM algorithm

Sepal length	Sepal width	Petal length	Petal width
5.004	3.406	1.481	0.249
5.893	2.760	4.377	1.409
6.794	3.059	5.665	2.060

Table 5 Defuzzification of centers found for lower and upper limits using the IT2FCM algorithm

Sepal length	Sepal width	Petal length	Petal width
5.004	3.406	1.480	0.250
5.892	2.758	4.376	1.409
6.794	3.060	5.667	2.060

Table 6 Results of the norm of the difference between the matrix of class centers data set of Iris flower and defuzzification of the centers found by IT2FCM and IT2FPCM algorithms

IT2FCM		IT2FPCM	
Parameters	Norm	Parameters	Norm
$m_1 = 1.5$	0.23521965	$m_1 = 1.5$	0.23512046
$m_2 = 2.5$		$m_2 = 2.5$	
		$\eta_1 = 1.5$	
		$\eta_2 = 2.5$	
$m_1 = 1.3$	0.235218752	$m_1 = 1.3$	0.235193466
$m_2 = 2.3$		$m_2 = 2.3$	
		$\eta_1 = 1.1$	
		$\eta_1 = 2.5$	

$$\|CM - CPCM\| \tag{22}$$

where

- CM: Mean by class of the Iris flower data set.
- CFCM: Defuzzification of centers found for lower and upper limits using IT2FCM algorithm.
- CFPCM: Defuzzification of centers found for lower and upper limits using IT2FCM algorithm.

6 Conclusions

IT2FPCM is an extension of the FPCM algorithm implemented with type-2 fuzzy logic tools concepts this in order that this is capable of handling uncertainty and is less susceptible to noise, this algorithm was tested using the Iris flower data set to observe whether the algorithm proposed was better than the IT2FCM and a comparison between both algorithms was performed.

The comparison was performed based on the norm calculated between the real centers of the Iris flower data set, and the defuzzification of the centers found by the ITFCM and IT2FPCM algorithms, where in Table 6 we can observe that the proposed algorithm is slightly better than IT2FCM, worth noting that the parameters used are not the optimal for both algorithms.

To find the optimal parameters for both algorithms used in this work we can use optimization algorithms like PSO, GSA, and GA; these were used in order to improve the performance of the interval type-2 clustering algorithms.

References

1. Bezdek, J.: Pattern Recognition with Fuzzy Objective Function Algorithms. Plenum, Berlin (1981)
2. Hirota, K., Pedrycz, W.: Fuzzy computing for data mining. Proc. IEEE **87**(9), 1575–1600 (1999)
3. Iyer, N.S., Kendel, A., Schneider, M.: Feature-based fuzzy classification for interpretation of mammograms. Fuzzy Sets Syst. **114**, 271–280 (2000)
4. Philips, W.E., Velthuinzen, R.P., Phuphanich, S., Hall, L.O., Clark, L.P., Sibiger, M.L.: Aplication of fuzzy C-means segmentation technique for tissue differentiation in MR images of hemorrhagic glioblastoma multiforme. Magn. Reson. Imaging **13**(2), 277–290 (1995)
5. Yang, M.-S., Hu, Y.-J., Lin, K.C.-R., Lin, C.C.-L.: Segmentation techniques for tissue differentiation in MRI of ophthalmology using fuzzy clustering algorithms. Magn. Reson. Imaging **20**, 173–179 (2002)
6. Chang, X., Li, W., Farrell, J.: A C-means clustering based fuzzy modeling method. In: Fuzzy Systems, 2000. The Ninth IEEE International Conference on FUZZ IEEE 2000, vol. 2, pp. 937–940 (2000)
7. Krishnapuram, R., Keller, J.M.: A possibilistic approach to clustering. IEEE Trans. Fuzzy Syst. **1**(2), 98, 110 (1993)
8. Pal, N.R., Pal, K., Bezdek, J.C.: A mixed C-means clustering model. In: Proceedings of the Sixth IEEE International Conference on Fuzzy Systems, 1997, vol. 1, pp. 11, 21, 1–5 Jul 1997
9. Pal, N.R., Pal, K., Keller, J.M., Bezdek, J.C.: A possibilistic fuzzy C-means clustering algorithm. IEEE Trans. Fuzzy Syst. **13**(4), 517–530 (2005)
10. Karnik, N., Mendel, M.: Operations on type-2 set. Fuzzy Set Syst. **122**, 327–348 (2001)
11. Mendel, J.: Uncertain Rule-Based Fuzzy Logic Systems: Introduction and new directions. Prentice-Hall Inc., Upper Saddle River (2001)
12. Rhee, F.C., Hwang, C.: A type-2 fuzzy C-means clustering algorithm. In: Annual Conference of the North American Fuzzy Information Processing Society, vol. 4, pp. 1926–1929 (2001)
13. Hwang, C., Rhee, F.C.-H.: Uncertain fuzzy clustering: interval type-2 fuzzy approach to C-means. IEEE Trans. Fuzzy Syst. **15**(1), 107, 120 (2007)
14. Zarandi, M.H.F., Zarinbal, M., Türksen, I.B.: Type-II fuzzy possibilistic C-mean clustering. In: IFSA/EUSFLAT Conference, pp. 30–35 (2009)
15. Choi, B., Rhee, F.: Interval type-2 fuzzy membership function generation methods for pattern recognition. Inf. Sci. **179**(13), 2102–2122 (2009)
16. Rubio, E., Castillo, O.: Interval type-2 fuzzy clustering for membership function generation. In: 2013 IEEE Workshop on Hybrid Intelligent Models and Applications (HIMA), pp. 13, 18, 16–19 Apr 2013
17. Ceylan, R., Özbay, Y., Karlik, B.: A novel approach for classification of ECG arrhythmias: type-2 fuzzy clustering neural network. Expert Syst. Appl. **36**(3), 6721–6726 (2009) (Part 2)
18. Tlig, L., Sayadi, M., Fnaeich, F.: A new descriptor for textured image segmentation based on fuzzy type-2 clustering approach. In: 2010 2nd International Conference on Image Processing Theory Tools and Applications (IPTA), pp. 258–263, 7–10 July 2010
19. Zarandi, M.H.F., Zarinbal, M.: A new image enhancement method type-2 possibilistic c-mean approach. In: IFSA World Congress and NAFIPS Annual Meeting (IFSA/NAFIPS), 2013 Joint, pp. 1131, 1135, 24–28 June 2013

Part VI
Data Analysis and Its Applications

Fuzzy-Based Mechanisms for Selection and Recommendation Processes

Ronald R. Yager and Marek Z. Reformat

Abstract Everyday, the users use the Web for things of their interest. They expect to find items that precisely, to the highest possible degree, match the items they are looking for. Quite often this is not enough, they would like to be exposed to things that provide them with some novelty. Systems that support users in their search activities provide them with some kind of variation, but it is not a controlled process. Diversity is accidental—the systems try to estimate what items users may like based on similarities between users, users' activities, or on explicitly specified preferences. The users do not have any influence on conditions governing formation of lists of suggested items. In this paper, we assert that application of fuzziness in systems supporting users in their search activities will allow the users to overlook and control mechanisms that identify alternatives and options suggested to them, as well as to influence selection of individuals that constitute groups providing suggestions. We focus on two applications of fuzzy methods that ensure controllable selection processes and illustrate benefits of fuzzy-based processing of available information. Firstly, we concentrate on social networks. A methodology for selecting groups of individuals that satisfy linguistically described requirements regarding the degree of matching between users' interests and collective interests of groups is presented. Secondly, we offer a novel recommending approach that provides users with a fuzzy-based process aiming at construction of lists of suggested items. This is accomplished via explicit control of requirements regarding rigorousness of identifying users who become a reference base for generating suggestions. A new way of ranking items rated by multiple users based on Pythagorean fuzzy sets (PFS) and taking into account not only assigned rates but also their number is described.

R.R. Yager (✉)
Iona College, New Rochelle, NY, USA
e-mail: yager@panix.com

M.Z. Reformat
University of Alberta, Edmonton, Canada
e-mail: Marek.Reformat@ualberta.ca

© Springer International Publishing Switzerland 2016
L.A. Zadeh et al. (eds.), *Recent Developments and New Direction in Soft-Computing Foundations and Applications*, Studies in Fuzziness and Soft Computing 342, DOI 10.1007/978-3-319-32229-2_15

1 Introduction

Users utilize the Web for multiple purposes. In the context of social networks, they hope to find other individuals or groups of individuals that have similar interests. In the context of entertainment or retail, they hope to find items of interest or potential interest. To support their activities, they use multiple methods and tools including search engines and recommender systems [1, 2].

Social-oriented systems allow their users to be involved in the process of adding to the content of Web and sharing it with others. The users can easily leave their observations and opinions on the Web and allow others to see them and express their own options about the same digital items, called hereafter Web resources. Some social services, such as Delicious (*del.isio.us*), Furl (*furl.net*), Flickr (*flickr.com*), and CiteULike (*citeulike.org*), allow users to annotate and categorize Web resources. The process of labeling—annotating—resources performed by users is called *tagging* [3]. At the same time, users' activities on the Web are easier to monitor. Analysis and processing of these activities can allow for determining users' characteristic features and building users' profiles.

Recommender systems, for the other hand, provide the users with items that represent a best possible match to their interests. The two most popular recommender schemas are collaborative and knowledge-based approaches [4, 5]. In the former approach, items are being suggested based on the items seen, bought, or simply rated by other users who seen, bought or rated, in the past, the same items as the user who is looking for recommendations. In the knowledge-based approach, the user provides her preferences and constrains, and recommended items match these requirements to the highest possible degree.

However, if users want to find other users or groups of users with similar sets of interests, or want to be provided with different items, the existing supporting systems are not very satisfying. Any deviation from "perfectly" matched users or items seems to be accidental. The user does not have any control over "a degree of matching," "extensions," or "additions." The processes of search and recommendation are controlled directly via the user's preferences, or by history of the user's activities. For recommender systems, an important issue is processing of multiple ratings associated with a single item. It is always a question how to balance the ratings themselves with their number, and how to combine them to obtain a single score identifying goodness of an item.

The chapter focuses on the application of fuzzy sets [6] to construct users' signatures and to aggregate them in order to create groups' signatures. Further, these signatures constitute the basis of the proposed methodology to support the user in her process of finding compatible groups of users. The methodology is designed in a way that the user who is looking for groups with similar interests governs the selection process in a human-like way using linguistic terms representing her requirements and precision of the search process. We also propose application of fuzziness to address the issue of the user's control of a selection process in a collaborative recommender system, as well as to determine a single score describing an item based on multiple ratings. The user controls a selection process via explicitly shaping composition of a group of users whose items and ratings constitute a reference base for building a list of suggested items. The process of calculating a single score is based on Pythagorean fuzzy sets (PFS) proposed in [7, 8].

The outline of the chapter is as follows. Section 2 contains a short introduction to important aspects required for finding users and groups of users: cardinality of fuzzy sets; and building a ranking of items: PFS. Section 3 is dedicated to a process of identifying pertinent individuals or groups. It begins with a short description of tagging systems and then continues with an explanation of creating user and group signatures. The process of identifying suitable individual and groups of users is fully presented. The section ends with a real-world example. An overview of the process of building a list of suggestions is presented in Sect. 4. This section also describes the user-controlled process of building a group of users, and a process of creating a ranking of items. A real-world example using the Netflix competition database is presented too. The paper is summed up with a conclusion section.

2 Fuzzy Mechanisms Required for Selection Processes

The proposed methodologies for identifying the most suitable friend and item are based on cardinality of fuzzy sets and PFS. A method of determining cardinality of a fuzzy set includes an input from the user in a form of a statement indicating the user's perception of minimum membership value that an element should have to be treated as a member of set. This concept allows the user to control "specificity" of a set, i.e., up to what degree elements should belong to a set to be recognized as members of this set by the user. The concept of PFS is described next. It introduces a measure of commitment together with its strength and direction. A value of strength could represent a number of inputs, while direction could be used to determine overall polarity of input—should it be positive or negative. Such interpretation is used in the process of building a ranking of items.

2.1 Fuzzy Sets Cardinality

Cardinality of fuzzy sets can be determined in a number of different ways [9]. Let us assume we want to determine cardinality of a fuzzy set A. Because elements belong to the set to a degree the cardinality of this set is done based on α-cuts of A for different values of α. For example, if we preform α-cut for $\alpha = 0.5$, then we can determine cardinality of the obtained set in a standard way counting elements with the membership value not less than 0.5.

Here, we propose a modification of this simple approach. Our modification involves an input from the user—a quantity called **membership requirement**. This is a fuzzy set, called hereafter *MemberReq*, defined on the universe of discourse

Fig. 1 Examples of fuzzy
sets for membership
requirement: *Relaxed,
Moderate,* and *Strict.* These
two sets of membership
requirements are used in the
case study (Sect. 3.6)

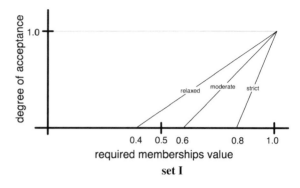

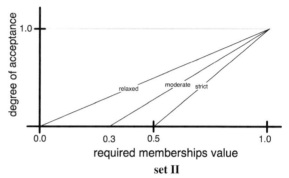

⟨0,1⟩ representing the user's perception of what membership values an element
should have to be treated as a member of a set. Examples of different sets: *Strict,
Moderate,* and *Relaxed* are shown in Fig. 1. The user does not need to understand
the concept of fuzzy sets; she provides only a linguistic term representing her level
of "strictness" in determining the level of similarity—the so-called similarity
regime.

For the purpose of determining cardinality the above-defined fuzzy set
MemberReq is "interpreted" in the following way: values from the universe of
discourse of *MemberReq* are treated as values of α used for α-cuts. So, right now the
set *Strict* indicates the user's rigorous requirements regarding α values, i.e., the user
considers as valid only α-cuts for α of 0.8 and higher. For example, the $\alpha = 0.9$ has
a membership value of 0.5 and this means the user's level of acceptance of an α-cut
for $\alpha = 0.9$ is equal to 0.5.

This allows us to create a fuzzy set, called *Cardinality*, representing user's
perception of cardinality in the context of user's acceptable levels of α's for α-cuts.
The set is

$$Cardinality_{memberReq}(A) = \frac{\mu(\alpha_i)}{card_{\alpha_i}}$$

where $\mu(\alpha_i)$ represents the acceptance of value α determined by the user (via *MemberReq* set), and *card* is the cardinality obtained for α-cut of the set A. In other words, this set represents user's degrees of acceptance of cardinalities calculated for different α-cuts performed on a given fuzzy. The value of cardinality is obtained via defuzzification [10, 11].

2.2 Pythagorean Fuzzy Sets

The PFS have been proposed by Yager in [7, 8]. It is a class of new non-standard fuzzy sets. A membership grade of PFS can be expressed as follows: $r(x)$ called the strength of commitment, and $d(x)$ called the direction of commitment, for each x from the domain X. Both of them are in the range from 0.0 to 1.0. Here, $r(x)$ and d (x) are linked with a pair of membership grades $A_Y(x)$ and $A_N(x)$. $A_Y(x)$ represents support for membership of x in A, while $A_N(x)$ represents support against membership of x in A. The relations between all these measures are follows:

$$A_Y^2(x) + A_N^2(x) = r^2(x)$$
$$A_Y(x) = r(x) * \cos(\theta)$$
$$A_N(x) = r(x) * \sin(\theta)$$

where

$$\theta = \arccos\left(\frac{A_Y(x)}{r(x)}\right)$$

and

$$\theta = (1 - d(x)) * \frac{\pi}{2}$$

The value of $r(x)$ allows for some lack of commitment. Its value is in the unit interval and the larger $r(x)$ is the stronger the commitment is, and the less the uncertainty it represented. The value of direction of commitment $d(x)$, on the other hand, can provide an interesting insight into relations between $A_N(x)$ and $A_Y(x)$. For $\theta = 0$, i.e., when there is not $A_N(x)$ part, the value of $d(x) = 1$ and this means that there is no "negative" comments and the commitment direction is fully positive. When both $A_Y(x)$ and $A_N(x)$ are equal, the angle $\theta = \pi/4$ and the value of $d(x) = 0.5$ indicating a natural direction of commitment. The other boundary condition is when only $A_N(x)$ is present, and there is no $A_Y(x)$—in this case, $d(x) = 0.0$ representing a lack of positive direction of commitment. As you can see the value of $d(x)$ is "detached" from the number of comments, it only depends on the ratio of comments. It is a very interesting feature of PFS that is used in the proposed approach.

2.3 Fuzzy Sets and Aggregation

Fuzzy set theory [6] aims at handling imprecise and uncertain information in various domains. Let D represents a universe of discourse. A fuzzy set F with respect to D is defined by a membership function $\mu_F: D \rightarrow [0,1]$, assigning a membership degree $\mu(d)$ to each $d \in D$. This membership degree represents the level of belonging of d to F. The fuzzy set can be represented as pairs:

$$F = \left\{ \frac{\mu(d_1)}{d_1}, \frac{\mu(d_2)}{d_2}, \ldots \right\}$$

For more information on fuzzy sets and systems, please consult [10, 11].

Aggregation of different pieces of information is a common aspect of any system that has to infer a single outcome from multiple facts. An interesting class of aggregation, called ordered weighted averaging (OWA) [12] operators, is a weighted sum over ordered pieces of information. In a formal representation, the OWA operator, defined on the unit interval I and having dimension n (n arguments), is a mapping $F_w: I^n \rightarrow I$ such that:

$$F_w(a_1, \ldots, a_n) = \sum_{j=1}^{n} (w_j * b_j)$$

where b_j is the jth largest of all arguments $a_1, a_2, \ldots, a_n$, and w_j is a weight such that w_j is in [0,1] and the sum of all w_j is equal to 1. For more on OWA, please go to [13].

3 Finding Pertinent Individuals or Groups of Interests in Social Networks

The users of social networks provide information regarding their likes, as well as opinions about items the network is dedicated to. Therefore, a lot of information is available to create a description of users and groups of users. The approach presented below is based on application of fuzzy sets and relations as a means of defining users' profiles based on such information.

3.1 Social Network and Tagging

Importance and popularity of social networks result in analysis of different aspects of the networks such as recommending friends or finding coherent groups of individuals. The basic approaches to determine similarity between users are (based on [14]): link matching—analysis of graphs created based on social networks; content matching—similarities of contents posted by the users; and content and link matching—the content matching algorithm equipped with social links derived from the structure of social networks.

For the content matching, the topic very much related to this chapter, we can distinguish three ways of identifying a community a given user can join [15]. The first one is via checking a directory of communities manually organized by topics. The second one is based on a set of social ties or via recommendations, while the third one is performed via a keyword and/or metadata search. One of the content matching techniques is the topological tree method used to automatically identify and organize social networking groups based on a simple query [15]. Detection of compatibility between users based on "semantic" and "complementary" relations is presented in [16]. Another content matching algorithm is fully based on the comparison of contents posted by users on the Web [14]. The approach is closely related to finding documents of similar content in the information retrieval field. An interesting approach to determine similarity between users is presented in [17]. The authors look for "clones" within a large, sparse social graph, as well as patterns of shared interests. A different approach to find similar users is considered in [18]. To model similarity, the authors assume that each user has a vector of preferences, and two users are similar if their preference vectors differ in only a few coordinates.

A growing involvement of users in building repositories of digital items and being responsible for their maintenance brings a new approach to describing resources. This new approach is called *tagging*. All items are described by anyone who "sees" them and wants to provide his/her description and/or comment. This approach becomes quite popular due to its simplicity, effectiveness, and enjoyableness [19]. Tagging becomes a source of information that is used for a number of research topics. For example, discovering changes in behavioral patterns of users [20], or discovering regularities in user activities, tag frequencies, and kinds of tags used [21], just to mention a few. Formally, a tagging system can be represented as a tuple [22]:

Definition 1 A tagging system is a tuple TS = (U, T, R, Y) where U, T, and R are finite sets with elements *users*, *tags* and *resources*, respectively, while Y is a relation between the sets, i.e., Y is subset of $U \times T \times R$. A *post* is a triple (u, T_{ur}, r) with $u \in U$, and $r \in R$, and a non-empty set $T_{ur} = \{t \in T | (u, t, r) \text{ from } Y\}$.

The relation (u_k, t_i, r_j) means that the user u_k assigned the tag t_i to the resource r_j.

3.2 User Signature

Social network users constantly interact with their peers, add new items and tags, as well as tag items they experienced. One of the ways of looking at this activity is a fragment of the $\langle U, R, T \rangle$ matrix that is related to a single user u_i. Such a fragment could look like the one presented in Table 1. This two-dimensional matrix

Table 1 A slice of $\langle U, R, T \rangle$ matrix representing activates of the user u_i, as well as a fuzzy relation—her signature (values in the brackets)

Tag	Res							Tag freq	TagPop
	r1	r2	r3	r4	r5	r6	r7		
t1	x (1/2)	x (1/2)						2	2/4
t2			x (1/4)					1	1/4
t3			x (3/4)	x (1/2)	x (3/4)		x (1/4)	4	4/4
t4	x (1/2)			x (1/2)				2	2/4
t5					x (1/2)	x (1/2)		2	2/4
t6	x (1/2)	x (1/2)						2	2/4
t7	x (1/2)		x (3/4)		x (3/4)			3	3/4
t8						x (1/4)		1	1/4
t9		x (1/2)						1	1/4
Tag/res	4	3	3	2	3	2	1		
Res Attract	4/4	3/4	3/4	2/4	3/4	2/4	1/4		

represents all information about activities of the user u_i. Any time, she uses a tag to describe an item (resource), a trace is left in the matrix. A mark "x" is put at the crossing of a specific tag and recourse to indicate user's interest in this tag and resource. If we sum up the cells (treating "x" as one) over a single row, we obtain frequency of usage of a given tag—the column "tag frequency"; if we sum them up over a column, we obtain a number of tags used by the user to describe a resource—the row "tags/resource." See Table 1 for sample values.

A closer look at the row "tags/resource" in Table 1 allows us to think about these numbers as indicators of importance/attractiveness of resources for the user. The idea of building a fuzzy set "Resource Attractiveness"—*ResAttract* for short—seems to be natural. This is a set with membership functions representing a degree of attention the user commits to each of the resources. There are multiple ways of building such a fuzzy set. One of them is proposed below:

$$ResAttract_{u\_k}(r) = \left\{ \frac{b_1}{r_1}, \frac{b_2}{r_2}, \rightleftharpoons, \frac{b_m}{r_m}, \rightleftharpoons, \frac{b_M}{r_M} \right\}$$

where

$$b_m = \frac{\#\ of\ tags\ used\ for\ r_m\ by\ u_k}{\max \#\ of\ different\ tags\ used\ for\ a\ single\ recource\ by\ u_k}$$

The values of the proposed membership function are determined as a ratio of a number of tags the user used for a given resource to a maximum number of tags the user used to label a single resource among all resources labeled by the user. In other words, the value assigned to a single resource expresses the user's degree of attention dedicated to this resource when compared to the most "popular" resource, i.e., the resource with the highest number of labels given by the user.

A very similar procedure can be applied to the column "tag frequency" in Table 1. Here, we create a fuzzy set "Tag Popularity"—*TagPop* for short. The purpose of this fuzzy set is to assign to each tag a degree of its utilization, i.e., a value that represents how often the user uses it among all tags. The membership function is

$$TagPop_{u\_k}(r) = \left\{ \frac{a_1}{t_1}, \frac{a_2}{t_2}, \rightleftharpoons, \frac{a_n}{t_n}, \rightleftharpoons, \frac{a_N}{t_N} \right\}$$

where

$$a_n = \frac{\#\ of\ times\ t_n\ was\ used\ by\ u_k}{\max \#\ of\ resources\ tagged\ by\ u_k\ with\ a\ single\ tag}$$

The procedure for calculating the values of a membership function a_n uses the maximum number of resources tagged with a single tag. In general, it is possible to use any reasonable number as the reference—for example, an average number of resources tagged by a single tag. In such a case, we assume that any tag that is used above average will have a membership value of one. This would represent an "optimistic" approach to estimating tag popularity. In the example presented in the paper, we use the formula above—more "pessimistic" view.

Both fuzzy sets defined *TagPop* and *ResAttract* describe user's activities—degrees of her interest in resources (higher degree of membership of a fuzzy set *ResAttract* more attention she dedicates to a resource), and degrees of her usage of labels (higher degree of membership of a fuzzy set *TagPop* more often a given tag is used). In order to capture both aspects of the user's typical usage of resources and tags we propose to build a fuzzy relation based on both sets. We will call this relation *UserSignature*:

$$UserSignature_{u_k}(r, t) = ResAttract_{u_k}(r) \times TagPop_{u_k}(t)$$

For a single resource r_i and a single tag t_j, the relation value is

$$UserSignature_{u_k}(r_i, t_j) = \min(ResAttract_{u_k}(r_i) \times TagPop_{u_k}(t_j))$$

This relation is interpreted in the following way: high values of the relation indicate resources that are of interest for the user labeled with tags that the user likes. In other words, this relation has the highest values for the case of the most distinctive—unique for the user—resource/tag combinations. Based on the definition presented above and the examples of fuzzy sets we can build such a fuzzy relation—values in the brackets in Table 1.

In general, we can perform some simple analysis of this user's signature. For example, we can project it on one of the dimensions—resources or tags—and study user's interest in resources and tags' popularity once the influence of tags and resources has been taken care of (via building the relation). If additionally, we make α-cuts [10, 11] for different values of α, we can look at subsets of resources representing different degrees of user's interests, and subsets of tags representing different degrees of popularity.

3.3   Finding Pertinent Individual

The relation *UserSignature* created for each user constitutes the user's description. Therefore, we assert that comparison of users is equivalent to comparison of their *UserSignature* relations. The relations *UserSignature* are two-dimensional fuzzy sets. We want to compare these fuzzy sets based on their elements. The similarity measure we use here, analogous to the Jaccard index, is

$$intrstBsdComp(u_i, u_j) = \frac{|T(UserSignature_{u\_i}(r, t), UserSignature_{u\_j}(r, t))|}{|UserSignature_{u\_i}(r, t)|}$$

where $|.|$ represents set's cardinality, a T-norm is used to determine the union of sets [9, 23]. The presented equation requires cardinality of a fuzzy set (Sect. 2.1). As it can be observed, the *intrstBsdComp* between *UserSignatures* is done in the reference to the *UserSignature* of the user who is trying to determine compatibility of other users to her. Such a notion is quite natural—in reality if we try to determine our similarity to another person we always compare everything to what we know about ourselves.

3.4   Group Signature

The process of determining interest-based compatibility between users can be also applied to the process of determining interest-based compatibility between a single user and a group, as well as between two groups.

A group G contains a number of users (we assume here a binary membership, i.e., a user belongs to a group or not, in general we can have a degree of membership to a group, any method suitable for determining users cliques can be applied here). When each user is represented with her *UserSignature*, we can aggregate all *UserSignatures* and obtain a *GroupSignature*:

$$GroupSignature(r, t) = AggregationOper_{u\_iG}(UserSignature_{u\_i}(r, t))$$

where *AggregationOper* represents any aggregation operation. The operation that is considered in the chapter is the *OWA* operator. The attractiveness of *OWA* comes from its ability to combine pieces of information using linguistic quantifiers (*LQ*) defining both range and degree of contribution of individual pieces toward the overall value. Utilization of different *LQ* allows us to control the degree of exactness of group's description (the depth of analysis of users' activities):

$$GroupSignature(r, t) = OWA^{LQ}_{u\_i \in G}(UserSignature_{u\_i}(r, t))$$

and is performed on a single pair $\langle resource, tab \rangle$ basis.

For example, if we use the linguistic quantifier "**max**" we construct *GroupSignature* in a very forgiving way—even resources/tags that are popular among a few users or used by a very few users become a part of this group's signature. This means a broad and full description of the group—even pairs $\langle resource, tag \rangle$ with small interests (tiny values of membership) become a part of signature representing the group.

The quantifier "**min**," on the other hand, means we want to represent the users' activities only via the most frequently used labels and most popular resources across all users. This means that a group signature contains only pairs that are the most popular among all members of the group. The signature is constituted only with core pairs.

The "in the middle" case is obtained with the quantifier "**most**." The group signature obtained with this quantifier contains $\langle resource; tag \rangle$ pairs that are most popular among the users.

3.5 Finding Pertinent Group

Once *GroupSignature* of a group is determined, we can compare it with the user's *UserSignature* in a similar way we do it for the case of two *UserSignatures*. The interest-based compatibility is determined with the following formula:

$$intrstBsdCompUG(u_i, G) = \frac{|T(UserSignature_{u\_i}(r, t), GroupSignature(r, t))|}{|UserSignature_{u\_i}(r, t)|}$$

The process of determining cardinality of fuzzy sets can be done with different membership requirements (different α-cuts). When this is combined with different approaches for determining a group signature, the user can identify different ways of comparing herself to a group.

Some possible variations of membership requirements and aggregation of signatures of group members give the following scenarios of identifying compatibility between the user and a group:

Relaxed and max (broad): The user wants to explore, she wants to find new things and looks for novelty, the user "compares" herself to all members of the group; however, the user does not require that majority of the group has to "fit" her interests, even single individuals who have very similar signatures would make the group look like a good match;

Strict and min (focus/core): The user is very focused and looks for a group that has interests very similar to her, a group signature reflects common interests of all users; therefore, a group the user is looking for is quite coherent in the interests that match hers to the highest possible level;

Moderate and most (majority): The user is open for novelty but does not want to "go" too far with newness; The user is okay if not all members are interested in the same things, but most of them are, at the same time, okay if only a few group members like things she does.

Presented above three possible scenarios describing different needs and requirements of the user shows flexibility of the proposed approach, as well as its ability to become a truly human-centric method of finding interest groups.

3.6 Case Study

The presented case study embraces a single scenario where the user is trying to find whether an existing group of users is a good fit for her. We use a small fraction of the tag cloud from the www.thelibrarything.com. The slices representing five "experienced" uses—A, B, C, D, and E, and a "novice" user—T—have been pulled out from the cloud. Figure 2 illustrates resources—authors—that are tagged by the experienced users. For example, the user A has tagged ten different authors: Borges, Cervantes, Calvino, Alighieri, Marquez, Neruda, Hamsun, Merwin, McEwan, and Rushdie. At the same time, we can say the users A, B, C, and D tagged the author Gabriel Garcia Marquez, while A, B, and D tagged Salman Rushdie. These five users constitute a group.

There are 28 different tags that have been used by the users. Some of them are as follows: fiction, twentieth–twenty-first-century novel, translation, and European literature. The user A is most active in the group, while the user T does not have a lot of tags and only few resources. A closer look at the user's T activities indicates her inclination toward fiction.

The determination process is performed for two sets of membership requirements: *Strict, Moderate*, and *Relaxed* (Sect. 2.1), Fig. 1. The **set I** represents a very

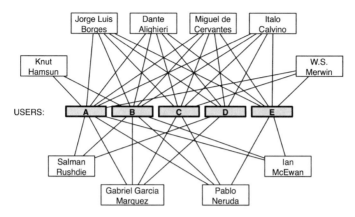

Fig. 2 Users *A, B, C, D,* and *E* together with resources representing authors tagged by them

demanding needs. Even for the *Relaxed* requirement, a membership value of 0.5 is accepted only to the degree 0.167. The acceptance level of 0.5 is set for the membership value of 0.7. The **set II** is different, for the *Relaxed* requirement membership values of 0.5 and 0.7 are accepted to the degrees of 0.5 and 0.7, respectively. The calculated values of interest-based compatibility for the user *T* for both sets of membership requirements are presented in Table 2.

For the group signature created with the operator **max**, and for the **set I** of membership requirements, the compatibility between *T* and the group is not higher than 0.542 (for the requirement *Relaxed*). These values are similar to the ones obtained for the compatibility between the user *T* and the user *A* (not reported in the paper). This seems justified by the fact that the user *A* is the most active one and the group signature reflects the fact that interests of all other users are "subsumed" by the interests of the user *A*. For the **set II**, the values for *Strict* and *Moderate* membership requirements are similar to the ones obtained for the user-to-user comparison (*T-to-A*, not shown here), but for the *Relaxed* requirements the value of compatibility equals 0.940 and is higher than for the user-to-user. This indicates that when the group is represented by the signature taking into account any book or tag used by its members, the group "subsumes" the user *A*.

The results for the comparison of the user *T* with the group when the group's signature is obtained using **min**-operator are quite different. For the **set I**—the values of interest-based compatibility are all zero except the *Relaxed* requirement,

Table 2 Interest-based compatibility between a user and a group

User *T*	Membership requirements					
	Strict		Moderate		Relaxed	
	Set I	Set II	Set I	Set II	Set I	Set II
Versus group (max)	0.000	0.438	0.300	0.617	0.542	0.940
Versus group (min)	0.000	0.000	0.000	0.118	0.042	0.400

but even then its value is only 0.042. The compatibility looks a bit better when the **set II** is used. In this case, the value for the *Relaxed* requirement is equal to 0.400.

All this means that the user T and the group are interest-based compatible in the broad sense. The group would fit the needs of the user T when she looks for novelty and wants to explore, as well as find new books to read. On the other hand, when the user T is focused and is looking for a group with core interests similar to hers, the group is not a very good fit.

4   Construction of Item Suggestion Lists

One of the most common utilization of the Web is a search for variety of items. Users look for suggestions what to buy, see or where to go, just to name a few of types of items for which they seek suggestions. In this section, we propose a new approach for ranking items evaluated by multiple users. The fuzzy-based method allows the user to determine in a linguistic way a group of users whose evaluations are taken into consideration when a ranking list is created.

4.1   Recommending or Suggesting

When a list of suggested items is provided to the user, it would be good to see some variety of items shown to the user. This would resemble a true browsing process—the user "roughly" knows what she wants and tends to wander around looking for things that could be of interest for her. Of course, the user is doing it with all her "background" knowledge regarding what she likes, but at the same time, she lets herself to be adventurous. In order to mimic such behavior, we propose the following procedure that ensures a more human-like experience in finding variety of items that could have potential interest to the user.

The **first step** is to determine a group of users that have evaluated the same items (for example movies) in a similar way as the user who looks for suggestions. The similarity is controlled by the user and reflects her way of determining a level of matching between her evaluations and evaluations of other individuals who experienced (watched) the same items (movies) as she. So far, we envision three possible ways of identifying similarity of items (movie) evaluations: (a) *strict*—evaluations have to match to the highest possible degree; (b) *strict on high*—positive evaluations have to match to the highest possible degree; however for the negative evaluations, the user is forgiving, in other words, the user wants to see a list of suggestions created based on users who agree with her in the case of the positive evaluations and could have different opinions in the case of negative evaluations; and (c) *strict on low*—the opposite to the above one, i.e., the user wants to see a list of suggestions created based on users with whom she agrees regarding the negative evaluations, and relaxes her requirements for matching positive evaluations. This process will lead to creation of a list of users with similar—up to the user's requirements—preferences regarding items (movies). Each user's preference pattern is determined by aggregation of all matching scores across of all items (movies) overlapping between the user and a potential member of a group.

The **second step** is to identify a list of items that have not been experienced by the user who looks for suggestions, but common to all the users from the group. It is the simplest step—it involves finding a common set of items among all items experienced by the members of the group identified in the first step.

The **third step** is to rank all selected items. The ranking of items has to take into consideration two components: quantitative—a number of ratings for each item; and qualitative—a distribution of positive and negative ratings.

4.2 Building Base of Related Users

In order to provide any suggestions, we need to have a pool of users who somehow resembles or have similar interests to the user who is looking for suggestions. Selection of such a group can be done in multiple ways, and in majority of situations such as process happens without any influence of the user. The approach we propose here provides the user with a way of indicting how "close" members of created group should be when compared with the user. The user governs this process via providing linguistic terms identifying strictness of the comparison between items rated by the user and items rated by members of a group being created.

Any items rated by two users can be compared using different approaches. The simplest one would be to check whether both ratings are the same. In such a case, we would obtain a binary result—a perfect match or total mismatch. Quite often, such a comparison is useful and it leads to finding a person who is very much like the user, kind of "a mirror image of the user." However, in the situation the user would like to "expend" her set of items and "go beyond" its own comfort zone, i.e., find something more diverse—a different type of comparison is needed. The approach presented here addresses such a need.

A proposed approach for comparison of ratings uses linguistic terms to control flexibility of this process, i.e., how differences between ratings should be treated. There are a number of different rating schemas, but without loss of generality we assume for the rest of the paper that the ratings are in the range from 1 to 5, where 1 represents "do not like" and 5 "like very much," with 3 indicating "neutral." To make the comparison of ratings controllable, we identify three terms describing strictness of the comparison process:

– *two-side-bounded* evaluation: the comparison is very strict, both ratings—positive and negative—have to be matched to the highest degree;
– *positive-side-bounded* evaluation: the comparison is strict for positive evaluations and relaxed for negative evaluations; in other words, the user is okay when

her negative ratings are not considered very "seriously," but positive ratings have to be matched to the highest degree;

– *negative-side-bounded* evaluation: the comparison is opposite to the above one— positive ratings do not have to match perfectly, but the negative ones have to be respected.

Let us assume we have two items to compare: an item to which the comparison is done—we called it a *reference_item*, and an item being compared—we called it an *other_item*. The comparison procedure is as follows: (1) the rating R_{ref} of the *reference_item* is subtracted from the rating R_{other} of the *other_item*, (2) the obtained difference is modified by an appropriate mapping representing the user's evaluation requirement, and the resulting value is a level of compatibility of two ratings.

An implementation of terms controlling the evaluation process is done using fuzzy sets. Let us define a universe of discourse **D** that is a range of differences between ratings. A fuzzy set is a mapping

$$\mu_{LTerm} : D \rightarrow [0, 1],$$

and the compatibility of ratings is

$$RateComp = \mu_{LTerm}(R_{\text{other}} - R_{\text{ref}})$$

where μ_{LTerm} is a fuzzy set associated with an appropriate linguistic term. Three possible sets representing the described above terms are shown in Fig. 3.

Figure 3a represents a fuzzy set for the *two-side-bounded* evolution. It allows for a small deviation of R_{other} from R_{ref}: the perfect match gives the score of 1.0, while a difference of 1.0 or −1.0 leads to the score of 0.5. Any other difference gives the

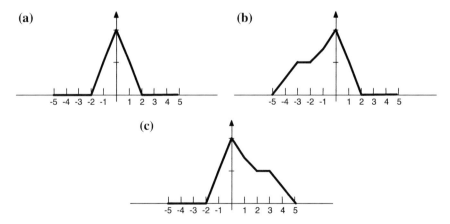

Fig. 3 Examples of fuzzy sets representing three linguistic terms: *two-side-bounded* (**a**), *positive-side-bounded* (**b**), *negative-side-bounded* (**c**)

score of 0.0. A fuzzy set for the *negative-side-bounded* evaluation, Fig. 3b, gives the score of 1.0 for the perfect match. However, it "penalizes" the situation when the difference $R_{other} - R_{ref}$ is positive, what ensures that the user's low ratings are well matched. On the other hand, when the difference $R_{other} - R_{ref}$ is negative, i.e., when R_{other} is smaller than R_{ref}, the evaluation is more forgiving—there is no requirement for a good match for the user's high ratings. A fuzzy set presented in Fig. 3c for the *positive-side-bounded* evaluation reflects the inverse behavior to the *negative-side-bounded* case.

A group is built in a process of selecting users who have a similar pattern of rating items to the user who looks for suggestions. Using different comparison fuzzy sets, we obtain different groups of users. Each of these groups contains users who match the user (our reference user) differently. This effect is desirable because this means that each group contains a slightly (not totally) different set of users. And a different set of users leads to a different set of items—therefore, suggestions provided by each group can be a bit different.

In a nutshell, the process of creating a group of users can be described via the following algorithm:

INPUT: M_{ref} movies rated by the user U_{ref}
 $\quad\quad\quad\quad$ UL_{all} a list of users
 $\quad\quad\quad\quad$ $\{M_i\}$ a set of movies rated by each user (U_i) from the list UL_{all}
OUTPUT: G $\quad$ a list of users constituting a group

1 Select Users who ranked the Same Movies as U_{ref}, i.e., M_i contains M_{ref}
2 Create a list of Selected Users: UL_{other}
3 **for** each user U_i from UL_{other}
4 $\quad$ **for** each movie from M_{ref}
5 $\quad\quad$ Compare Ratings: calculate score and store it
6 $\quad$ **rof**
7 $\quad$ Create Rating Compatibility measure: aggregate scores
8 **rof**
9 Sort Users from UL_{other} based on Rating Compatibility measures
10 Create group G: select top 25 percent from sorted UL_{other}

Another human-like aspect of creating a group is a linguistic-based control of an aggregation process (line 7). In the proposed approach, we use linguistic-based OWA [23]. Any linguistic quantifiers can be applied to calculate a rating compatibility measure.

Example To illustrate the process, let us take a look at a very simple situation where we have the U_{ref} and another user—U_i, and both of them ranked the same five items. Their ratings are in Table 3 column 1 and 2, respectively.

The difference between ratings is shown in column 3. This difference is used as the input to three different linguistic terms introduced in the previous section: *two-side-bounded*—the obtained values are in column 4; *positive-side-bounded*—column 5; and *negative-side-bounded*—column 6. It can be easily observed that the

Table 3 Illustration of a proposed method for composing a group of users

U_{ref} Ratings	U_i	Difference R_i-R_{ref}	Two-side	Positive-side	Negative-side
			Rating compatibility		
5	1	−4	0	0.25	0
4	2	−2	0	0.5	0
3	3	0	1	1	1
2	4	2	0	0	0.25
1	5	4	0	0	0.5

two-side-bounded term is very strict, the *positive-side-bounded* term "forgives" when there is a mismatch for items ranked high by U_{ref}, and the *negative-side-bounded* term "forgives" when the mismatch happens for items which the U_{ref} ranked low.

The application of OWA with the linguistic quantifier MOST leads to a small compatibility score of 0.04 in the case of the *two-side-bounded* evaluation. For the *negative-* and *positive-side-bounded* evaluation, the score is the same 0.15.

4.3 Creating List of Possible Items

The group generation process ends up with a group of users who rank items in a way that resembles the user's way, i.e., it could "tightly" resemble the user's rankings for both positive and negative rankings, or just one of them. At the same time, many of these users ranked items that have not been experienced/ranked by the user yet, and a set of these items is a starting point for building a suggestion list.

In general, multiple users from the group can rank, in the scale from 1 (worst) to 5 (best), a single item. It means that a single item can have multiple rankings. The proposed approach is taking into consideration two aspects in order to determine a single score representing a degree of attractiveness to the user. These two aspects are as follows: overall combined ratings provided by the users from the group; and a number of users who provided these ratings. Such calculated degree of attractiveness is used to sort items and provide a list of suggestions to the user. In the proposed method, we use PFS to identify the final ranking of suggestions.

4.4 Ranking of Items

A process of building a ranking starts with determining a score, called degree of attractiveness, for a single item. The process comprises a number of steps.

The first step is to build PFS for a single item. In order to do this, we perform a mapping from the users' ratings into $A_Y(x)$ and $A_N(x)$ values. It happens in the following way:

- Ratings 5 and 4 are mapped into 1.0 and 0.5, and the sum of them is assigned to $A_Y(x)$.
- Ratings 1 and 2 are mapped into 1.0 and 0.5, and the sum of them is assigned to $A_N(x)$.
- Rating 3 is natural and is not mapped at all.

Based on obtained values of $A_Y(x)$ and $A_N(x)$, we calculate values $r(x)$—strength of commitment, and $d(x)$—direction of commitment. This is performed for all movies. We find r_{max} and normalize all $r(x)$. Now, the maximum $r(x)$ is 1.0, and all other ones are in the range from 0.0 to 1.0.

The next step is to calculate a score—an attractiveness value. And this is done using a single transformation of $d(x)$ and $r(x)$. If we represent a PFS in polar coordinates (Fig. 4), we can think of an area bounded by the axe $A_N(x)$, $r(x)$, and a fragment of the circle circumference connecting $A_N(x)$ and the tip of the $r(x)$—a thick line in Fig. 4a. Such defined fragment of the circle represents simultaneously two things: a level of commitment (normalized $r_{norm}(x)$) and a direction of commitment—in this particular case more commitment means smaller angle θ and larger the circle fragment. The formula representing this relationship is

$$\text{score}_1 = r_1 * \alpha_1 = r_1\left(\frac{\pi}{2} - \theta_1\right)$$

and knowing that

$$\theta_1 = (1 - d_1) * \frac{\pi}{2}$$

so

$$\text{score}_1 = r_1 d_1 \frac{\pi}{2}$$

As we can see the attractiveness is simply a product of $r(x)$ (strength commitment) and $d(x)$ (direction of commitment), the $\pi/2$ is a simple constant that can be omitted for in the comparison process. If we calculate score for another PFS, Fig. 4b, we can compare both PFSs.

With such defined attractiveness of an item, ranking of items rated by the users for a determined group is a simple sorting process. As the result, we obtain a list of suggestions.

Example As in the case of building a group of users, also here we present a simple example that explains how five items with multiple ratings are compared and eventually ranked. Let us assume the ratings as shown in Table 4, column 1.

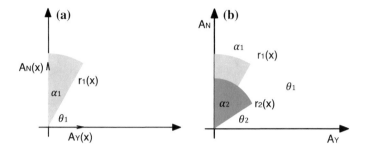

Fig. 4 PFSs in polar coordinates: a single PFS (**a**), and two PFS with different $r(x)$ and $d(x)$ (**b**)

Table 4 Illustration of a proposed method for ranking items with multiple ratings

Ratings		A_Y	A_N	r	r_{norm}	d	Score	Rank
Item-1	5, 4, 3, 2, 1	1.5	1.5	2.1213	0.6975	0.50	0.3486	3
Item-2	5, 3, 1	1.0	1.0	1.4142	0.4650	0.50	0.2325	4
Item-3	5, 4	1.5	0	1.5000	0.4932	1.00	0.4932	2
Item-4	5, 4, 4, 4, 3, 2	2.5	0.5	2.5495	0.8383	0.87	0.7293	1
Item-5	4, 3, 2, 2, 1, 1	0.5	3.0	3.0414	1.0000	0.11	0.1100	5

Table 4 contains all intermediate results; however, we focus on the last four columns: r_{norm}—normalized strength of commitment, d—direction of commitment, as well as score and rank. As we can see the strongest commitment has Item-5, the weakest has Item-2. In the case of direction of commitment—the most positive has Item-3 and Item-4, and the most negative is Item-5. Overall, the score, a product of r_{norm} and d, indicates Item-4 as the most suitable suggestion, with Item-3 as the second (even if it has only two ratings but both positive), and Item-1 as the third (quite a number of ratings but evenly split between positive and negative).

4.5 Netflix Example

In order to illustrate the proposed method in a real-world scenario, we have applied it to building a list of suggested movies. For this purpose, we use Netflix data that have been used in Netflix competition [24]. The database of movies and ratings contains a total of 17,770 movies and 480,190 users.

We have created a user U_{ref} who wants a suggestion list created from Netflix movies. We assume that the U_{ref} have seen the fifteen movies. A sample of these movies and their rating are presented in Table 5.

Group Building Process The first step is to determine a group of users who "fit" the U_{ref} requirements regarding compatibility of movie ratings. We used all three terms introduced in Sect. 4.2. As the result, we obtained three groups of users:

Table 5 A sample of movies and their ratings

Movie		Rating
Title	ID	
Star Trek: Voyager: season 7	10141	5
Star Trek: insurrection	12513	5
Star Trek: the next generation: season 1	10666	5
The sound of music	12074	2
The Exorcist 2: the heretic	9387	2
The Exorcist	16793	1
Night of the living dead	9940	1

- **group T-SB**: obtained with the *two-side-bounded* term; it contains 2 users (their userIDs 1578801 and 647979), and a number of movies rated by the users is 4972;
- **group N-SB**: obtained with the *negative-side-bounded* term: a number of users in the group is 6, and a number of different rated movies is 11034;
- **group P-SB**: obtained with the *positive-side-bounded* term: the group is the largest, it includes 8 users, and a number of different movies to be selected from is 17540.

Generation of Recommendations: The movies associated with each of the groups, and not seen by the U_{ref}, have been used to create the suggestion lists. Due to the space limitations, only top three movies, their scores, and some statistics are shown:

- **group T-SB**: top movies are Akira Kurosawa's Dreams, The Good, and the Bad and the Ugly, Chariots of Fire; a total of 562 movies have a score above 0.5, including 235 with a perfect score of 1.0;
- **group N-SB**: top movies are The Lord of the Rings: The Fellowship of the Ring: Extended Edition, Lord of the Rings: The Return of the King, and Star Wars: Episode IV: A New Hope; in this case 421 movies obtained a score above 0.5, including 8 with 1.0;
- **group P-SB**: top movies are Lord of the Rings: The Return of the King, Lord of the Rings: The Fellowship of the Ring, and Lord of the Rings: The Two Towers; a total of 130 movies with a core above 0.5, only 3 with 1.0 score.

Please note that the recommendations generated based on the group T-SB contain the highest number of movies, and almost half of them have a perfect score. The fact that the U_{ref} is very strict here shows that only two users from 480190 watched the same movies and rated them in the same way. Therefore, the suggestions do not have a huge "base"—only two users have rated each movie. This explains a high number of movies with a perfect score, and somehow diverse set of recommendations.

For the recommendations built using the other two groups, N-SB and P-SB, the situation is different: more users match the movie ratings made by U_{ref}, so the lists of suggestions seem to include movies that appear to be a good fit to U_{ref} interests. There are a much smaller number of movies with the perfect score, but this can be explained by the fact that it is more difficult to find consensus among a larger number of users.

5   Conclusion

The increasing perception of the Web as a new social platform has introduced a need for methods and tools that support users' search for other users or user groups that share, at least partially, their interests. In the paper, we propose to use two fuzzy sets as representations of users. One of these sets is built out of all tags used by the user to annotate resources. It represents popularity of tags and is called *TagPop*. Another fuzzy set is built out of all resources that are annotated by the user. This set represents importance or attractiveness of resources to the user. This set is called *ResAttract*. If both fuzzy sets are created for a single user and combined as a relation, a fuzzy-based description of the user is generated. These descriptions of users—*UserSignatures*—can be used to determine similarity between them.

We propose a measure of interest-based compatibility of a single user to another user. This measure is a ratio of cardinalities of two fuzzy sets. In order to make this feasible, we introduce the concept of cardinality of a fuzzy set that involves an input from the user—a quantity called a membership requirement. This membership requirement is a fuzzy set representing user's perception of what membership values an element should have to be treated as a member of a set. Based on these membership requirements, cardinality values of fuzzy sets are determined. The chapter also proposes the method to aggregate signatures of all users that are members of a single group and create a group signature. The aggregation process is governed via the OWA operator with linguistic quantifiers provided by the user. Using the proposed approach for determining compatibility between users, a level of interest-based compatibility between the user and a group is determined.

The paper reports on preliminary experimental studies. It includes a real-world-based case study using the data from the Web site www.librarything.com. This Web site contains a collection of books and a multitude of users who annotate the books with variety of tags. From this system, we have picked a number of users and performed an experiment for determining interest-based compatibility between a single user and a group.

We envision a series of experiments addressing the construction of fuzzy sets *ResAttract* and *TagPop*; however, the limited space and a need for a thorough explanation of the proposed approach have forced us to leave it for the future work.

The users' anticipations regarding what can be found on the Web exceeds initial expectations of finding items that represent a perfect match to what the users already know. On many occasions, the users want to be exposed to more variety of items, items that go beyond their core interests. The recommendation systems try to provide the user with diversity of suggested items, but it is done without the users' input.

Further, we introduce two novel techniques that lead to construction of alternative versions of lists of recommended items. One of them allows the user to control a process of identification of groups of users that are used as the reference for created an initial list of possible suggestions. The other technique is applied to a process of generating rankings of times. It combines multiple ratings and generates a single score that reflects both a number of ratings and their values.

The new methods have been used to generate lists of suggested movies using Netflix database.

References

1. Bobadilla, J., Ortega, F., Hernando, A., Gutierrez, A.: Recommender systems survey. Knowl.-Based Syst. **46**, 109–132 (2013)
2. Brin, S., Page, L.: The anatomy of a large-scale hypertextual Web search engine. Comput. Netw. ISDN Syst. **30**, 107–117 (1998)
3. Hammond, T., Hannay, T., Lund, B., Scott, J.: Social bookmarking tools (I): a general review. D-Lib Magazine, vol. 11, www.dlib.org/dlib/april05/hammond/04hammond.html (2005)
4. Burke, R.: Knowledge-based recommender systems. In: Encyclopedia of Library and Information Systems (2000)
5. Terveen, L., Hill, W.: Beyond Recommender Systems: Helping People Help Each Other. HCI New Millenn. 487–509 (2001)
6. Zadeh, L.A.: Fuzzy sets. Inf. Control **8**, 338–353 (1965)
7. Yager, R.R.: Pythagorean fuzzy subsets. IFSA/NAFIPS Joint World Congress 57–61 (2013)
8. Yager, R.R.: Pythagorean membership grades in multi-criteria decision making. IEEE Trans. Fuzzy Syst. **22**, 958–965 (2014)
9. Wygralak, M.: On the best scalar approximation of cardinality of a fuzzy set. Int. J. Uncertain. Fuzziness Knowl. Based Syst. **56**, 681–687 (1997)
10. Klir, G., Yuan, B.: Fuzzy Sets and Fuzzy Logic: Theory and Applications. Prentice Hall, Upper Saddle River (1995)
11. Pedrycz, W., Gomide, F.: Fuzzy Systems Engineering: Toward Human-Centric Computing. Wiley-IEEE Press (2007)
12. Yager, R.R.: On ordered weighted averaging aggregation operators in multi-criteria decision making. IEEE Trans. Syst. Man Cybern. **18**, 183–190 (1988)
13. Yager, R.R.: Families of OWA operators. Fuzzy Sets Syst. **59**, 125–148 (1993)
14. Chen, J., et al.: Make new friends, but keep the old—recommending people on social networking sites. In: Proceedings of the ACM CHI, Boston, MA, USA (2009)
15. Freeman, R.T.: Topological tree clustering of social network search results. In: Yin, H. et al. (Eds.) IDEAL 2007, pp. 760–769. LNCS 4881 (2007)
16. Kazemi, A., Nematbakhsh, M. (2011) Finding compatible people on social networking sites: a semantic technology approach. In: Second International Conference on Intelligent Systems, Modelling and Simulation (ISMS), pp. 306–309
17. Tanner, C., Litvin, I., Joshi, A.: Social networks: finding highly similar users and their inherent patterns. Soc. Netw. (2008)
18. Nisgav, A., Patt-Shamir, B.: Finding similar users in social networks. Theor. Comput. Syst. **49**, 720–737 (2011)
19. Smith, G.: Tagging: People-Powered Metadata for the Social Web. New Riders (2008)
20. Fu, W.-T., Kannampallil, T., Kang, R.: A semantic imitation model of social tag choices, 2009. In: International Conference on Computational Science and Engineering, pp. 66–73 (2009)

21. Golder, S., Huberman, B.: The structure of collaborative tagging systems. J. Inf. Sci. **32**, 198–208 (2006)
22. Hotho, A., Jaschke, R., Schmitz, C., Stumme, G.: Information Retrieval in Folksonomies: Search and Ranking, LNAI 4011: The Semantic Web: Research and Applications, pp. 411–426. Springer, Berlin (2006)
23. Yager, R.R.: On ordered weighted averaging aggregation operators in multicriteria decision making. IEEE Trans. Syst. Man Cybern. **18**, 190–193 (1988)
24. http://www.netflixprize.com

Association Measures on Sets
with Involution and Similarity Measure

I. Batyrshin

Abstract The methods of construction of non-statistical association measures on the sets with involution operation and similarity measure are proposed. The Pearson's correlation coefficient is obtained as a particular case of the class of association measures associated with Lukasiewicz t-conorm. Examples of association measures on [0, 1] and on the set of fuzzy sets are considered.

1 Introduction

The association measures are widely used in data analysis. They can measure direct or inverse relationships between data. The inverse relationships can have different nature: "high" versus "small" values of variables, "increasing" versus "decreasing" patterns of time series, similarity with a statement or with its negation in fuzzy logic. The examples of association measures are the Pearson's correlation coefficient [1, 2] playing fundamental role in statistical data analysis, the local trend association measure and its variations [3, 4] used in analysis of time series shape associations and a measure of correlation of membership functions considered in [5]. Recently, a problem of construction of time series shape association measures defined as functions satisfying some reasonable properties was discussed in [6] and the methods of construction of such measures using distance-based similarity measures have been proposed in [7]. The results of [7] have been extended in [8] on a set with involution, and the methods of generation of association measures proposed in [7] have been extended here using the pseudo-difference operations associated with t-conorms. It was shown [7, 8] that the sample Pearson's correlation coefficient can be constructed as a particular case of the family of association measures related to Lukasiewicz t-conorm.

I. Batyrshin (✉)
Centro de Investigación en Computación (CIC-IPN),
Instituto Politécnico Nacional, DF, Mexico
e-mail: batyr1@gmail.com

© Springer International Publishing Switzerland 2016
L.A. Zadeh et al. (eds.), *Recent Developments and New Direction in Soft-Computing Foundations and Applications*, Studies in Fuzziness and Soft Computing 342, DOI 10.1007/978-3-319-32229-2_16

221

This work is based on the paper [9] with the same name presented on 4th World Conference on Soft Computing, Berkeley, California, 25–27 May 2014. It extends the results of [7, 8] on general case of non-statistical association measures as functions defined on a set with involution and similarity measure and satisfying the properties similar to the properties of Pearson's correlation coefficient. It considers some problems appearing in unified definition of non-statistical association measures on such different domains such as the set of time series, the set of truth or probability values [0, 1] and the set of fuzzy sets. Some new results in this area can be found in [10–15].

The paper has the following structure. Sect. 2 discusses the desirable properties and possible definitions of association measures on the set with involution. Section 3 considers some definitions of basic concepts of fuzzy logic used further in the paper. Section 4 considers the general methods of construction of association measures by means of similarity measures and pseudo-difference operations associated with t-conorms. The methods proposed in this section are used in Sect. 5 and in Sect. 6 for the construction of examples of association measures on [0, 1] and on the set of fuzzy sets with discrete domain. The conclusions and discussions are given in the last section.

2 Definition of an Association Measure

As a prototype in the definition of association measure, consider the sample Pearson's correlation coefficient:

$$\mathrm{corr}(x, y) = \frac{\sum_{i=1}^{n}(x_i - \bar{x})(y_i - \bar{y})}{\sqrt{\sum_{i=1}^{n}(x_i - \bar{x})^2 \cdot \sum_{i=1}^{n}(y_i - \bar{y})^2}}, \tag{1}$$

taking values in [−1, 1]. Note that this function does not defined for $x = const$ and $y = const$ and one can restrict himself calculating corr(x, y) only when $x, y \neq const$. For the completeness of the definition, corr(x, y) can be redefined properly for x, $y = const$. The correlation coefficient has the following properties fulfilled for all x, $y \neq const$:

$$\mathrm{corr}(x, y) = \mathrm{corr}(y, x), \tag{2}$$

$$\mathrm{corr}(x, x) = 1, \tag{3}$$

$$\mathrm{corr}(x, -y) = -\mathrm{corr}(x, y), \tag{4}$$

$$\mathrm{corr}(-x, -y) = \mathrm{corr}(x, y), \tag{5}$$

Table 1 Example of three synthetic time series with corr $(x, y) = corr(x, -y) = 0$

I	1	2	3	4	5	6	7	8	9	10
X	10	8	5	6	9	15	20	25	18	14
Y	20	12	1	2	4	5	6	7	4	2
−y	−20	−12	−1	−2	−4	−5	−6	−7	−4	−2

$$corr(x, -x) = -1, \tag{6}$$

$$corr(x, x) \geq corr(x, y). \tag{7}$$

(5) follows from (4) and (2); (6) follows from (4) and (3); and (7) follows from (3) taking into account corr$(x, y) \in [-1, 1]$. Hence, the properties (5)–(7) follow from (2)–(4). These properties can be considered as basic properties of the correlation coefficient. Note that (6) and (3) [and hence, (4) and (3)] are contradictory when $x = -x$, so we need to avoid such contradictions between possible axioms of association measures.

The correlation coefficient plays fundamental role in statistical data analysis. However, in time series data mining [3, 4, 16–20] in the analysis of similarity and associations between shapes of time series, the correlation coefficient is not a best choice for association measure. For example, suppose that X is a set of real-valued time series $x = (x_1, ..., x_n)$ with $-x = (-x_1, ..., -x_n)$. Consider the modified benchmark example of synthetic time series from [4, 14, 21] given in Table 1. Figure 1 depicts the charts of these time series with all values multiplied by 10.

The time series x and y have the similar shapes, and it can be supposed that the reasonable shape association measure should have sufficiently high positive $A(x, y) > 0$ and negative $A(x, -y) < 0$ values, but we have for them corr$(x, y) = $ corr$(x, -y) = 0$. The better choice for a measure of time series shape associations gives the local trend association measure [3, 14] based on moving approximation transform (MAT). MAT transforms time series $x = (x_1, ..., x_n)$ into a sequence $MAT_k(x) = (a_1, ..., a_{n-k+1})$ of slope values (local trends) of simple linear regressions, $f_i = a_i t + b_i$, of time series values $(x_i, ..., x_{i+k-1})$, $i \in \{1, ..., n - k + 1\}$ in sliding window of size k. Suppose that $x = (x_1, ..., x_n)$ and $y = (y_1, ..., y_n)$ are the two time series and $MAT_k(x) = (a_{x1}, ..., a_{xm})$, $MAT_k(y) = (a_{y1}, ..., a_{ym})$, $k \in \{2, ..., n - 1\}$, $m = n - k + 1$ are their MATs. The following function is called a measure of local trend associations (LTA):

$$lta_k(x, y) = \frac{\sum\limits_{i=1}^{m} a_{xi} \cdot a_{yi}}{\sqrt{\sum\limits_{i=1}^{m} a_{xi}^2 \cdot \sum\limits_{j=1}^{m} a_{yj}^2}}, \quad k \in \{2, ..., n - 1\}. \tag{8}$$

This measure, for example, for time series given in Table 1 and Fig. 1 for window size $k = 2$ gives value lta$_2(x, y) = 0.53$ that corresponds to our perception about positive association between the shapes of time series x and y, because they

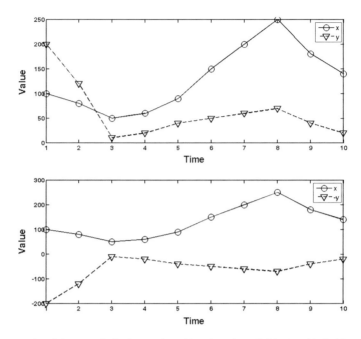

Fig. 1 Example of three synthetic time series with values from Table 1 multiplied by 10. Time series: x and y (*from above*), x and $-y$ (*from below*). We have: corr(x, y) = corr(x, $-y$) = 0

are synchronously moving up and moving down. For time series x and $-y$, this measure gives dually negative association value $\text{lta}_2(x, -y) = -0.53$. In [4], it was considered a modification of this method when slope values a_{xi} in (8) are replaced by $\text{sign}(a_{xi}) \in \{-1, 0, 1\}$ values. The modified in such form association measure (8) for benchmark example from Table 1 gives values as $\text{lta}_2(x, y) = 1$, $\text{lta}_2(x, -y) = -1$. Both measures satisfy the properties (2)–(4):

$$\text{lta}_k(x, y) = \text{lta}_k(y, x), \tag{9}$$

$$\text{lta}_k(x, x) = 1, \tag{10}$$

$$\text{lta}_k(x, -y) = -\text{lta}_k(x, y). \tag{11}$$

These measures are not defined if all slope values are equal to 0.

In [6], a time series shape association measure is defined for all x, $y \neq$ const, as a function taking values in $[-1, 1]$ and satisfying the properties:

$$A(x, y) = A(y, x), \tag{12}$$

$$A(x, x) = 1, \tag{13}$$

$$A(x, -x) = -1, \tag{14}$$

$$A(x, -y) = -A(x, y). \tag{15}$$

$$A(x + q, y) = A(x, y) \text{ for all real values } q. \tag{16}$$

In many applications, also the following property can be required:

$$A(px, y) = A(x, y), \quad \text{if } p > 0. \tag{17}$$

The general method of construction of shape association measures satisfying these properties has been proposed in [7]. It contains the following components: (1) a time series standardization $F(x)$; (2) a dissimilarity measure $D(F(x), F(y))$; (3) a transformation of D into a similarity measure SIM; and (4) a transformation of SIM in an association measure A. Each component should satisfy some conditions to obtain finally a shape association measure A satisfying considered above properties. This approach gives possibility to define a wide range of association measures. For example, the correlation coefficient (1) can be obtained using the following components:

$$F(x)_i = \frac{x_i - \bar{x}}{\sqrt{\sum_{j=1}^{n} (x_j - \bar{x})^2}}, \quad \bar{x} = \frac{1}{n} \sum_{j=1}^{n} x_j \tag{18}$$

$$D(x, y) = \sqrt[2]{\sum_{i=1}^{n} |F(x)_i - F(y)_i|^2}, \tag{19}$$

$$SIM(x, y) = 1 - D^2(x, y)/4 = 1 - \frac{1}{4} \sum_{i=1}^{n} |F(x)_i - F(y)_i|^2, \tag{20}$$

$$A_{SIM}(x, y) = SIM(x, y) - SIM(x, -y). \tag{21}$$

In [8], it was shown that (21) can be considered as a pseudo-difference operation associated with Lukasiewicz t-norm, and the sample Pearson's correlation coefficient can be considered as a member of the family of association measures related to Lukasiewicz t-conorm.

In [5], it was considered a measure of correlation of membership functions f (in the case of finite domain):

$$c(f_1, f_2) = \begin{cases} 1 - \frac{4}{H} \sum_{x \in U} [f_1(x) - f_2(x)]^2, & \text{if } H \neq 0 \\ 1, & \text{if } H = 0 \end{cases} \tag{22}$$

$$H = \sum_{x \in U} [2f_1(x) - 1]^2 + \sum_{x \in U} [2f_2(x) - 1]^2.$$

This measure was constructed as a measure satisfying some properties similar to (2)–(7) where instead of $-f_1$ it is used $1 - f_1$. The properties (2)–(4) have the following form:

$$c(f_1, f_2) = c(f_2, f_1), \tag{23}$$

$$c(f_1, f_1) = 1, \tag{24}$$

$$c(f_1, 1 - f_2) = -c(f_1, f_2). \tag{25}$$

The Pearson's correlation coefficient, the local trend association measure and the measure of correlation of fuzzy sets aimed to measure different things: the relationships between the values of two variables, the relationships between the local trends of time series and the relationships between the membership functions. But all these measures satisfy similar properties and it would be interesting to study general methods of construction of functions satisfying these properties.

In [8], the methods of construction of an association measure proposed in [7] were extended on a set X with involution N and the transformation of SIM into A was based on pseudo-difference operation associated with some t-conorm. The involution N depending on the set X can be defined as involutive negation, an involutive complement, the multiplication on -1, etc. The association measure was defined in [8] as a function taking values in $[-1, 1]$ and satisfying the properties:

$$A(x, y) = A(y, x), \tag{26}$$

$$A(x, x) = 1, \tag{27}$$

$$A(N(x), x) \in \{1, -1\}, \tag{28}$$

$$A(N(x), y) = -A(x, y), \quad \text{if } A(N(x), x) \neq 1. \tag{29}$$

The last two properties have been introduced to avoid the contradiction similar to the contradiction between properties (3) and (6) discussed above for the correlation coefficient. For example, for the fixed points of involution N, we have $N(x) = x$ and $A(N(x), x) = A(x, x) = 1$, whereas the desirable property of the association measure is $A(N(x), x) = -1$. Below we propose corrected and simplified version of this definition.

Consider a set X with an **involution** $N : X \to X$, such that for all x from X it is fulfilled:

$$N(N(x)) = x. \tag{30}$$

$N(x)$ will be referred to as **a reflection** of x. We suppose that N is not an identity function, i.e. $N(x) \neq x$ for all x from X. Denote FP a set of fixed points of N, i.e. the elements of X such that $N(x) = x$. If FP is not empty, denote its elements by x_{FP}.

Consider some examples:

1. The set $X = [0, 1]$ with the involutive negation $N(x) = 1 - x$ containing one fixed point $x_{FP} = 0.5$.
2. The finite chain $X = (a_0, a_1, ..., a_m)$, with negation $N(a_k) = a_{m-k}$, for all $k = 0, 1, ..., m$. FP is empty if m is an odd number.
3. X is a set of fuzzy sets $x:U \rightarrow [0, 1]$ with involutive complement N defined by $N(x(u)) = 1 - x(u)$ for all u from U. We have one fixed point x_{FP} such that $x_{FP}(u) = 0.5$ for all u from U.
4. X is a set of real-valued n-tuples $x = (x_1, ..., x_n)$ with involution $N(x) = -x$, where $-x = (-x_1, ..., -x_n)$, $x_{FP} = (0, ..., 0)$.
5. De Morgan's lattice can have more than one fixed point.

The following properties will be considered as the desirable properties of the association measure:

$$A(x, y) = A(y, x). \quad \text{(symmetry)} \tag{31}$$

$$A(x, x) = 1. \quad \text{(reflexivity)} \tag{32}$$

$$A(N(x), y) = -A(x, y). \quad \text{(inverse relationship)} \tag{33}$$

As it was mentioned for correlation coefficients, the properties (32) and (33) are contradictive, in this case for fixed points of N. For fixed point x_{FP} from (33), it follows: $A(N(x_{FP}), y) = A(x_{FP}, y) = -A(x_{FP}, y)$ that gives

$$A(x_{FP}, y) = 0 \quad \text{for all } y \in X \text{ and all } x_{FP} \in \text{FP}, \tag{34}$$

$$A(x_{FP}, y_{FP}) = 0 \quad \text{for all } x_{FP}, y_{FP} \in \text{FP}, \tag{35}$$

$$A(x_{FP}, x_{FP}) = 0 \quad \text{for all } x_{FP} \in \text{FP}. \tag{36}$$

Anti-reflexivity (36) of fixed points contradicts with reflexivity (32). Several definitions of association measure can be proposed to avoid this contradiction.

Definition 1 Suppose X is a set with an involution N and FP is a set of fixed points of N. A function $A:X \times X \rightarrow [-1, 1]$ will be called:

1. **an association measure of type 1** if (31) and (33) are fulfilled for all x, y from X and (32) is fulfilled for all $x \notin \text{FP}$.
2. **an association measure of type 2** if (31) and (32) are fulfilled for all x, y from X and (33) is fulfilled for all $y \in X$ and all $x \notin FP$.

Alternatively, as an association measure we will consider further in this paper a function defined only on the set without fixed points $X \backslash FP$. (The more general approach considered in [15]). If it will be necessary, the values $A(x_{FP}, y)$ of association measure for all $x_{FP} \in \text{FP}$ and all $y \in X$ can be additionally defined depending on an agreement or an application. Such association measure, considered below, can be extended on association measure of type 1 or type 2.

Definition 2 Suppose X is a set with an involution N, FP is a set of fixed points of N and $X_N = X \backslash FP$. An association measure on X_N is a function $A{:}X_N \times X_N \rightarrow [-1, 1]$ satisfying (31)–(33) for all $x \in X_N$.

Note that for all $x \in X_N$, it is fulfilled $N(x) \neq x$.

An association measure defined by definition 2 has the following properties.

Proposition 1 *The association measure satisfies for all x, $y \in X_N$, the following properties:*

$$A(N(x), x) = -1, \tag{37}$$

$$A(N(x), N(y)) = A(x, y), \tag{38}$$

$$A(N(x), y) = A(x, N(y)). \tag{39}$$

In the following section, we consider association measures related to similarity measures defined on X.

Definition 3 A similarity measure on X is a function $\text{SIM}{:}X \times X \rightarrow [0, 1]$ satisfying for all x, y from X the properties:

$$\text{SIM}(x, y) = \text{SIM}(y, x), \tag{40}$$

$$\text{SIM}(x, x) = 1. \tag{41}$$

Note that from the definitions given above, it follows that an association measure of type 2 is a similarity measure on X.

We will say that an association measure A is related to a similarity measure *SIM* if the following relationships between these measures are fulfilled:

$$A(x, y) > 0, \quad \text{if } \text{SIM}(x, y) > \text{SIM}(x, N(y)), \tag{42}$$

$$A(x, y) < 0, \quad \text{if } \text{SIM}(x, y) < \text{SIM}(x, N(y)). \tag{43}$$

The condition (42) means that the association between x and y is positive if x is more similar to y than to its reflection $N(y)$. Conversely, (43) means that the association between x and y is negative if x more similar to the reflection $N(y)$ of y than to y.

Section 4 discusses the methods of construction of association measures (defined by Definition 2) using similarity measures such that relations (42) and (43) are fulfilled. Generally, these methods are the modified versions of the methods proposed in [7, 8]. The novel methods of construction of similarity measures that can be used in the generation of association measures are considered. The following section gives a short introduction to known results from fuzzy logic that will be used further.

3 Some Basic Definitions

Remember some definitions and results from [22–31] that will be used further.

A *t*-**conorm** is a function $S:[0, 1]^2 \rightarrow [0, 1]$ satisfying for all $a, b, c \in [0, 1]$ the properties of commutativity, associativity monotonicity and the boundary condition:

$$S(a, 0) = a. \tag{44}$$

From the definition of *t*-conorm, it follows for all $a \in [0,1]$:

$$S(1, a) = S(a, 1) = 1, \quad S(0, a) = a. \tag{45}$$

An element $a \in \,]0, 1[$ will be referred to as **a nilpotent element** [24] of S if there exists some $b \in \,]0, 1[$ such that $S(a, b) = 1$. A *t*-conorm S has no nilpotent elements if and only if from $S(a, b) = 1$ it follows $a = 1$ or $b = 1$.

Consider simplest *t*-conorms:

$$S_M(a, b) = \max\{a, b\}, \tag{46}$$

$$S_L(a, b) = \min\{a + b, 1\}, \tag{47}$$

$$S_P(a, b) = a + b - ab. \tag{48}$$

The maximum (46) and the probabilistic sum (48) have no nilpotent elements but the Lukasiewicz *t*-conorm (47) has.

Let S be a *t*-conorm. A **S-difference** is defined by [26]:

$$a \overset{S}{-} b = \inf\{c \in [0, 1] | S(b, c) \geq a\} \tag{49}$$

for any a, b in $[0, 1]$.

Let S be a *t*-conorm. The **pseudo-difference** associated with S is defined by [26]:

$$a(-)_S b = \begin{cases} a \overset{S}{-} b, & \text{if } a > b \\ -\left(b \overset{S}{-} a\right), & \text{if } a < b \\ 0, & \text{if } a = b \end{cases} \tag{50}$$

for any a, b in $[0, 1]^2$. Equivalently

$$a(-)_S b = \text{sign}(a - b)(\max(a, b) \overset{S}{-} \min(a, b)). \tag{51}$$

The following pseudo-differences are associated with t-conorms S_M, S_L, respectively:

$$a(-)_M b = \begin{cases} a, & \text{if } a > b \\ -b, & \text{if } a < b \text{ ,} \\ 0, & \text{if } a = b \end{cases} \tag{52}$$

$$a(-)_L b = a - b. \tag{53}$$

For S_P, we have:

$$a(-)_P b = \begin{cases} \frac{a-b}{1-\min(a,b)}, & \text{if } a \neq b \\ 0, & \text{if } a = b \end{cases}. \tag{54}$$

An **aggregation function** of 2 arguments is a function $M:[0, 1]^2 \to [0, 1]$ that is non-decreasing in each variable and that satisfies the boundary conditions:

$$M(0,0) = 0, \tag{55}$$

$$M(1,1) = 1. \tag{56}$$

4   Similarity and Association Measures

Proposition 2 *A similarity measure* SIM *satisfies for all* x, y *from* X *the property:*

$$\text{SIM}(N(x), N(y)) = \text{SIM}(x, y). \tag{57}$$

if and only if it satisfies:

$$\text{SIM}(N(x), y) = \text{SIM}(x, N(y)). \tag{58}$$

The properties (57) and (58) will be referred to as a cancellation of reflections (CR for short) and permutation of reflections (PR for short), respectively.

Theorem 1a *Suppose* SIM *is a similarity measure satisfying the cancellation of reflections property* (57) *and* S *is a t-conorm then the function* $A_{\text{SIM},S}$: $X_N \times X_N \to [-1, 1]$ *defined for all* x, $y \in X_N$ *by*

$$A_{\text{SIM},S}(x, x) = 1, \tag{59}$$

$$A_{\text{SIM},S}(x, N(x)) = -1,$$

$$A_{\text{SIM},S}(x,y) = \text{SIM}(x,y)(-)_S\text{SIM}(x,N(y)), \quad \text{if } y \neq x, y \neq N(x) \qquad (60)$$

is an association measure on X_N.

Definition 4 A similarity measure SIM will be called **strict reflexive** if

$$\text{SIM}(x,x) > \text{SIM}(x,y) \quad \text{for all } x \neq y. \qquad (61)$$

Theorem 1b *Suppose SIM is a strict reflexive similarity measure satisfying the cancellation of reflections property* (57) *and S is a t-conorm then the function $A_{\text{SIM, }S}:X_N \times X_N \rightarrow [-1, 1]$ defined for all $x, y \in X_N$ by*

$$A_{\text{SIM},S}(x,y) = \text{SIM}(x,y)(-)_S\text{SIM}(x,N(y)), \qquad (62)$$

is an association measure if one of the following is fulfilled:

1.
$$\text{SIM}(N(x),x) = 0, \quad \text{for all } x \in X_N \qquad (63)$$

2.
$$\text{the } t-\text{conorm S has no nilpotent elements.} \qquad (64)$$

Proposition 3 *Suppose M is an aggregation function of two arguments and SIM is a similarity measure, then the following function is defined for all x, y from X by*

$$\text{SIM}_M(x,y) = M(\text{SIM}(x,y), \text{SIM}(N(x),N(y))) \qquad (65)$$

is a similarity measure satisfying the CR property (57). *SIM_M is strict if SIM is strict and if M satisfies:*

$$M(a,b) < 1 \text{ if } \min(a,b) < 1. \qquad (66)$$

In (65), one can use the simplest aggregation functions, for example:

$$\text{SIM}_M(x,y) = \text{Min}(\text{SIM}(x,y), \text{SIM}(N(x),N(y))), \qquad (67)$$

$$\text{SIM}_M(x,y) = (\text{SIM}(x,y) + \text{SIM}(N(x),N(y)))/2. \qquad (68)$$

In [7, 8], the distance-based methods of construction of similarity measures satisfying PR and CR properties that can be used in construction of association measures have been introduced. The Proposition 3 gives the method of construction of such similarity measures directly from any similarity measure by means of suitable aggregation function. This proposition together with Theorem 1 gives possibility to build association measures from similarity measures defined by experts, from fuzzy equivalence or indistinguishability relations, etc. [24, 27–30].

5   Association Measures on [0, 1]

An **involutive negation** [27, 31] on [0, 1] is a strictly decreasing function neg: [0, 1] $\rightarrow$ [0, 1] that satisfies the boundary conditions:

$$\text{neg}(0) = 1, \ \text{neg}(1) = 0. \tag{69}$$

and involutivity:

$$\text{neg}(\text{neg}(a)) = a, \quad \text{for all } a \in [0, 1]. \tag{70}$$

A strictly increasing function g from the unit interval onto itself is called an **automorphism** of the unit interval [27]. Any automorphism satisfies boundary conditions

$$g(0) = 0, \quad g(1) = 1. \tag{71}$$

neg is an involutive negation if and only if there exists an automorphism g of [0,1], such that for all $a \in [0,1]$, it is fulfilled [27, 31]:

$$\text{neg}(a) = g^{-1}(1 - g(a)). \tag{72}$$

Such g is called a **generator** of *neg*. It is easy to see that the following functions are strict similarity measures satisfying the cancellation of reflections property (57):

$$\text{SIM}(x, y) = 1 - |g(x) - g(y)|, \tag{73}$$

$$\text{SIM}(x, y) = 1 - g^{-1}|g(x) - g(y)|. \tag{74}$$

Applying Theorem 1, we can construct from these similarity measures the association measures defined on the set of non-fixed points of negation *neg*. For fixed point x_{FP}, we have $g(x_{\text{FP}}) = 0.5$. As it was mentioned in Sect. 2, the values of association measure $A(x_{\text{FP}}, y)$ can be defined depending on application or by agreement.

For example, from Theorem 1b using pseudo-difference (54) for similarity measures (73) and (74), we obtain:

$$A(x, y) = \frac{|g(x) + g(y) - 1| - |g(x) - g(y)|}{\max\{|g(x) + g(y) - 1|, |g(x) - g(y)|\}}, \tag{75}$$

$$A(x, y) = \frac{g^{-1}(|g(x) + g(y) - 1|) - g^{-1}(|g(x) - g(y)|)}{\max\{g^{-1}(|g(x) + g(y) - 1|), g^{-1}(|g(x) - g(y)|)\}}. \tag{76}$$

For Zadeh negation $\text{neg}(x) = 1 - x$ generated by $g(x) = x$ from (75), we obtain the association measure:

$$A(x, y) = \frac{|x + y - 1| - |x - y|}{\max\{|x + y - 1|, |x - y|\}}. \tag{77}$$

For Yager negation

$$\text{neg}(x) = \sqrt[p]{1 - x^p}, \quad p > 0$$

generated by generator $g(x) = x^p$ from (76), we obtain the following association measure:

$$A(x, y) = \frac{\sqrt[p]{|x^p + y^p - 1|} - \sqrt[p]{|x^p - y^p|}}{\max\left\{\sqrt[p]{|x^p + y^p - 1|}, \sqrt[p]{|x^p - y^p|}\right\}}$$

For similarity measure (73) for Zadeh negation with $g(x) = x$ applying pseudo-difference (52) in Theorem 1b, we obtain:

$$A(x, y) = \begin{cases} 1 - |x - y|, & \text{if } |x + y - 1| > |x - y| \\ |x + y - 1| - 1, & \text{if } |x - y| > |x + y - 1| \end{cases}. \tag{78}$$

For this association measure, the condition

$$|x + y - 1| = |x - y|, \tag{79}$$

is fulfilled only when x or y is the fixed point. For this reason, the third line in (52) when $A(x, y) = 0$ does not shown in (78).

For Zadeh negation applying pseudo-difference (53) associated with Lukasiewicz t-conorm from Theorem 1a, we obtain the following association measure:

$$A(x, y) = \begin{cases} 1, & \text{if } x = y \\ -1, & \text{if } x = N(y) \\ |x + y - 1| - |x - y|, & \text{otherwise} \end{cases} \tag{80}$$

Another way to build association measures on [0, 1] is to use one of the dozens of similarity or equivalence relations defined on [0, 1] (see, e.g. [30] for the different types of fuzzy equivalences on [0, 1]). If it does not satisfy to CR property (57), then it can be transformed to similarity measure satisfying (57) by means of suitable aggregation function as it was proposed in Proposition 3.

Consider for example the biresiduation of product t-norm T_P [30]:

$$E_P(x, y) = \frac{\min(x, y)}{\max(x, y)}. \tag{81}$$

Transform this similarity measure into similarity measure satisfying CR property (57) as follows:

$$\mathrm{SIM}_P(x, y) = \frac{1}{2}(E_P(x, y) + E_P(N(x), N(y))). \tag{82}$$

Using in (82) the negation of Zadeh $N(x) = 1 - x$, we can represent (82) in the form:

$$\mathrm{SIM}_P(x, y) = \frac{x(1 - x) + y(1 - y)}{2(\max(x, y) - xy)}. \tag{83}$$

To build by methods given in Theorem 1 an association measure related to this similarity measure, we need to use $\mathrm{SIM}_P(x, N(y))$ that can be represented as follows:

$$\mathrm{SIM}_P(x, N(y)) = \frac{x(1 - x) + y(1 - y)}{2(\max(x, 1 - y) - x(1 - y))}. \tag{84}$$

The similarity measure (83) is strict, so we can use Theorem 1b and build an association measure by (62) using pseudo-difference operation $(-)_S$ associated with some t-conorm S without nilpotent elements. We can use in (62) the pseudo-difference operation (52) associated with t-conorm S_M (46) and we will obtain the following association measure:

$$A(x, y) = \frac{x(1 - x) + y(1 - y)}{2(\max(x, y) - xy)}, \quad \text{if } \max(x, 1 - y) - x(1 - y) > \max(x, y) - xy,$$

$$A(x, y) = -\frac{x(1 - x) + y(1 - y)}{2(\max(x, 1 - y) - x(1 - y))},$$
$$\text{if } \max(x, 1 - y) - x(1 - y) < \max(x, y) - xy.$$

Note that the condition $\max(x, 1 - y) - x(1 - y) = \max(x, y) - xy$ is fulfilled when x or y is the fixed point of negation $N(x) = 1 - x$, i.e. $x = 0.5$ or $y = 0.5$, and by definition 2 we do not define association value in these points. But if we agree, we can redefine the values $A(0.5, y) = 0$ for all y from [0,1] that proposed by pseudo-difference operation $(-)_S$.

6 Association Measures of Fuzzy Sets

The value of association between two sets can be considered as a normalized integration of association values between elements of these sets. Consider a set of fuzzy sets $x:U \to [0, 1]$ defined on a finite domain $U = \{u_1, \ldots, u_n\}$. Fuzzy set x can be considered as a sequence $x = (x_1, \ldots, x_n)$, $x_i \in [0, 1]$, $i = 1, \ldots, n$. Based on the methods of construction of association measures proposed in [7], we can use in generation of association measure on the set of fuzzy sets with finite domain the following similarity measure:

$$SIM(x, y) = 1 - \frac{1}{n} \sum_{i=1}^{n} |g(x_i) - g(y_i)|^2$$

where $g(x)$ is a generator of involutive negation on $[0, 1]$. From Theorem 1 for $A(x, y) = A_{SIM,P}(x, y)$ and generator $g(x) = x$, we obtain:

$$A(x, y) = \frac{\sum_{i=1}^{n}(2x_i - 1)(2y_i - 1)}{\max(\sum_{i=1}^{n}|x_i - y_i|^2, \sum_{i=1}^{n}|x_i + y_i - 1|^2)}.$$

As it is supposed from definitions, we have $A(x, y) = 1$ if $y = x$, and $A(x, y) = -1$, if $y = N(x)$, i.e. the association measure can be considered as a measure of relationship between fuzzy set and their complement.

7 Conclusions

The paper discusses the possible definitions of an association measure on the set with involution, and, as an example, on $[0, 1]$ with involutive negation. It is proposed to consider association measures on the sets without fixed points of involution and to define the values of association measures in these points by additional agreement depending on application. The general methods of construction of association measures based on given similarity measure satisfying the cancellation of reflections (CR) property are considered in Theorem 1 that extend the methods first introduced in [7, 8]. In this paper, we propose also the simple method (Proposition 3) of construction of similarity measures satisfying CR from similarity measures not satisfying this property. It gives possibility to construct dozens of association measures from fuzzy equivalences and similarity measures studied in literature (see, e.g. [24, 27–30]). Another method to construct similarity measures satisfying CR is to obtain them from a dissimilarity measure (or metric) used together with data transformation [7, 8]. This condition gives possibility to extend the set of transformations often applied in data standardization [7] and to use generators of negations (on [0, 1]) in construction of association measures. Some examples of association measures constructed by proposed methods on [0, 1] and on a set of fuzzy sets are considered.

Acknowledgment The results have been partially supported by the projects SIP 20151589 and 20162204 of Instituto Politécnico Nacional, Mexico.

References

1. Hair Jr, J.F., Anderson, E.R., Tatham, R.L.: Multivariate Data Analysis with Readings. Macmillan Publishing Co., Inc., New York (1986)
2. Anscombe, F.J.: Graphs in statistical analysis. Am. Stat. **27**, 17–21 (1973)
3. Batyrshin, I., Herrera-Avelar, R., Sheremetov, L., Panova, A.: Moving approximation transform and local trend associations in time series data bases. In: Perception-based Data Mining and Decision Making in Economics and Finance, Studies in Computational Intelligence, vol. 36, pp. 55–83. Springer Physica Verlag, Berlin (2007)
4. Batyrshin, I.: Up and down trend associations in analysis of time series shape association patterns. In: MCPR 2012. LNCS, vol. 7329, pp. 246–254. Springer, Berlin (2012)
5. Murthy, C.A., Pal, S.K., Dutta Majumder, D.: Correlation between two fuzzy membership functions. Fuzzy Sets Syst. **17**(1), 23–38 (1985)
6. Batyrshin, I., Sheremetov, L., Velasco-Hernandez, J.X.: On axiomatic definition of time series shape association measures. In: ORADM 2012, Workshop on Operations Research and Data Mining, Cancun, pp. 117–127 (2012)
7. Batyrshin, I.: Constructing time series shape association measures: Minkowski distance and data standardization. In: BRICS CCI 2013, Brasil, Porto de Galhinas (2013). http://arxiv.org/pdf/1311.1958v3
8. Batyrshin, I.: Association measures and aggregation functions. In: Advances in Soft Computing and Its Applications. Lecture Notes in Computer Science, vol. 8266, pp. 194–203. Springer, Berlin (2013)
9. Batyrshin, I.: Association measures on sets with involution and similarity measure. In: Proceedings of 4th World Conference on Soft Computing, Berkeley, California, May 25–27, 2014
10. Batyrshin, I., Kreinovich, V.: One more geometric interpretation of Pearson's correlation. Thai. Stat. **13**(1), 125–126 (2015). ISSN 1685-9057
11. Batyrshin, I.: Measures of association of plausible events. Fuzzy Sets Soft Comput. **10**(1), 23–34 (2015) (In Russian). ISSN 1819-4362
12. Batyrshin, I.: Association measures on [0,1]. J. Intell. Fuzzy Syst. **29**(3), 1011–1020 (2015)
13. Batyrshin, I., Villa-Vargas, L.A., Solovyev, V.: Association measures on the set of subintervals of [0, 1]. In: Fuzzy Information Processing Society (NAFIPS) Held Jointly with 2015 5th World Conference on Soft Computing (WConSC), 2015 Annual Conference of the North American, pp. 1–3. IEEE (2015)
14. Batyrshin, I., Solovyev, V., Ivanov, V.: Time series shape association measures and local trend association patterns. Neurocomputing **29**, 924–934 (2016)
15. Batyrshin, I.: On definition and construction of association measures. J. Intell. Fuzzy Syst. **29**(6), 2319–2326 (2015)
16. Das, G., Gunopulos, D.: Time series similarity and indexing. In: Handbook on Data Mining, pp. 279–304. Lawrence Erlbaum Associates, Hillsdale (2003)
17. Fu, T.-C.: A review on time series data mining. Eng. Appl. Artif. Intell. **24**, 164–181 (2011)
18. Gavrilov, M., Anguelov, D., Indyk, P., Motwani, R.: Mining the stock market: which measure is best? In: Proceedings of the Sixth ACM SIGKDD International Conference on Knowledge Discovery & Data Mining, KDD 2000, Boston, MA, USA, pp. 487–496 (2000)

19. Goldin, D.Q., Kanellakis, P.C.: On similarity queries for time-series data: constraint specification and implementation. In: 1995 International Conference on the Principles and Practice of Constraint Programming, pp. 137–153. Springer, Berlin (1995)

20. Rafiei, D., Mendelzon, A.: Similarity-based queries for time series data. In: Proceedingds of the ACM SIGMOD International Conference on Management of Data (SIGMOD '97), Tucson, Arizona, pp. 13–24 (1997)

21. Batyrshin, I., Solovyev, I.: Positive and negative local trend association patterns in analysis of associations between time series. In: MCPR 2014, Lecture Notes in Computer Science, Vol. 8495, pp. 92–101. Springer, Berlin (2014)

22. Zadeh, L.A.: Fuzzy sets. Inf. Control 8(3), 338–353 (1965)

23. Klement, E.P., Mesiar, R., Pap, E.: Triangular Norms. Kluwer, Dordrecht (2000)

24. Fodor, J., Roubens, M.: Fuzzy Preference Modelling and Multi-criteria Decision Support. Kluwer, Dordrecht (1994)

25. Beliakov, G., Pradera, A., Calvo, T.: Aggregation Functions: A Guide for Practitioners. Springer, Berlin (2008)

26. Grabisch, M., Marichal, J.-L., Mesiar, R., Pap, E.: Aggregation Functions. Cambridge University Press, Cambridge (2009)

27. Ovchinnikov, S., Roubens, M.: On strict preference relations. Fuzzy Sets Syst. 43, 319–326 (1991)

28. Recasens, J.: Indistinguishability Operators. Modelling Fuzzy Equalities and Fuzzy Equivalence Relations. Springer, Berlin (2010)

29. Wagenknecht, M.: On some relations between fuzzy similarities and metrics under archimedian t-norms. J. Fuzzy Math. 3, 563–572 (1995)

30. Li, Y., Qin, K., He, X.: Some new approaches to constructing similarity measures. Fuzzy Sets Syst. 234, 46–60 (2014)

31. Trillas, E.: Sobre funciones de negacion en la teoria de conjuntos difusos. Stochastica 111, 47–59 (1979)

Two-Phase Memetic Modifying Transformation for Solving the Task of Providing Group Anonymity

Oleg Chertov and Dan Tavrov

Abstract Nowadays, it has become a common practice to provide public access to various kinds of primary non-aggregated statistical data. Necessary precautions ought to be taken in order to guarantee that sensitive data features are masked, and data privacy cannot be violated. In the case of protecting information about a group of people, it is important to protect intrinsic data features and distributions. To do so, it is obligatory to introduce a certain level of distortion into the dataset. The problem of minimizing this distortion is a complex optimization task, which can be successfully solved by applying appropriate heuristic procedures, e.g., memetic algorithms. The task of determining whether a particular solution masks sensitive data features is an ill-defined one and often can be solved only by expert evaluation. In the paper, we propose to apply two-phase memetic algorithm to solve such tasks of providing group anonymity, for which it is not always possible to define appropriate constraints.

1 Introduction

A man is a social animal [1]. Most of our everyday actions depend on or are based on how people treat them, especially the ones closest to us (or whose opinion is most relevant to us). Such people constitute what may be called a "close circle".

At the same time, a person may not always be eager to disclose information about members of such a close circle. This reluctance may originate either from some subjective views or from the nature of the circle (religious identity, professional community, income level, LGBT community, radical group, etc.).

O. Chertov (✉) · D. Tavrov
Applied Mathematics Department, National Technical University of Ukraine
"Kyiv Polytechnic Institute", Kiev, Ukraine
e-mail: chertov@i.ua

D. Tavrov
e-mail: dan.tavrov@i.ua

© Springer International Publishing Switzerland 2016
L.A. Zadeh et al. (eds.), *Recent Developments and New Direction in Soft-Computing Foundations and Applications*, Studies in Fuzziness and Soft Computing 342, DOI 10.1007/978-3-319-32229-2_17

239

What we face here is a problem of concealing membership of a given respondent in a certain group. This problem can be formulated as the task of masking certain characteristics of a given respondent [2–4]. This task is usually called the task of providing *individual anonymity*, where anonymity means [5] the property of a subject to be unidentifiable within a set of other subjects. It is possible to set a complementary task of providing *group anonymity*, where we need to conceal information not about a single respondent, but about a group of respondents (e.g., we need to mask regional, age, or other kinds of distributions of a certain group).

The procedure for providing data anonymity should meet the following conditions [6, p. 399]:

1. Disclosure risk is low or at least adequate to protected information importance.
2. Both original and protected data, when analyzed, yield close or even equal results.
3. The cost of transforming the data is acceptable.

When providing group anonymity, preserving data utility in the sense of the second requirement is an optimization task at least of the same complexity [7, p. 12] as the well-known k-anonymization problem in the field of statistical disclosure control, which is NP-hard [8]. Moreover, providing group anonymity can be viewed as a constrained optimization problem, in which the search space consists of both feasible and unfeasible solutions. The feasibility of a solution is interpreted as the ability to mask sensitive data features in the modified dataset, and its optimality is determined by the level of modified data utility.

In this paper, we propose to use memetic algorithms (MAs) [9], which are usually implemented as evolutionary algorithms with incorporated local search procedures [10, p. 173], sometimes called *memes* after the term introduced in [11, p. 192]. A review of memetic algorithms for dealing with constrained optimization problems presented in [12] shows the benefits of using MAs instead of conventional heuristics.

There can be distinguished four [10] commonly used ways to handle the constraints in any evolutionary algorithm:

1. Using penalty functions that reduce fitness of infeasible solutions [13].
2. Using repair functions that convert infeasible solutions into feasible ones [14].
3. Restricting search to a feasible subspace of the search space by using specific alphabet for problem representation [10, pp. 215–216].
4. Using decoder functions that map infeasible solutions into feasible ones, thus transforming initial search space into another one [15].

In some cases, constraints on the solutions can be deduced from additional considerations. For instance, the solutions might be restricted to the one preserving specific data features such as high frequency [16] or periodic [17] components. It is therefore possible to restrict search space to a well-defined subspace by using a special form of solution representation in the memetic algorithm [18]. In general, however, there are no specific requirements for the solution to meet, and using penalty functions seems to be the most appropriate alternative.

In practice, more often than not quality of a solution heavily depends on quality of constraints. If they are too severe (the desired data distribution that masks sensitive data features is known), the level of distortion introduced into the dataset might become too high. On the other hand, if constraints are too mild, the solution may not mask sensitive data features, even though the penalties are relatively small.

In general, when it is not clear what constraints should be introduced (different constraints guide evolution in different directions, not necessarily leading to obtaining optimal solutions), we propose to use a two-phase memetic algorithm for solving the TPGA. At the first phase, it helps define appropriate constraints for the solution, and at the second phase, it optimizes data utility loss caused by the solution.

2 Theoretic Background

2.1 The Task of Providing Group Anonymity

Let the data for anonymizing be gathered in a depersonalized *microfile* $\mathbf{M}$. Each record r_i, $i = \overline{1, \rho}$, of this microfile contains values of attributes w_j, $j = \overline{1, \eta}$. The set of all the values of w_j is denoted as $\mathbf{w}_j$.

Let w_{v_j}, $j = \overline{1, t}$, denote *vital* microfile attributes. Then, a *vital value combination* V can be defined as an element of Cartesian product $\mathbf{w}_{v_1} \times \mathbf{w}_{v_2} \times \ldots \times \mathbf{w}_{v_t}$. We will denote a set of vital value combinations by $\mathbf{V} = \{V_1, \ldots, V_{l_v}\}$. We call microfile records whose attribute values belong to $\mathbf{V}$ *vital records*.

Let w_p denote a *parameter* microfile attribute. Then, a *parameter value P* can be defined as a value of this attribute, $P \in \mathbf{w}_p$. We will denote a set of parameter values by $\mathbf{P} = \{P_1, \ldots, P_{l_p}\}$. Parameter values can be used to divide the microfile $\mathbf{M}$ into *submicrofiles* $\mathbf{M}_1, \ldots, \mathbf{M}_{l_p}$.

Let us denote by $G(\mathbf{V}, \mathbf{P})$ a *group* of respondents whose distribution needs to be protected when providing group anonymity. The group is thereby determined by appropriately defined values of *parameter* and *vital* microfile attributes.

The *task of providing group anonymity* (TPGA) [19] lies in performing modification of $\mathbf{M}$ in order to mask sensitive data features. The generic scheme of providing group anonymity according to single-stage approach to solving the TPGA goes as follows:

1. Construct a (depersonalized) microfile $\mathbf{M}$ representing statistical data to be processed.
2. Define groups of respondents to be protected $G_i(\mathbf{V}_i, \mathbf{P}_i)$, $i = \overline{1, k}$.
3. For each i from 1 to k:

(a) Choose a dataset of arbitrary structure $\Omega_i(\mathbf{M}, G_i)$ called *goal representation* that represents features of G_i in a way appropriate for their masking;

(b) Define a transformation A: $\Omega_i(\mathbf{M}, G_i) \rightarrow \Omega_i^*(\mathbf{M}^*, G_i)$ called *modifying transformation* and obtain both a *modified goal representation* and *modified microfile*.

4. Prepare the modified microfile $\mathbf{M}^*$ for publishing.

In this work, we will illustrate the two-phase modifying transformation based on the most widely used goal representation, the *quantity signal*, in the form $\mathbf{q} = (q_1, q_2, \ldots, q_{l_p})$, where q_i is a total count of respondents in submicrofile $\mathbf{M}_i$, $i = \overline{1, l_p}$, whose vital attribute values belong to $\mathbf{V}$.

Any modifying transformation A has to provide two kinds of data modification. On the one hand, the quantity signal has to be altered in order to mask its sensitive features according to *restrictions* imposed on all or some of its values. On the level of microfile modification, this can be achieved by swapping the vital and non-vital records between different submicrofiles. Records should be swapped in a pairwise fashion to preserve the number of records in each submicrofile.

This brings us to the second kind of data modification. Swapping two records between submicrofiles obviously introduces certain amount of distortion into the microfile, which can be measured by the *influential metric* [7]

$$\mathrm{InfM}(r, r^*) = \sum_{p=1}^{n_{\mathrm{ord}}} \omega_p \left(\frac{r(I_p) - r^*(I_p)}{r(I_p) + r^*(I_p)} \right)^2 \tag{1}$$
$$+ \sum_{k=1}^{n_{\mathrm{cat}}} \gamma_k \chi^2(r(J_k), r^*(J_k)),$$

where I_p (J_k) stands for the pth ordinal (kth categorical) *influential attribute* (attribute whose distribution over parameter values is of interest for researchers), $r(\cdot)$ stands for the operator returning the attribute value of the record r, $\chi(v_1, v_2)$ stands for the operator, which is equal to a certain number χ_1 if its arguments belong to the same category, otherwise χ_2, and ω_p and γ_k are nonnegative weights (the more important is the attribute, the greater is the weight).

In other words, properly defined modifying transformation has to modify the quantity signal to mask sensitive data features and, at the same time, minimize the distortion introduced into the microfile in terms of (1).

2.2 Memetic Modifying Transformation and Individual Representation

Since the task that needs to be solved by modifying transformation seems to be the one that can be solved only by exhaustive search, we propose to use the modifying transformation in the form of memetic algorithm introduced in [19] (memetic modifying transformation, MMT):

1. Create initial population $P = \{U_i\}$ of μ randomly generated individuals, $i = \overline{1, \mu}$. Apply local search operator $S(U_i) \ \forall i = \overline{1, \mu}$.
2. Calculate fitness function $f(U_i) \ \forall i = \overline{1, \mu}$.
3. If termination condition holds, stop. Continue otherwise.
4. Select λ pairs of parents. Put them into set P'.
5. Apply recombination operator $R(U_{i_1}, U_{i_2})$ to each pair $\langle U_{i_1}, U_{i_2} \rangle$ from P', $i_1 \in \overline{1, \lambda}$, $i_2 \in \overline{1, \lambda}$, and $i_1 \neq i_2$. Put the resulting offspring into population P''.
6. Apply mutation operator $M(U_j) \ \forall U_j \in P''$, $j = \overline{1, \lambda}$.
7. Apply local search operator $S(U_j) \ \forall j = \overline{1, \lambda}$.
8. Calculate fitness function $f(U_j) \ \forall j = \overline{1, \lambda}$.
9. Select among individuals from $P \cup P''$ μ fittest ones. Put them into P in place of the current ones.
10. Go to step 3.

Each individual is a matrix U with Q rows and four columns with the following elements:

1. Element of the first column $u_{i1} \ \forall i = \overline{1, Q}$ is an index of a submicrofile to remove vital records from. The set of such submicrofiles needs to be defined by the user.
2. Element of the third column $u_{i3} \ \forall i = \overline{1, Q}$ is an index of a submicrofile to add vital records to. The set of such submicrofiles also needs to be defined by the user.
3. Element of the second column $u_{i2} \ \forall i = \overline{1, Q}$ is an index of the record from $\mathbf{M}_{u_{i1}}$ to be removed.
4. Element of the fourth column $u_{i4} \ \forall i = \overline{1, Q}$ is an index of the record from $\mathbf{M}_{u_{i3}}$ to be swapped with the one defined by u_{i2}.

Number of rows can vary from individual to individual.

Let us denote by $|U|^{(i)}$ the total count of occurrences of index i in the first and the third column of U. Then, for any index i in the first column, $|U|^{(i)} \leq q_i$.

Each particular pair $\langle u_{i1}, u_{i2} \rangle$ $(\langle u_{i3}, u_{i4} \rangle) \forall i = \overline{1, Q}$ can occur in U only once.

Requirements mentioned above cannot be violated during the whole run of the algorithm.

Each individual U uniquely defines both a modified quantity signal $\mathbf{q}^*$ and a precise sequence of pairwise swaps to be performed in order to modify the microfile, thereby defining a solution to the TPGA.

2.3 Fitness Function for the MMT

In this work, we propose to use the fitness function as a product of three independent terms:

$$f(U) = \Upsilon(U) \cdot \Phi(U) \cdot \Psi(U), \tag{2}$$

where $\Upsilon(U)$ represents estimation of a TPGA solution quality from the minimizing microfile distortion point of view, $\Phi(U)$ is a penalty function representing estimation of TPGA solution quality from the masking sensitive quantity signal features point of view, and $\Psi(U)$ is a penalty term against obtaining individuals with too many rows. Since all three terms are of equal importance, their values lie in the interval $[0, 1]$.

We propose to use the following expression for the first term of (2):

$$\Upsilon(U) = \frac{C_{\max} - \sum_{i=1}^{Q} \mathrm{InfM}(\mathbf{M}_{u_{i1}}(u_{i2}), \mathbf{M}_{u_{i3}}(u_{i4}))}{C_{\max}}, \tag{3}$$

where $C_{\max}$ is the greatest possible value of the cumulative influential metric and $\mathbf{M}_i(j)$ is the operator yielding the jth vector of the submicrofile $\mathbf{M}_i$.

We also propose to use the following expression for the penalty function:

$$\Phi(U) = \prod_{j=1}^{l_p} \mu_j \left(|U|^{(j)} \right), \tag{4}$$

where $\mu_j(x), j = \overline{1, l_p}$, is a *restriction function* that takes values in the interval $[0, 1]$ and expresses the degree of compatibility of the current value q_j with the corresponding restriction.

2.4   Other MMT Components

Operator $R(U_{i_1}, U_{i_2})$ should be defined as a proper recombination operator applied to two parent individuals U_{i_1} and U_{i_2} that yield two offspring individuals U_{j_1} and U_{j_2}. It should be applied with a high probability p_c. In this work, we propose to use recombination operator introduced in [19] based on the "cut" and "splice" operator [20]. It randomly generates two *crossover points* $k_1 \in [0, Q_{i_1}]$ and $k_2 \in [0, Q_{i_2}]$, then splits each parent at appropriate points, and exchanges the tails between them, thus creating the offspring.

Operator $M(U)$ should be defined as a proper mutation operator applied to the individual U that yields the mutated one U'. In this work, we propose to use operator introduced in [19], which is a superposition $M = M_4 \circ M_3 \circ M_2 \circ M_1$ of the following operators:

1. Operator M_1 is applied with small probability p_{m_1} to the first column of U as to the permutation. Each pair $\langle u_{i1}, u_{i2} \rangle$ needs to be preserved $\forall i = \overline{1, Q}$.
2. Operator M_2 is applied with small probability p_{m_2} to the third column of U as to the permutation. Each pair $\langle u_{i3}, u_{i4} \rangle$ needs to be preserved $\forall i = \overline{1, Q}$.

3. Operator M_3 is applied with small probability p_{m_3} to the second column of U as to the vector of categorical values.
4. Operator M_4 is applied with small probability p_{m_4} to the fourth column of U as to the vector of categorical values.

Operator $S(U)$ in this work is defined as an operator applied to the individual U that yields the modified one U' according to the following procedure [19]:

1. Carry out steps 2–4 $\forall i = \overline{1, Q}$.
2. Generate a uniformly distributed random number $r \in [0, 1]$.
3. If $r \leq p_{mem}$, assign to u_{i4} the index of a record from $\mathbf{M}_{u_{i3}}$ closest to the record u_{i2} from $\mathbf{M}_{u_{i1}}$ in terms of (1).
   If $r > p_{mem}$, assign to u_{i2} the index of a record from $\mathbf{M}_{u_{i1}}$ closest to the record u_{i4} from $\mathbf{M}_{u_{i3}}$ in terms of (1).
4. Go to step 2.

Other MMT components, such as selection, initialization, termination, and population size, can be chosen individually for each TPGA at hand.

3 Two-Phase Memetic Algorithm

In this section, we will discuss the two-phase memetic algorithm for the TPGA, in which sensitive data features to be masked are maximum values of the quantity signal.

For a certain element of the quantity signal, two types of restriction functions are defined:

1. *Decreasing restriction functions*, which are monotonically non-increasing functions that tend to unity as the corresponding quantity signal value decreases to a particular value.
2. *Increasing restriction functions*, which are monotonically non-decreasing functions that tend to unity as the corresponding quantity signal value increases to a particular value.

In most cases, we can determine only decreasing restriction functions for sub-microfiles to remove vital records from. The choice of submicrofiles to add records (without explicitly defining corresponding increasing restriction functions) can be left to the evolutionary process itself [19]. However, the quality of the solution heavily depends on the quality of the decreasing restriction functions. If the restrictions are too severe (too many swaps need to be performed), the cumulative metric (1) might become too great. On the other hand, if they are too mild, the solution may not be feasible, since maximums may remain greater than other signal values, even though their absolute values have decreased.

In some cases, with the help of information from external sources, it may be possible to choose submicrofiles to add vital records too, and define appropriate increasing restriction functions. However, in general, when there is no additional information other than present in the data at hand, it is not clear what submicrofiles to choose and what restrictions to impose, because different restrictions guide evolution in different directions, not necessarily leading to obtaining optimal solution. In this work, we propose to apply MMT according to the following procedure:

1. Based on analyzing the quantity signal **q** representing microfile **M**, define suitable decreasing restriction functions for those signal elements that violate the requirement of masking maximum signal values.
2. Apply the memetic algorithm as described in Sect. 2.2.
3. Classify individuals obtained into *feasible solutions* (compatible with decreasing restrictions to a high degree and mask maximum signal values), *subfeasible solutions* (compatible with decreasing restrictions to a high degree and do not mask maximum signal values), and *infeasible solutions* (not compatible with decreasing restrictions to a high degree).
4. Group all subfeasible solutions, for which it is possible to define the same increasing restriction functions, in clusters. One solution may belong to several clusters.
5. Choose the cluster with the smallest mean value of cumulative metric (1). If the cluster contains less than μ solutions (μ is the population size in the algorithm), then increase its size to μ by duplicating solutions at random. If the cluster contains more than μ solutions, decrease its size to μ by removing solutions at random.
6. Apply memetic algorithm of step 2 to the set of solutions obtained on step 5 as to the initial population.

The first two steps of this procedure constitute the *first phase* of the MMT, and the other four ones constitute the *second phase* of the MMT.

4 Practical Results

4.1 General Description of the Task

To illustrate the application of the MMT to the real data based on task of providing group anonymity, we decided to consider the following problem. Let us mask the regional distribution of military active personnel in the state of Massachusetts (the USA) according to the 5-Percent Public Use Microdata Sample File of the 2000 US census [21]. The total of 141,838 records was taken for analysis.

To define the group of military active personnel distributed by place of work, we took "Military Service" as the vital attributes, its value "Active Duty" as the only vital value, "Place of Work PUMA" ("Public Use Microdata Area") as the parameter attribute, and its values 25,010–25,120 with the step 10 (codes of Massachusetts statistical areas) as parameter values.

Quantity signal **q** corresponding to the group is shown in Fig. 4 (solid line). Signal elements 1, 2, …, 12 correspond to statistical areas 25,010, 25,020, …, 20,120, respectively.

4.2 The First Phase of the MMT

As we can see from the graph of the quantity signal (Fig. 4, solid line), anonymity can be provided by reducing the value of the second, the seventh, the ninth, and the twelfth signal elements. This leads us to the following decreasing restriction functions (Fig. 1):

$$\mu_2(x) = \begin{cases} 1, & x \leq 20 \\ 1 - 2\left(\frac{x-20}{47}\right)^2, & 20 \leq x \leq 43.5 \\ 2\left(\frac{x-67}{47}\right)^2, & 43.5 \leq x \leq 67 \\ 0, & x \geq 67 \end{cases}$$

$$\mu_7(x) = \begin{cases} 1, & x \leq 25 \\ 1 - 2\left(\frac{x-25}{5}\right)^2, & 25 \leq x \leq 27.5 \\ 2\left(\frac{x-30}{5}\right)^2, & 27.5 \leq x \leq 30 \\ 0, & x \geq 30 \end{cases}$$

$$\mu_9(x) = \begin{cases} 1, & x \leq 25 \\ 1 - 2\left(\frac{x-25}{3}\right)^2, & 25 \leq x \leq 26.5 \\ 2\left(\frac{x-28}{3}\right)^2, & 26.5 \leq x \leq 28 \\ 0, & x \geq 28 \end{cases}$$

$$\mu_{12}(x) = \begin{cases} 1, & x \leq 25 \\ 1 - 2\left(\frac{x-25}{13}\right)^2, & 25 \leq x \leq 21.5 \\ 2\left(\frac{x-38}{13}\right)^2, & 21.5 \leq x \leq 38 \\ 0, & x \geq 38 \end{cases}$$

Indices of all the signal elements other than those restricted by functions from Fig. 1 were chosen to appear in the third column of individuals in the MMT population.

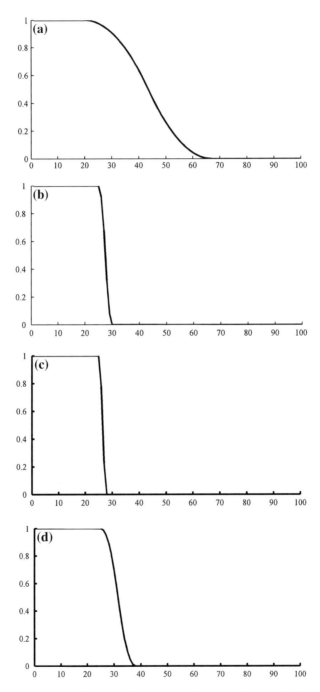

Fig. 1 Decreasing restriction functions for the example: **a** for the second element, **b** for the seventh element, **c** for the ninth element, and **d** for the twelfth element

Fig. 2 Penalty term heavily discriminating from obtaining individuals with more than 100 rows

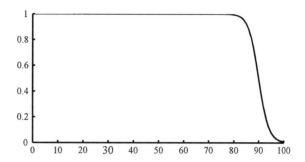

To minimize the distortion introduced into the microfile, we took "Sex," "Age," "Hispanic or Latino Origin," "Marital Status," "Educational Attainment," "Citizenship Status," and "Person's Total Income in 1999" as the influential attributes. We considered all these attributes to be categorical ones. To simplify the matter, we chose the following parameters of (1): $\gamma_k = 1 \ \forall k = \overline{1,7}$, $\chi_1 = 1$, $\chi_2 = 0$. In this case, the metric (1) shows the number of attribute values to be altered during one swap of the records between the submicrofiles.

To prevent individuals in the MMT from growing indefinitely, we used the following penalty term (Fig. 2):

$$\Psi^{ex}(U) = \frac{1}{1 + e^{0.5(Q-90)}} \ .$$

The fitness function (2), for example, is as follows:

$$f^{ex1}(U) = \frac{1099 - \sum_{i=1}^{Q} \sum_{k=1}^{7} \text{sign}|\mathbf{M}_{u_{i1}}(u_{i2}, A_k) - \mathbf{M}_{u_{i3}}(u_{i4}, A_k)|}{1099} \times$$
$$\times \prod_{j \in \{2,7,9,12\}} \mu_j \left(|U|^{(j)} \right) \cdot \Psi^{ex}(U),$$

where $\text{sign}(\cdot)$ is a function yielding -1 if its argument is negative, 0 if it equals 0, and 1 if it is positive; A_k, $k = \overline{1,7}$, is the kth influential attribute; $\mathbf{M}_j(i, A_k)$ returns the value of the attribute A_k of the ith record in submicrofile $\mathbf{M}_j$.

We chose the swap mutation [22] as mutation operators M_1 and M_2, and the random resetting mutation [10, p. 43] as mutation operators M_3 and M_4. We decided to apply tournament selection [23] as an efficient and easy-to-implement selection operator, with the tournament size 5.

The population was initialized by randomly generating matrices with different numbers of rows. Elements of the first column were generated with probabilities proportional to the values of the corresponding elements of $\mathbf{q}$. Elements of the third column were generated with probabilities proportional to the total numbers of records in corresponding submicrofiles.

Table 1 Clusters obtained after the first phase of MMT

Quantity signal elements to increase	Cluster size	Mean metric
1 and 6	78	45.436
3 and 6	84	46.048
3 and 10	26	46.269
4 and 6	43	48.488
6 and 8	183	46.519
8 and 10	101	44.238

During the MMT run, we multiplied the mutation probabilities by the factor of 10 whenever the standard deviation of the population fitness values dropped below 0.03 [24].

Other MMT parameters were chosen to be $\mu = 100$, $\lambda = 40$, $p_c = 1$, $p_{m_1} = p_{m_2} = p_{m_3} = p_{m_4} = 0.001$, $p_{mem} = 0.75$.

We performed 30 independent runs of the MMT, terminating each run after having obtained 1000 generations.

4.3   The Second Phase of the MMT

Among 3000 solutions obtained as the result of the first phase of applying MMT, only 754 (25.133 %) are feasible ones. Two solutions with the lowest cumulative metrics (1) are given in Fig. 4a (dashed–dotted and dotted lines). The mean cumulative metric (1) is 57.901.

The majority of solutions are subfeasible ones (1837, or 61.233 %). We divided them into several clusters, and the most prominent ones are shown in Table 1.

As deduced from Table 1, it is reasonable to choose solutions, for which values of elements 8 and 10 should be increased, for the second phase. This leads us to the following increasing restriction functions (Fig. 3):

$$\mu_8(x) = \mu_{10}(x) = \begin{cases} 0, & x \le 15 \\ 2\left(\frac{x-15}{12}\right)^2, & 15 \le x \le 21 \\ 1 - 2\left(\frac{x-27}{12}\right)^2, & 21 \le x \le 27 \\ 1, & x \ge 27 \end{cases}$$

The fitness function (2) for the second phase is as follows:

$$f^{ex2}(U) = \frac{1099 - \sum_{i=1}^{Q} \sum_{k=1}^{7} \text{sign}|\mathbf{M}_{u_{i1}}(u_{i2}, A_k) - \mathbf{M}_{u_{i3}}(u_{i4}, A_k)|}{1099} \times$$
$$\times \prod_{j \in \{2,7,8,9,10,12\}} \mu_j\left(|U|^{(j)}\right) \cdot \Psi^{ex}(U).$$

Fig. 3 Increasing restriction function for the eighth and tenth signal elements from the example

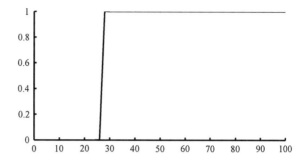

Fig. 4 Initial (*solid line*) and modified quantity signals: **a** feasible one with the metric 40 (*dashed–dotted line*), feasible one with the metric 43 (*dotted line*), and subfeasible one (*dashed line*), and **b** the one with the metric 37 (*dashed–dotted line*) and the one with the metric 38 (*dotted line*)

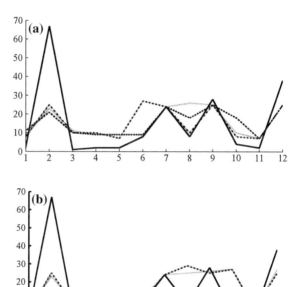

Among 3000 solutions obtained as the result of the second phase of applying MMT, 2693 ones (or 89.767 %) are feasible ones. Two solutions with the lowest cumulative metrics (1) are given in Fig. 4b (dashed and dotted lines). The mean cumulative metric (1) is 47.873. In other words, it is sufficient to alter as few as 0.005 % of the microfile attribute values in order to provide group anonymity.

This result is better than the one obtained in [19], even though restrictions imposed on the solution are stricter here.

5 Conclusions

Combining local search techniques with evolutionary algorithms to increase efficiency of the latter ones has become a widely accepted practice. This idea can be applied to solve real-life problems in quite diverse ways yielding results of varying quality.

In the paper, we proposed the two-phase approach to applying memetic algorithms that can provide results of a significantly better quality. During the first phase, feasible solutions to the optimization task are obtained, and potential ways of their improvement are discovered. During the second phase, more concise constraints can be formulated, leading to obtaining solutions of a better quality.

Experimental results presented in the paper prove the two-phase memetic algorithm to be worthy of practical interest.

However, several issues need further investigation, for example, automatic clustering of the results obtained after the first phase, enhancing algorithm efficiency by choosing appropriate components and analyzing algorithm efficiency as a function of its parameters.

References

1. Brooks, D.: The Social Animal: The Hidden Sources of Love, Character, and Achievement. Random House Trade Paperbacks, New York (2011)
2. Aggarwal, C.C., Yu, P.S.: A general survey of privacy-preserving data mining: models and algorithms. In: Aggarwal, C.C., Yu, P.S. (eds.) Privacy-Preserving Data Mining: Models and Algorithms. Advanced in Database Systems, vol. 34, pp. 11–52. Springer Science+Business Media, LLC, New York (2008)
3. Fung, B., Wang, K., Chen, R., Yu, P.: Privacy-preserving data publishing: a survey of recent developments. ACM Comput. Surv. 42(4), 1–53 (2010)
4. Sowmyarani, C.N., Srinivasan, G.N.: Survey on recent developments in privacy preserving models. Int. J. Comput. Appl. 38(9), 18–22 (2012)
5. Phitzmann, A., Hansen, M.: A terminology for talking about privacy by data minimization: anonymity, unlinkability, undetectability, unobservability, pseudonymity, and identity management. Version v0.34 [Online]. Available: http://dud.inf.tu-dresden.de/Anon_Terminology.shtml (2010)
6. Chertov, O., Pilipyuk, A.: Statistical disclosure control methods for microdata. In: 2009 International Symposium on Computing, Communication, and Control. Proceedings of CSIT, vol. 1, pp. 339–343. IACSIT Press, Singapore (2011)
7. Chertov, O. (ed.): Group Methods of Data Processing. Raleigh, Lulu.com (2010)
8. Meyerson, A., Williams, R.: General k-anonymization is hard. Carnegie Mellon School of Computer Science, Technical Report, CMU-CS-03-113 (2003)
9. Moscato, P.: On evolution, search, optimization, genetic algorithms and martial arts: toward memetic algorithms. Caltech Concurrent Computation Program, Caltech, CA, C3P Report 826 (1989)
10. Eiben, A.E., Smith, J.E.: Introduction to Evolutionary Computing, 2nd edn. Springer, Berlin (2007)
11. Dawkins, R.: The Selfish Gene: 30th Anniversary Edition. Oxford University Press, Oxford (2006)

12. Ray, T., Sarker, R.: Memetic algorithms in constrained optimization. In: Neri, F., Cotta, C., Moscato, P. (eds.) Handbook of Memetic Algorithms, pp. 135–151. Springer, Berlin (2012)
13. Smith, A.E., Coit, D.W.: Penalty functions. In: Bäck, T., Fogel, D.B., Michalewicz, Z. (eds.) Evolutionary Computation 2. Advanced Algorithms and Operators, pp. 41–48. Institute of Physics Publishing, Bristol (2000)
14. Michalewicz, Z.: Repair algorithms. In: Bäck, T., Fogel, D.B., Michalewicz, Z. (eds.) Evolutionary Computation 2. Advanced Algorithms and Operators, pp. 56–61. Institute of Physics Publishing, Bristol (2000)
15. Michalewicz, Z.: Decoders. In: Bäck, T., Fogel, D.B., Michalewicz, Z. (eds.) Evolutionary Computation 2. Advanced Algorithms and Operators, pp. 49–55. Institute of Physics Publishing, Bristol (2000)
16. Chertov, O., Tavrov, D.: Providing group anonymity using wavelet transform. In: MacKinnon, L.M. (ed.) Data Security and Security Data, LNCS, vol. 6121, pp. 25–36. Springer, Berlin (2012)
17. Tavrov, D., Chertov, O.: SSA-caterpillar in group anonymity. In: Presented at the World Conference in Soft Computing, San Francisco, CA (2011)
18. Chertov, O.R., Tavrov, D.Y.: Memetic algorithm for microfile modification with distortion minimization while providing group anonymity. Bull. Volodymyr Dahl East Ukrainian Nat. Univ. 8(179), 256–262 (2012). (in Ukrainian)
19. Chertov, O., Tavrov, D.: Memetic algorithm for solving the task of providing group anonymity. In: Jamshidi, M., Kreinovich, V., Kacprzyk, J. (eds.) Advanced Trends in Soft Computing. Studies in Fuzziness and Soft Computing, vol. 312, pp. 281–292. Springer, Switzerland (2014)
20. Goldberg, D.E., Korb, B., Deb, K.: Messy genetic algorithms: motivation, analysis, and first results. Complex Syst. 3, 493–530 (1989)
21. U.S. Census 2000.: 5-percent public use microdata sample files [Online]. Available: http://www.census.gov/main/www/cen2000.html (2000)
22. Syswerda, G.: Schedule optimization using genetic algorithms. In: Davis, L. (ed.) Handbook of Genetic Algorithms, pp. 332–349. Van Nostrand Reinhold, New York (1991)
23. Brindle, A.: Genetic algorithms for function optimization. Doctoral dissertation, Department of Computer Science, Technical Report, TR81-2, University of Alberta (1981)
24. Goldberg, D.E.: Genetic Algorithms in Search, Optimization, and Machine Learning. Addison-Wesley, New York (1989)

Querying Cyber-Networks Using Words

John T. Rickard and Allen E. Ott

Abstract Cyber-networks are characterized by two distinct types of nodes—devices and users—and by voluminous interactions in real time. All computational intelligence approaches to the analysis of these networks must deal with complexities of these interactions and their high data rates. This paper describes a computationally feasible approach to querying large cyber-networks using word-based queries. The attribute memberships and connection strengths in these cyber-networks are described granularly using appropriate vocabularies of words, where the words themselves are modeled using interval type-2 (IT2) fuzzy membership functions (MF) on an appropriate scale. By employing precomputation and storage of these word representations and queries, automated monitoring functions in large cyber-networks can be performed in real time via simple arithmetic calculations. We provide an illustrative example using data from a real cyber-network.

1 Introduction

The ability to model and query cyber-network activities is an important tool in cyber security. Typical approaches to this problem describe these networks using crisp attribute memberships and connection strengths between nodes, which may consist either of individual users or network devices. In previous work [1], we described an approach to perceptual computing in *social* networks, based on Yager's "Paradigm for Intelligent Social Network Analysis" (PISA) [2, 3]. This work extended Yager's results to the use of interval type-2 (IT2) membership function (MF) representations of word vocabularies to describe attribute memberships and relationship strengths, which enable us to granulate this information to an

J.T. Rickard (✉)
Distributed Infinity, Inc., 4637 Shoshone Drive, Larkspur, CO, USA
e-mail: terry.rickard@reagan.com

A.E. Ott
Distributed Infinity, Inc., 1382 Quartz Mountain Drive, Larkspur, CO, USA
e-mail: aott@distributedinfinity.com

© Springer International Publishing Switzerland 2016   255
L.A. Zadeh et al. (eds.), *Recent Developments and New Direction
in Soft-Computing Foundations and Applications*, Studies in Fuzziness
and Soft Computing 342, DOI 10.1007/978-3-319-32229-2_18

adequate degree of precision for purposes of analysis while substantially reducing the computational requirements.

In this paper, we extend this approach to the cyber security domain, which involves some unique network constructs, and present examples to illustrate its application. The extensions proposed herein uniquely explore the connections with identifying both people and devices as nodes, and provide a means of addressing the real time volume of link and attribute data within a cyber-network.

2 Type-1 Fuzzy Cyber-Networks

We first review some of the technical background for this paper [1–3]. In Yager's PISA model, a set X of nodes of a network is imbued with a collection of attributes U on which a vector fuzzy set $F(U)$ is defined. For example, two such attributes might be the experience level and the computer science skills of a system user, and elements of F might then include fuzzy sets for "amateur" and "high." A fuzzy relationship $R(x,y) : X \times X \rightarrow [0,1]$ represents the strength of the connection between nodes $x, y \in X$, where a node can be either a person or a device. Such a relationship might describe, for example, the frequency of logging into a device through an account on that device. Typically, many different relationships will be of interest in characterizing a cyber-network, each of them having its own descriptive matrix function $R(x,y)$.

The triple $\langle X, F, R \rangle$ is a fuzzy graph in which the nodes have fuzzy degrees of membership with respect to different attributes and the links have fuzzy strengths of relationship. Thus, an overall relationship $R.F$ is defined such that

$$R.F(x,y) = R(x,y) \wedge F(x) \wedge F(y), \tag{1}$$

where the symbol $\wedge$ denotes a t-norm conjunction operator, typically the min operator. $R(x,y)$ represents the direct link strength between a pair of nodes without any intervening links. This is generalized through the composition operator defined by

$$R^2(x,z) = R(x,z) \blacklozenge R(x,z) \triangleq \max_{\substack{y \in X \\ y \neq x,z}} [R(x,y) \wedge R(y,z)], \tag{2}$$

where by convention, $R^0(x,z) = 0$ and $R^1(x,z) = R(x,z)$. From these definitions, we can recursively define

$$R^k(x,z) = R(x,z) \blacklozenge R^{k-1}(x,z) \tag{3}$$

as the k-fold composition of $R(x,y)$. Note that $R^j(x,y) \leq R^k(x,y)$ for $j < k$; $\forall x, y \in X$, since from (2) and (3), $R^j(x,y)$ represents the greatest strength of connection between x and y involving paths of at most j links. Thus, the larger the j, the more the possible paths between x and y are considered, including all paths of length less than j.

The above definitions are extensible to type-1 fuzzy memberships for the elements of F and R, which are equivalently defined in terms of their corresponding α-cuts F_α and R_α, e.g.,

$$R_\alpha(x, y) = \{(x, y) : R(x, y) \geq \alpha\}, \tag{4}$$

and similarly for F_α. $R_\alpha^k(x, y)$ is computed by applying (3) to the left and right endpoints of the corresponding α-cuts of $R(x, y)$ and $R^{k-1}(x, z)$. This enables us to represent imprecise knowledge of node memberships and link strengths, as compared to crisp singleton values.

Next, we consider a path $\rho = \{x_0, x_1, \ldots, x_n\}$ between a set of nodes on X. The path *strength* for each α-cut is defined by

$$ST_\alpha(\rho) = \min_i R_\alpha(x_{i-1}, x_i), \quad x_{i-1}, x_i \in \rho, \quad \alpha \in [0, 1], \tag{5}$$

while the *path length* is defined by

$$L_\alpha(\rho) = \sum_{i=1}^{n} \frac{1}{R_\alpha(x_{i-1}, x_i)}, \quad x_{i-1}, x_i \in \rho, \quad \alpha \in [0, 1]. \tag{6}$$

Denoting α_ℓ as the left endpoint of an α-cut, we have $ST_{\alpha_\ell}(\rho) = 0$ and similarly $L_{\alpha_r}(\rho) = \infty$ if $R_{\alpha_\ell}(x_{i-1}, x_i) = 0$ for any α-cut of any node pair (x_{i-1}, x_i) in ρ.

The measures $ST_\alpha(\rho)$ and $L_\alpha(\rho)$ can in turn be used to describe such terms as "strong path" or "short path" from which we define additional network features and descriptions. Of particular interest in cyber-networks are a *work group* and the *centrality* of a particular node. A work group Q might, for example, be defined as a subset of users and network devices in X such that (1) all elements in Q are connected by short strong paths and (2) no element not in Q is connected to any element in Q by a strong path.

The kth-order centrality of a node x_i may be defined as the normalized sum of the strengths of the strongest paths between this node and all other nodes in the network, i.e.,

$$C_\alpha(x_i) = \frac{1}{n-1} \sum_{\substack{j=1 \\ j \neq i}}^{n} R_\alpha^k(x_i, x_j). \tag{7}$$

Note that (7) provides a measure both for the centrality of a given user in a cyber-network and for a given device on the network. Furthermore, one can restrict the sums in (7) to calculate four different degrees of centrality appropriate to cyber-networks: (1) between a given user and all other users; (2) between a given user and all network devices; (3) between a given network device and all users; and (4) between a given network device and all other network devices. Each of these centrality measures has its own importance relative to queries and automated monitoring functions for cyber-networks.

The type-1 formulations of (4)–(7) can be extended to IT2 fuzzy sets, where these sets are used to represent words that describe the attribute memberships and relationship strengths. The use of IT2 fuzzy sets to describe these words captures the imprecision in the meanings of such words. This computing with words (CWW) approach originally was proposed by Zadeh [4] and advanced by Mendel and Wu [5]. The latter authors make extensive use of the IT2 weighted average aggregation operator, which they refer to as a "linguistic weighted average." This operator was later extended by Rickard et al. [6, 7] to the more general "linguistic weighted power mean" (LWPM), where both the inputs x_i and weights w_i are specified as IT2 membership functions MFs. Expressively, the LWPM is defined by

$$\text{LWPM}(\mathbf{x}, \mathbf{w}, p) = \left(\sum_{i=1}^{n} w_i x_i^p \middle/ \sum_{i=1}^{n} w_i \right)^{\frac{1}{p}}, \tag{8}$$

where $\mathbf{x}$ and $\mathbf{w}$ are vector variables with each element described by an IT2 MF, and we use the term "expressively" to denote that (8) cannot be computed directly via standard interval arithmetic on the α-cuts of the variables, since the w_i appear in both the numerator and the denominator.

A modified version of the enhanced Karnik–Mendel (EKM) algorithm for the computation in (8) is provided in [6]. This EKM algorithm can be applied to any mean-type operator, not just the LWPM. The algorithm yields an interval MF for each value of α, so that the resulting IT2 MF can be calculated via the α-cuts of the corresponding upper and lower MFs. A further extension to non-mean, thresholding-type aggregation operators is described in a forthcoming pair of papers [8, 9]. These aggregation operators account for "all-in" and "all-out"-type decisions at threshold values controlled by a single parameter.

Particular choices of the exponent p $(-\infty \leq p \leq +\infty)$ in the LWPM (8) produce some commonly used weighted mean-type aggregation operators, including the min $(p = -\infty)$, harmonic mean $(p = -1)$, geometric mean $(p = 0)$, arithmetic mean $(p = 1)$, root mean square $(p = 2)$ and max $(p = +\infty)$. The LWPM is a t-norm only for the two cases $p = -\infty$ (min) and $p = 0$ (product). However, for $p < 1$, it provides an aggregation operator with a continuously variable degree of conjunctiveness greater than 0.5, and for $p \leq 0$,

$$\text{LWPM}(\mathbf{x}, \mathbf{w}, p) = 0 \quad \text{if any } x_i = 0 \tag{9}$$

in (8). The analogous case holds for selecting the degree of disjunctiveness for $p > 1$, while the case $p = 1$ (i.e., the weighted arithmetic average) is equally conjunctive and disjunctive with degree 0.5. Thus, we may generalize the "∧"

operator in (1)–(3) to an importance-weighted conjunctive operator $\wedge_p(x, y)$ defined using (8) as

$$\wedge_p(\mathbf{x}, \mathbf{w}) \triangleq \mathrm{LWPM}(\mathbf{x}, \mathbf{w}, p), \quad p \leq 1, \tag{10}$$

where $p \leq 0$ preserves the desirable property in (9) for a conjunctive operator in this application. Similarly, we denote the generalized disjunctive operator $\vee_p(x, y)$ by

$$\vee_p(\mathbf{x}, \mathbf{w}) \triangleq \mathrm{LWPM}(\mathbf{x}, \mathbf{w}, p), \quad p \geq 1. \tag{11}$$

LWPM operators can also be used to construct conjunctive or disjunctive *partial absorption operators* [10, 11]. Conjunctive partial absorption operators enable us to designate some of the input variables as *mandatory*, in the sense that a good score in *each* of these inputs is necessary to produce an aggregate good score, while the other inputs are simply *desirable*, in the sense that a good score in any of these enhances the output score but is not necessary to achieve a good overall score. Similarly, disjunctive partial absorption operators enable us to designate some of the input variables as *sufficient*, in the sense that a good score in any one of these variables produces an aggregate good score.

The general idea behind this style of CWW is to specify a vocabulary of descriptive input and importance words, along with their corresponding IT2 MFs defined, for example, on a scale of [0, 10] and then specify the type of inference engine desired. Given this set of input and importance words, the engine computes a corresponding output IT2 MF, which may then be used directly as computed, or it may be granularly decoded (i.e., classified) into a word from an output vocabulary via similarity computations with respect to the IT2 MFs representing these output words.

A commonly used scalar similarity measure is the Jaccard similarity [12], for which the similarity of an IT2 output set A to the ith output vocabulary word B_i in a vocabulary B is defined by

$$\mathrm{sim}(A, B_i) = \frac{\int_X dx \min\left(\bar{\mu}_A(x), \bar{\mu}_{B_i}(x)\right) + \int_X dx \min\left(\underline{\mu}_A(x), \underline{\mu}_{B_i}(x)\right)}{\int_X dx \max\left(\bar{\mu}_A(x), \bar{\mu}_{B_i}(x)\right) + \int_X dx \max\left(\underline{\mu}_A(x), \underline{\mu}_{B_i}(x)\right)}, \tag{12}$$

where $\bar{\mu}_A(x)$ and $\underline{\mu}_A(x)$ are the upper- and lower-bounding MFs (UMF and LMF, respectively) of the IT2 MF of A and similarly for each B_i. The integrals in (12) are approximated by summations in practice. Given this similarity measure, we then define the matching function

$$\omega(A, B) = \arg \max_i \mathrm{sim}(A, B_i) \tag{13}$$

as the index of the vocabulary word in B having the highest similarity to A.

In other instances, we may wish to determine the degree to which an output IT2 MF satisfies a certain property, e.g., path shortness in a cyber-network. In these cases, we use the *subsethood* measure to determine this degree of satisfaction. Theoretically, the subsethood $ss(A, B)$ of one IT2 fuzzy set A in another such set B is described by an *interval* whose endpoints are computed algorithmically [13]. However, for the sake of simplicity, we employ a computationally simpler scalar measure of this quantity as proposed by Vlachos and Sergiadis [14] and defined by

$$ss(A, B) = 1 - \frac{\int_X dx \, \max\left(0, \underline{\mu}_A(x) - \underline{\mu}_B(x)\right) + \int_X dx \, \max(0, \bar{\mu}_A(x) - \bar{\mu}_B(x))}{\int_X dx \, \underline{\mu}_A(x) + \int_X dx \, \bar{\mu}_A(x)}$$

(14)

Note that similarity is a symmetric relationship between its arguments, whereas subsethood is not. Further note that the domain X in (12) and (14) need not necessarily be one-dimensional. For example, we could compute the similarity between two network processes in a three-dimensional workflow space consisting of team attributes, connections to servers, and data flow between team members, where the subject entities may be defined only imprecisely via IT2 fuzzy sets.

3   Querying Cyber-Networks Using Words

As pointed out in [1], it is relatively straightforward to extend the type-1 fuzzy constructs described in Section II to the IT2 case by careful application of (4)–(7) to both the UMF and LMF of the IT2 MFs describing the attributes and relationship strengths of interest. There are a number of advantages to doing this.

First, it allows us easily to specify both attributes and relationship strengths in a cyber-network using relatively static vocabularies of words rather than dealing with a continuum of numerical values, which often change frequently with time and vary widely across global network sites. While words clearly are less precise than crisp quantitative values, human interpretation of numerical values is almost always in terms of qualitative expressions that are considerably more granular. In such cases, we might simply specify a set of overlapping IT2 MFs appropriate to a given vocabulary, or alternatively construct vocabularies having word MFs derived from polling of corresponding interval data from human subjects or historical network record logs [15].

A second and major advantage to using word descriptions for attributes and relationships is that they allow us to precompute and store many of the quantities of interest in a cyber-network. For example, in place of computing the path strength of a given path ρ as in (5), the suitably discretized α-cuts of the various combinations of $F.R(x_{i-1}, x_i)$ for specified vocabularies for F and R can be precomputed and stored and are then available for recall to select the IT2 MF having the minimum defuzzified centroid for any adjacent nodes (x_{i-1}, x_i) along the path ρ. The latter becomes the IT2 MF for the path strength, and its defuzzified value provides a scalar value if desired. Then, the only computation required is to select the minimum of n scalar values, where the number of links in the path is n. All else is simple recall processing.

This strategy can be extended to precomputing and storing these results using the operators $\wedge_p(x, y)$ and $\vee_p(\mathbf{x}, \mathbf{w})$ defined in (10) and (11) for a range of incremental degrees of conjunction/disjunction as defined by the *andness* of the operator as a function of p [11]:

$$
\text{andness}(p) = \begin{cases} 2 - 3 \int_0^1 \int_0^1 \mathrm{d}x\mathrm{d}y \left(\frac{x^p + y^p}{2}\right)^{\frac{1}{p}}, & p \neq 0 \\ 2 - 3 \int_0^1 \int_0^1 \mathrm{d}x\mathrm{d}y \sqrt{xy}, & p = 0 \end{cases} \tag{15}
$$

Consider the case where 7-word vocabularies are used for both F and R. Then, we precompute $7 \times 7 \times 7 = 343$ IT2 MFs for the possible combinations of $F.R$ relationship strengths between any pair of nodes and store the α-cuts, centroid intervals, and defuzzified scalar values corresponding to these MFs. Extending this to say 11 incremental degrees of andness between 0 and 1.0 in increments of 0.1, with corresponding p values computed from (15), results in 3773 cases that must be precomputed one time and stored. Given the flow of event records describing the dynamics of a virtual cyber-network, this provides a highly desirable trade-off, even in cases where relatively large vocabularies may be required.

Even greater simplification can be accomplished by granulating the results of the above computations to a discrete vocabulary of node pair attribute/relationship strengths, where we match each initially computed $F.R(x_{i-1}, x_i)$ IT2 MF to a word from this vocabulary by selecting the word whose MF has the highest similarity computed using (12). The k-fold composition ($k \geq 2$) of relationship strengths defined recursively for singleton values in (3) can similarly be granulated recursively for IT2 MFs by matching $R^{k-1}(x, z)$ to its most similar word in the vocabulary used for $R(x, z)$ and then selecting the vocabulary word for the composition that is most similar to precomputed word pairs representing $R^2(x, z)$.

So for example, if we use a 7-word vocabulary for $R(x, z)$, we would precompute $7 \times 7 = 49$ IT2 MFs representing each possible combination of word pairs and then match each pair computation to the word in the vocabulary for $R(x, z)$ whose MF is most similar. An additional desirable feature of this approach is that the MFs resulting from these recursive computations do not "broaden," as often happens in successive computations involving IT2 MFs due to the compounding of imprecision, since at each stage they are matched back to one of the original vocabulary MFs.

A similar approach can be applied to path length calculations as in (6). In this case, we precompute and store the α-cuts of the IT2 MFs corresponding to the inverses of all possible combinations of $F.R$ (and possibly andness values), along with their centroid intervals and defuzzified values. Then, the only computation required on the fly to compute the length of a given path ρ is summation of the left and right endpoints of the α-cuts for each node pair in the path (when the full IT2 MF for path length is desired), or summation of the centroid interval endpoints (when an interval path length is desired) or a simple summation of the defuzzified scalar values (when the fastest possible computation is desired).

By employing the above approach, we can exploit the generality of IT2 representations of attributes and relationships, which capture the inherent imprecision in describing the mapping of activities to cyber-networks, without paying a potentially infeasible price for on-the-fly computations involving these quantities in queries and/or dynamic modeling for large virtual networks. This is equivalent to "quantizing" the IT2 inference space into a rich, but discrete set of mappings, so that arbitrary IT2 MFs for attribute values and relationship strengths and their inference combinations in a cyber-network can be well approximated by a finite set of inputs and outputs.

4 Examples

In this section, we consider a cyber-network whose nodes consist of a set of users and a set of devices. The users are described by a set of attributes $f_i \in F$, with a relationship R that characterizes the strength of connection with other users, e.g., the frequency of communications via chat, e-mail, or sharepoint connections. In general, the ith attribute for the jth individual is described by an IT2 MF $\tilde{\mu}_{i,j}^F(x)$ corresponding to a vocabulary word $\beta_{ij} \in B_i^F$, where B_i^F is the vocabulary chosen for the ith attribute in F. Similarly, each element $r_{i,j} = [R]_{i,j}$ of the relationship R between user i and user j is described by an IT2 MF $\mu_{ij}^R(x)$ corresponding to a vocabulary word $\gamma_{ij} \in B^R$, where B^R is the vocabulary chosen for the relationship.

For illustrative purposes, suppose that f_{1i} is the attribute describing the *experience level* and f_{2i} is the attribute describing the *computer science skill* of the ith user and $r_{i,j}$ is the relationship strength of user i to user j, each described in words drawn from their respective vocabularies B_1^F, B_2^F, and B^R shown in Table 1.

The corresponding trapezoidal MFs for the UMFs and LMFs are shown in Table 2, where the five parameters of a trapezoidal function $\text{trap}(x, h, lb, lt, rt, rb)$ following its argument x are height, left bottom, left top, right top, and right bottom, respectively. For illustration, we use the same set of MFs for each vocabulary.

Thus, for example, the ith user might have the attribute values $f_{1i} = $ "Considerable" and $f_{2i} = $ "Average" in Table 2, represented by corresponding IT2 MFs $\tilde{\mu}^{(5)}(x)$ and $\tilde{\mu}^{(4)}(x)$ defined on the domain $[0, 1]$. The choice of $[0, 1]$ as the domain enables us to map the subjective judgments of experience level and computer science skill into unit ranges appropriate to users on a given network.

Suppose now that the ith user is attempting to identify a set of network resources capable of conducting a specific set of complex application tasks. He then may wish to identify teams with strong connections to individuals having high computer

Table 1 Vocabularies B_1^F, B_2^F, and B^R for the attributes experience level, computer science skill, and relationship strength

Experience level	Computer science skill	Relationship strength	Vocabulary MF
Novice (N)	Very low (VL)	None/very weak (NVW)	$\tilde{\mu}^{(1)}(x)$
Amateur (A)	Low (L)	Weak (W)	$\tilde{\mu}^{(2)}(x)$
Modest (M)	Modest (M)	Fairly weak (FW)	$\tilde{\mu}^{(3)}(x)$
Intermediate (I)	Average (A)	Casual (C)	$\tilde{\mu}^{(4)}(x)$
Considerable (C)	Moderately high (MH)	Moderately strong (MS)	$\tilde{\mu}^{(5)}(x)$
Substantial (S)	High (H)	Strong (S)	$\tilde{\mu}^{(6)}(x)$
Guru (G)	Very high (VH)	Very strong (VS)	$\tilde{\mu}^{(7)}(x)$

Table 2 Trapezoidal upper and lower mfs for vocabulary word mfs

Vocabulary MF	Trapezoidal upper and lower MFs
$\tilde{\mu}^{(1)}(x)$	$\bar{\mu}^{(1)}(x) = \text{trap}(x, 1, 0, 0, 0.126, 0.472)$
	$\underline{\mu}^{(1)}(x) = \text{trap}(x, 1, 0, 0, 0.01, 0.143)$
$\tilde{\mu}^{(2)}(x)$	$\bar{\mu}^{(2)}(x) = \text{trap}(x, 1, 0.001, 0.0114, 0.273, 0.434)$
	$\underline{\mu}^{(2)}(x) = \text{trap}(x, 0.03, 0.014, 0.176, 0.176, 0.2)$
$\tilde{\mu}^{(3)}(x)$	$\bar{\mu}^{(3)}(x) = \text{trap}(x, 1, 0.014, 0.269, 0.557, 0.823)$
	$\underline{\mu}^{(3)}(x) = \text{trap}(x, 0.36, 0.366, 0.429, 0.429, 0.461)$
$\tilde{\mu}^{(4)}(x)$	$\bar{\mu}^{(4)}(x) = \text{trap}(x, 1, 0.204, 0.402, 0.626, 0.827)$
	$\underline{\mu}^{(4)}(x) = \text{trap}(x, 0.36, 0.481, 0.512, 0.512, 0.563)$
$\tilde{\mu}^{(5)}(x)$	$\bar{\mu}^{(5)}(x) = \text{trap}(x, 1, 0.413, 0.613, 0.824, 0.997)$
	$\underline{\mu}^{(5)}(x) = \text{trap}(x, 0.39, 0.673, 0.718, 0.718, 0.765)$
$\tilde{\mu}^{(6)}(x)$	$\bar{\mu}^{(6)}(x) = \text{trap}(x, 1, 0.337, 0.774, 1, 1)$
	$\underline{\mu}^{(6)}(x) = \text{trap}(x, 1, 0.739, 0.98, 1, 1)$
$\tilde{\mu}^{(7)}(x)$	$\bar{\mu}^{(7)}(x) = \text{trap}(x, 1, 0.592, 0.902, 1, 1)$
	$\underline{\mu}^{(7)}(x) = \text{trap}(x, 1, 0.877, 0.991, 1, 1)$

science skills via at most k links. Let $\tilde{\mu}_{ST}(x)$ and $\tilde{\mu}_{HC}(x)$ be IT2 MFs for the classes "strong path" and "high computer science skill," defined on the [0, 1] domains of path strength and computer science skill, respectively. In particular, let these MFs be specified by

$$\begin{aligned}
\bar{\mu}_{ST}(x) &= \text{trap}(x, 1, 0.5, 0.85, 1, 1) \\
\underline{\mu}_{ST}(x) &= \text{trap}(x, 1, 0.7, 0.95, 1, 1) \\
\bar{\mu}_{HC}(x) &= \text{trap}(x, 1, 0.65, 0.85, 1, 1) \\
\underline{\mu}_{HC}(x) &= \text{trap}(x, 1, 0.75, 0.9, 1, 1)
\end{aligned} \tag{16}$$

The jth users' $(j \neq i)$ memberships $ST(i,j)$ and $HC(j)$ in these classes are calculated using (14) as

$$\begin{aligned} ST(i,j) &= ss\big(R^k(x_i, x_j), \tilde{\mu}_{ST}(x)\big) \\ HC(j) &= ss\big(\tilde{\mu}_{2j}(x), \tilde{\mu}_{HNW}(x)\big). \end{aligned} \tag{17}$$

Thus, the degree $c_1(i,j)$ to which individual j $(j \neq i)$ satisfies this criterion can be calculated using a singleton input LWPM in (8) and (10) with exponent $p \leq 1$ as

$$c_1(i,j) = \wedge_p \left(\begin{bmatrix} ST(i,j) \\ HC(j) \end{bmatrix}, \begin{bmatrix} 1 \\ 1 \end{bmatrix} \right). \tag{18}$$

The choice of p determines the degree of conjunction required for the satisfactions of the two criteria "strong connection" and "high computer science skill." Different weight values can be used if one of these criteria is more important than the other. Computing $c_1(i,j)$ for each $j \neq i$ and ranking the results provide a list of prospective network resources in order of their suitability.

Recall that the results for both $R^k(x_i, x_j)$ and $\tilde{\mu}_{2j}(x)$ in (17) have been precomputed and stored; therefore, it is only necessary at most to perform the subsethood calculations. However, if the classes "strong path" and "high computer science skill" are of general interest in queries, these subsethood calculations too can be precomputed and stored for each combination of (word) values for the first arguments, so that only recall processing and simple arithmetic calculations are required to compute the values in (18).

For illustration, suppose we consider a network of six users, where individual #1 is the potential task lead. Assume the attribute vectors for experience level f_1 and computer science skill f_2 are given by (see Table 1)

$$\begin{aligned} f_1 &= [\text{I} \quad \text{A} \quad \text{C} \quad \text{M} \quad \text{G} \quad \text{I}]^T \\ f_2 &= [\text{M} \quad \text{VL} \quad \text{H} \quad \text{L} \quad \text{VH} \quad \text{M}]^T, \end{aligned} \tag{19}$$

and the relationship strength matrix R is given by

$$R = \begin{bmatrix} 1 & W & C & FW & VS & NVW \\ W & 1 & MS & MS & C & FW \\ S & FW & 1 & W & NVW & NVW \\ S & MS & W & 1 & FW & C \\ C & W & C & VS & 1 & W \\ FW & MS & NVW & W & S & 1 \end{bmatrix}, \tag{20}$$

where the diagonal "1s" correspond to singleton MFs at unity value and are not involved in our computations. The corresponding vectors $ST(k)$ of degrees of "strong connection" between node 1 and the remaining nodes for $k = 1, 2$ and the vector HC of degrees of "high computer science skill" are calculated from (17) as:

$$ST(1) = \begin{bmatrix} 1 & 0 & 0.223 & 0.151 & 1 & 0 \end{bmatrix}^T$$
$$ST(2) = \begin{bmatrix} 1 & 0.151 & 0.223 & 1 & 1 & 0.151 \end{bmatrix}^T \qquad (21)$$
$$HC = \begin{bmatrix} 0.058 & 0 & 0.666 & 0 & 0.952 & 0.058 \end{bmatrix}^T.$$

We note that for this example, $ST(k) = ST(2)$ for $k > 2$. From (18) with $p = -0.72$ (corresponding to an andness of 0.75) in (15), the vectors of criterion satisfaction for node elements 2 through 6 for $k = 1, 2$ are given by

$$c_1(1) = \begin{bmatrix} * & 0 & 0.347 & 0 & 0.975 & 0 \end{bmatrix}^T$$
$$c_1(2) = \begin{bmatrix} * & 0 & 0.347 & 0 & 0.975 & 0.086 \end{bmatrix}^T. \qquad (22)$$

Thus, the potential task leader's best candidate would be user #5, with user #3 a distant second.

Now suppose that our potential task leader is leveraging a strategy that will benefit particularly from more experienced network users. He might then wish to query his cyber-network site with a desire for strong paths to more experienced users, for whom high computer science skill is mandatory. In this case, let $\tilde{\mu}_{ME}(x)$ be the IT2 MF for the class "more experienced user," where the membership of individual j in this class is given by

$$ME(j) = ss\left(\tilde{\mu}_{1j}(x), \tilde{\mu}_{ME}(x)\right). \qquad (23)$$

We then would use a partial absorption operator [11] with the mandatory input being $HC(j)$ and the desired inputs being a conjunction (of specified degree p') between $ST(j)$ and $ME(j)$. This partial absorption operator computation involves 1) a weighted partial disjunction (denoted ∇) of the mandatory and desired inputs (using $\vee_{p_d}(\mathbf{x}, \mathbf{w})$ with exponent $p_d \geq 1$) followed by 2) a weighted partial conjunction (denoted Δ) of the mandatory input with the result from 1) (using $\wedge_{p_c}(\mathbf{x}, \mathbf{w})$ with exponent $p_c \leq 1$).

In a type-1 fuzzy context, $x_1 \Delta(x_1 \nabla x_2)$ necessarily is zero if x_1 is zero, and for nonzero x_1, $x_1 \Delta(x_1 \nabla x_2) - x_1$ is positive (respectively negative) when x_2 is greater (respectively less) than x_1. The absolute difference $|x_1 \Delta(x_1 \nabla x_2) - x_1|$ is called the *reward* when x_2 is greater than x_1 and is otherwise called a *penalty* [11]. We construct the operator $x_1 \Delta(x_1 \nabla x_2)$ from nested weighted power means as:

$$c_2(j) = \wedge_{p_c}\left(\begin{bmatrix} HC(j) \\ \nabla_{p_d}\left(\begin{bmatrix} HC(j) \\ \wedge_{p'}\left(\begin{bmatrix} ST(j) \\ ME(j) \end{bmatrix}, \mathbf{w}'\right) \end{bmatrix}, \mathbf{w}_d\right) \end{bmatrix}, \mathbf{w}_c\right), \qquad (24)$$

where $\mathbf{w}'$ is the importance weight vector for the two desired inputs and $\mathbf{w}_d$, p_d, $\mathbf{w}_c$, and p_c are chosen to achieve the desired reward and penalty values as described in

Fig. 1 IT2 membership
function of "more
experienced user"

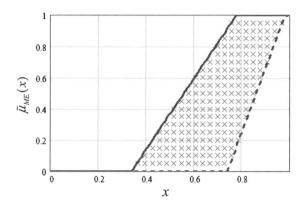

[11]. Again, since the variables in the weighted power means in (24) have been
precomputed, only arithmetic operations must be performed to arrive at this result.

Let the more experienced user class MF correspond to that of S in Table 1 (i.e.,
$\tilde{\mu}^{(6)}(x)$), as shown in Fig. 1.

Setting a penalty of -25% and a reward of $+15\%$, with $p' = -0.72$,
$p_d = 5.802$, and $p_c = 0.619$, we obtain from (24) the vectors of criterion satisfactions for $k = 1, 2$:

$$c_2(1) = [*\quad 0\quad 0.586\quad 0.007\quad 0.961\quad 0.043]^T$$
$$c_2(2) = [*\quad 0\quad 0.586\quad 0.012\quad 0.961\quad 0.071]^T. \tag{25}$$

Thus, while user #5 is still the best candidate, user #3 provides a reasonably
strong match to this query as well.

To further strategize his team composition, our potential task leader may wish to
know the strength of the centrality of the other users in the network. The IT2
centrality $C(i; k)$ of the ith user over a maximum of k links is calculated using (7)
on the upper and lower MFs of R. We then calculate the subsethood of $C(i; k)$ in
the set "strong path" having MF $\tilde{\mu}_{ST}(x)$ as in (17), i.e.,

$$ST(C(i; k)) = ss(C(i; k), \tilde{\mu}_{ST}(x)). \tag{26}$$

This yields the following vectors of centrality strengths $CS(k)$ for $k = 1, 2$, with
$CS(k) = CS(2)$ for $k > 2$:

$$CS(1) = [0.275\quad 0.308\quad 0.189\quad 0.35\quad 0.289\quad 0.306]$$
$$CS(1) = [0.505\quad 0.436\quad 0.397\quad 0.585\quad 0.559\quad 0.589] \tag{27}$$

Thus, user #3 has the lowest strength of centrality of any of the users, which
depending on the nature of the task assignment may prove to be an advantage or
disadvantage to the team leader.

As a final query, the task leader may wish to know the degree to which users #2 and #5 constitute a work group to themselves. As defined above, this would be the degree to which users #3 and #5 are connected by short strong paths, and no other users are connected to them by a strong path. Performing this calculation yields a degree of 0, indicating that these users are not an isolated workgroup.

5 Conclusion

We have extended the social network constructs of [1–3] to the perceptual computing domain for cyber-networks using word representations of attribute membership and relationship strengths between users. In particular, we have identified means for feasibly performing calculations using these constructs in large virtual networks by precomputing and storing the combinations of attribute/relationship words involved in the typically relatively small vocabularies required in cyber-networks. We are presently working on the applications of these approaches to dynamic network modeling using fuzzy cognitive maps.

Acknowledgements This material is based on research sponsored by the US Air Force Academy under agreement number FA7000-12-2-0020. The US Government is authorized to reproduce and distribute reprints for governmental purposes notwithstanding any copyright notation thereon.

The authors acknowledge the use of the USAF Academy Center for Cyberspace Research, including cadet teams to develop prototype software and simulation scenarios. Graph construction capability and scalability were explored and evaluated by Lieutenant Bryan Hall and Lieutenant Josiah Lane using simulated cyber-network records. Their work was extended to include the perception engine vocabulary attributes. The application of imprecise queries and graph similarity was evaluated by Lieutenant Elliot Unseth and Lieutenant Jon Beabout. The imprecise graph similarity research was extended with the perceptual computing IT2 fuzzy approach.

Disclaimer The views and conclusions contained herein are those of the authors and should not be interpreted as necessarily representing the official policies and endorsements, either expressed or implied of the US Air Force Academy or the US Government.

References

1. Rickard, J.T., Yager, R.R.: Perceptual computing in social networks. In: Proceeding of International Fuzzy System Association World Congress/North American Fuzzy Information Processing Society Annual Meeting, Paper #9, Edmonton, Alberta, Canada, June 2013
2. Yager, R.R.: Intelligent social network analysis using granular computing. Int. J. Intell. Syst. **23**(11), 1196–1219 (2008)
3. Yager, R.R.: Concept representation and database structures in fuzzy social relational networks. IEEE Trans. Syst. Man Cybern Part A Syst. Hum. **40**(2), 413–419 (2010)
4. Zadeh, L.: Fuzzy logic = computing with words. IEEE Trans. Fuzzy Syst. **4**(2), 103–111 (1996)

5. Mendel, J.M., Wu, D.: Perceptual Computing: Aiding People in Making Subjective Judgments, Piscataway. IEEE Press, NJ (2010)
6. Rickard, J.T., Aisbett, J., Yager R.R., Gibbon, G.: Fuzzy weighted power means in evaluation decisions. In: Proceeding World Symposium on Soft Computing, Paper #100, San Francisco, CA, May 2011
7. Rickard, J.T., Aisbett, J., Yager, R.R., Gibbon, G.: Linguistic weighted power means: comparison with the linguistic weighted average. In: Proceeding of FUZZ-IEEE 2011, 2011 World Congress on Computational Intelligence, pp. 2185–2192, Taipei, Taiwan, June 2011
8. Rickard, J.T., Aisbett, J.: New classes of threshold aggregation functions based upon the Tsallis q-exponential with applications to perceptual computing. IEEE Trans. Fuzzy Syst. **22** (3), 672–684 (2014)
9. Aisbett, J., Rickard, J.T.: Centroids of type-1 and type-2 fuzzy sets when membership functions have spikes. IEEE Trans. Fuzzy Syst. **22**(3), 685–692 (2014)
10. Dujmović, J.J.: Continuous preference logic for system evaluation. IEEE Trans. Fuzzy Syst. **15**(6), 1082–1099 (2007)
11. Dujmović, J.J.: Partial absorption function. J. Univ. Belgrade, EE Dept., ser. Mathematics and Physics, **659**, pp. 156–163 (1979)
12. Jaccard, P.: Nouvelles recherches sur la distribution florale. Bull de la Societe de Vaud des Sciences Naturelles **44**, 223 (1908)
13. Rickard, J.T., Aisbett, J., Gibbon, G.: Fuzzy subsethood for fuzzy sets of Type 2 and generalized Type n. IEEE Trans. Fuzzy Syst. **17**(1), 50–60 (2009)
14. Vlachos, I., Sergiadis, G.: Subsethood, entropy, and cardinality for interval-valued fuzzy sets —an algebraic derivation. Fuzzy Sets Syst. **158**, 1384–1396 (2007)
15. Wu, D., Mendel, J.M., Coupland, S.: Enhanced interval approach for encoding words into interval type-2 fuzzy sets and its convergence analysis. IEEE Trans. Fuzzy Syst. **20**(3), 499–513 (2012)

On the Concept of Big Data Analysis

A.B. Pashayev and E.N. Sabziev

Abstract The concept of analysis of the information in Big Data is offered. In the proposed concept, it is introduced some universal set of values with the limited number of words. All files (sources of information) projected into the universal set. The search purpose was formed in terms of universal set. Then, search process was performed in the universal set, i.e. in the set of projection of sources of information. Such technology reduces the localization of searching information. Such approach allows locate the required information within the framework of the traditional sizes and makes possible further application of methods and algorithms of the information processing for them.

1 Introduction

Intensive development of information technology and hardware systems led to accumulation of large amount of digital data, such as text, audio, video data and specific content.

The accumulated data are stored in files of different structure, e.g. doc, txt, bmp, jpg, wav, mp3, rtf, mpg, flv etc, and each type has their source and purpose. For instance, video data are mainly accumulated in the process of security monitoring of areas and objects, video monitoring of traffic, video recording of sports events, political events and mass festivities, as well as natural phenomena, progress of science experiments, aerospace imaging of Earth and extraterrestrial space etc. [1]. The amount of accumulated data are so large that its processing is associated with great difficulties. This is why new concept of Big Data was introduced in information theory.

A.B. Pashayev (✉)
Software Development Department of Kiber Ltd Company, Baku, Azerbaijan
e-mail: adalat@kiber.az

E.N. Sabziev
Department of Applied Researches of Kiber Ltd Company, Baku, Azerbaijan
e-mail: elkhan@kiber.az

© Springer International Publishing Switzerland 2016   269
L.A. Zadeh et al. (eds.), *Recent Developments and New Direction
in Soft-Computing Foundations and Applications*, Studies in Fuzziness
and Soft Computing 342, DOI 10.1007/978-3-319-32229-2_19

Big data [2] is a collection of data sets so large and complex that it becomes difficult to process using on-hand database management tools or traditional data processing applications. The challenges include capture, curation, storage, search, sharing, transfer, analysis and visualization. The trend to larger data sets is due to the additional information derivable from analysis of a single large set of related data, as compared to separate smaller sets with the same total amount of data, allowing correlations to be found to "spot business trends, determine quality of research, prevent diseases, link legal citations, combat crime and determine real-time roadway traffic conditions".

The challenge for large enterprises is determining who should own Big Data initiatives that straddle the entire organization.

Obviously, acquisition and storing of such large amounts of data cannot be justified without it being used for its designated purpose. In this regard, we should undoubtedly point out that while searching for any kind of useful information, it is important to know what data set can contain this information. For instance, if some man is searched for in video data, then such information as data of the progress of science experiments, aerospace Earth imaging data and scanning data for traffic on expressways must be neglected. Thus, the data set to be processed can be narrowed depending on the purpose in view.

Another side of the problem is that the same data can be used for different purposes. For instance, a video file of a hockey match broadcast can be of interest to the panel of judges investigating into the quality of the referee work in the match, to the coaches analysing the tactics of the rival team, to the reporters reviewing the most intriguing moments of the game, etc. The episodes of interest for different user categories may or may not cross in some ways. Moreover, different episodes can have different degrees of importance depending on the users interested in them. For instance, conflicts among players are important to the referees after the game is interrupted by a whistle, while for the coaches of the teams, those conflicts might hold no interest whatsoever.

The aim of this paper was to investigate the possibilities of Big Data preprocessing to facilitate further retrieval and processing of more useful information.

2 Problem Statement

As noted above, data retrieval and processing in Big Data are associated with certain difficulties. This brings up the question: can desired data be organized in a way such that it would be possible to use conventional processing methods afterwards?

With regard to this question, we set the following objective:

– *To develop data locating mechanism in Big Data based on search purpose*

Assume that we search among all available files for a file containing the information with the semantic context "young woman and man talking in the rain".

This can be a photograph, a fragment of a novel or a letter, a video clip, etc. To that end, let us narrow down the variety of the files, among which the search will be performed, by viewing all files with the following content: "young man", "young woman", "rain" and "conversation". Thus, the search area for the sought-for information is identified.

Investigating the set problem, we propose decomposing the original task into simpler ones as a possible solution. Thus, the original task is divided into the following subtasks:

Task 1—Definition of universal set of values;

Task 2—Developing the mechanism of great ion of the image of the files on the universal set of values;

Task 3—Developing the search mechanism based on the searching in the set of universal values.

So, proposed concept is to introduce some universal set of values with the limited number of words. All files (sources of information) projected into the universal set. The search purpose was formed in terms of universal set. Then, search process was performed in the universal set, i.e. in the set of projection of sources of information. Such technology reduces the localization of searching information.

3 Universal Set of Values

In the present paper, we propose the concept of primary locating of the desired data. For this purpose, the universal set of values is introduced. This set contains basic concepts of human cognition and is limited quantitatively. In doing so, we can formulate the purpose of search, using the elements of the universal set of values.

We should note that such limited word sets were introduced and successfully applied for initial study of daily use vocabulary of foreign languages. For instance [3], Basic English contains only 850 words.

This is a list of the 850 words grouped and listed by Ogden in The ABC of Basic English (1932) [4]. These words all denote simple concepts commonly used in everyday life.

Obviously, depending on the purpose of Big Data processing, this list of words can be shortened or supplemented.

We suppose that the universal set of values will be something like "Basic English core vocabulary", introduced for the purpose of data processing. Let us define the universal set of values more accurately.

Let the set M be so-called **universal set of values**. Thus, we have some set of finite number of elements—values. Labels for fast search will be composed of these elements later. Therefore, we believe that the general number of elements of the set M will be limited by some reasonable number n, e.g. $n \leq 1000$.

For convenience in operation in the set M, we can introduce the indexing system of elements numeration $i = 0, 1, \ldots, n$, where n is the number of the elements. Denote the elements of M by a_i.

The numbering is proposed to make it possible, while implementing the offered concept, to focus on operations on the elements of the set separated from the meanings of specific words. The numbering sequence does not affect the progress of further research. The elements can be numbered alphabetically, for instance.

4 Value Sentence

As noted above, different algorithms for data processing exist and are in development now, which allow recognizing subjects, actions, etc. in files of different structure. The features of such algorithms can certainly be expanded so that they could determine the degree of importance of the attributes found in files and expressed by the real number μ within the range [0, 1].

For instance, let us consider a case of a football match broadcast. The broadcast is recorded into a video file. It rains; the host team defeats the guest team. During half-time, the camera captures a young man and a young woman peacefully discussing the match. In that case, the ranging can be as follows:

- Host 1
- Guest 1
- Host wins 1
- Rain 1
- Discussion 1
- Conversation 1
- Fight 0
- Young man 0.9
- Young woman 0.1
- Accident 0
- Money 0
- Airplane 0

etc.

Thus, the fuzzy nature of degree of importance in further may allow performing operations above them that is appropriate to assess the importance of phrases or sentences.

Definition *A sequence of pairs of elements and the corresponding coefficients of belonging* $\{a_i \in M, \mu_i \in [0,1]\} i = 0, 1, \ldots, n$ *is called a value sentence.*

Note that the elements of M, which is the universal set of values, are determined and numbered in advance. Therefore, the index $i = 0, 1, \ldots, n$ uniquely determines the element of the set M. Taking into account this fact, we can claim that the sequence of numbers $\{\mu_i \in [0,1]\}$, $i = 0, 1, \ldots, n$, determines the value sentence.

While a file is created (saved), it can be preprocessed. The degree of μ_i has to be determined for each value a_i. Obviously, the values μ_i can be different for the same values a_i, depending on the expert method and determination algorithm. However, they can be taken as basic given a certain share of admissible deviation.

Therefore, each data group (each file), with a certain "admissible accuracy", can be matched with some value sentence, which is the sequence $\{\mu_i\}$ of the length n. In that case, $\mu_i = 0$ will mean that the value a_i is absent in this file.

5 Big Data Sources

The main sources of Big Data are text data (content of web pages, e-mail, forums, etc.), images and video files, as well as speech data. Signals of different nature, such as seismic signals, space surveillance signals, machinery vibration signals, can also act as data.

Different instruments designed for processing of all the aforementioned types of data are available these days. Information technology is developing rapidly, and new methods and tools for data processing emerge every day.

6 Cloud Computing

The analysis of the Big Data sources, as well as the instruments for processing of Big Data components, demonstrates that they do not originate from the same source but are rather distributed. Hence, it is reasonable to use Cloud technologies. In other words, the resources can be combined in a same Cloud.

Different units (packages) of data processing are developed and maintained by certain groups and stored on certain servers. Big Data sources are also distributed, and the data are stored on different servers.

In accordance with the proposed approach, in a Cloud environment, the incoming data will be preprocessed. In that case, the relevant software resources from the Cloud environment can be involved, depending on the nature of the incoming data. When a file is saved, a label to that file is created. The label is the sequence of degrees of importance of the elements within the universal set of values. Thus, alongside with every data carrier file, their labels will appear in the Cloud environment.

During Big Data processing, the sentence of processing purpose will be created. That sentence will be the sequence of degrees of importance of each element within the universal set of values. Useful data will be located by comparing the search sentence with the labels of the files. Such kind of search will tangibly reduce the required resources, including the time for retrieving the required data. Then, obtained data, which will be of usual size, can undergo further processing.

7   Search Sentence

As described above, each file is given an additional attribute, a label consisting of the sequence of words from the universal set of values. To form the purpose of search, we need to build a sentence from elements of the universal set of values. The search sentence can be formed using semantics or as a random sequence of elements of the universal set of values.

Let us consider a case when the purpose of search is formed semantically. Suppose that using elements of the universal set of values, we build a sentence in English, which indicates the purpose of search. We can assign some coefficient of importance v to each word, beginning with the first word of the sentence. The coefficient decreases uniformly from 1 to $1/k$, where k is the number of words in the sentence. In that case, if some words recur in the sentence, we can identify them, assigning the coefficient corresponding to the value of its first location in the sentence.

Lets us consider the above example of search of the file containing information on the young man and woman having a conversation in the rain. The search target can be formulated by the following set of words: rain, young man, young woman and conversation. The proposed algorithm will then assign the importance as follows:

$k = 4$;
rain $\leftarrow 4/4 = 1$;
young man $\leftarrow 3/4 = 0.75$;
young woman $\leftarrow 2/4 = 0.5$;
conversation $\leftarrow 1/4 = 0.25$.

Coefficient 0 is assigned to other words of the universal set M automatically.

Another way to form a search sentence is for the user to select a random sequence of words from the universal set of values, assigning some coefficient of importance v (within the range $[0, 1]$) to each word.

In order to search for some required information, the search target is usually formulated in a natural language. The proposed concept implies that a sentence formulated in common language can be projected into some search sentence contained in the universal set of values.

Take any element of the set M. Each element of the set M generates a class consisting of all sentences, the projection of which matches that element. Any sentence can be projected into the set M by identifying natural-language words with similar meaning to one another. The task of identification can be given to experts. Obviously, such identification is ambiguous and depends both on experts and ambiguousness of semantic values of words. Nevertheless, a certain operational projection can be determined.

Thus, any random set of words of a natural language can have a certain projection into the set M.

It should be noted that projection of all natural-language words into some set M does not in fact depend on the language and is a clustering. Such a projection can be built for the English language as well as for any other language.

Thus, the purpose of search will have the form of a sentence by means of elements of some subset of the universal set of values with the coefficients of importance v within the range $[0, 1]$.

8 Search Algorithm

Let us consider some search sentence. According to the definition of the universal set of values, each word (element) of that sentence has a unique number i and each word of the sentence has a coefficient of importance v_i. Thus, we have some sequences, $\{v_{i_1}, v_{i_2}, \ldots, v_{i_k}\}$. Data search (location) in the set of labels within Big Data will be based on this sequence.

Let us form a subset of search depending on the possible limitations such as file type and date of creationetc. Obviously, each file of the set will have its label of the proposed structure: $\{\mu_{i_1}, \mu_{i_2}, \ldots, \mu_{i_k}\}$.

Depending on the search target, different methods of data filtration can be determined. In the following paragraphs, we describe the task of determining such methods.

Assume that some element $\mu^0 = \{\mu_1^0, \mu_2^0, \ldots, \mu_n^0\}$, which is a search sentence, is given. The task is to find in the set $M \equiv [0, 1]^n$ a subset of elements $\mu = \{\mu_1, \mu_2, \ldots, \mu_n\}$ that are close in a certain sense to the given sentence μ^0. To that end, let us determine the binary relations in the set M. These relations can be regular or fuzzy. Depending on the introduced relation, we can obtain some or other subset of points close to μ^0. The definition of a relation (fuzzy relations) can be found, for instance, in [5].

Let us provide some examples of relations that express different sets of elements depending on the search target.

"Exact match". For a given set of positive numbers ε_i, $i = 1, 2, \ldots, n$, the set of "exact match" can be determined through the relation $E(\mu^0) \equiv \{|\mu_i - \mu_i^0| < \varepsilon_i, (\mu, \mu^0) \in M \times M\}$. A reasonable selection of numbers ε_i, $i = 1, 2, \ldots, n$ can be the subject of a separate discussion.

Local match. It should be noted that usually, when search conditions are formulated, mandatory filters are set. Therefore, "exact match" can significantly narrow down the range of the sought-for elements (files). From this point of view, it is reasonable to consider another relation determined in the set $M \times M$:

$$L(\mu^0) \equiv \{|\mu_i - \mu_i^0| < \varepsilon_i, \text{ for } \mu_i^0 > 0\}.$$

Bottom coverage. The relation

$$T\left(\mu^0\right) \equiv \left\{\mu_i \geq \mu_i^0, \left(\mu, \mu^0\right) \in \mathsf{M} \times \mathsf{M}\right\}$$

is a formalization of the "all greater than or equal to" request and generalizes the classical inequality that divides the set $\mathsf{M} \times \mathsf{M}$ into two parts. Similar to the local match, the relation "*local bottom coverage*" can be determined:

$$TL\left(\mu^0\right) \equiv \left\{\mu_i \geq \mu_i^0, \text{ for } \mu_i^0 > 0\right\}.$$

Top coverage, ***local top coverage***. Obviously, the following relations can be considered

$$B\left(\mu^0\right) \equiv \left\{\mu_i \leq \mu_i^0, \left(\mu, \mu^0\right) \in \mathsf{M} \times \mathsf{M}\right\}$$

and

$$BL\left(\mu^0\right) \equiv \left\{\mu_i \leq \mu_i^0, \text{ for } \mu_i^0 > 0\right\},$$

which will in a sense be in contrast to the relations $T(\mu^0)$ and $TL(\mu^0)$, respectively.

Thus, the file search will be performed in the set of all their shortcuts. The files will be highlighted, for which the relation generated in compliance with the search target holds true.

In the absence of any coefficients in the search sentence, they will be regarded as equal to unity.

9   Conclusion

Note, that there are various methods of searching information in Big Data. One of them is the Google algorithm based on the search of given word or sequences of words in the all Big Data and then arranged the sources by some mechanism [6]. Principally, Google algorithm focuses on finding data that contain the selected words in the text file or web page, or in the captions to the illustrations, charts, etc.

The proposed approach is based on the search for a concept similar in content (meaning) to the desired keyword. In addition, the search is process not direct in the file. Each file is assigned some label and search process in the set of labels. The labels can be created for any type of file (for both text and audio, photo and video files, etc.). The proposed of ranking and filtering of data can significantly narrow the range of the search, which will facilitate further search of the necessary information.

References

1. Michael, K., Miller, K.W.: Big data: New opportunities and new challenges. Computer. **46**(6), 22–24 (June 2013)
2. Big Data. (http://en.wikipedia.org/wiki/Big_data#cite_note-2)
3. Basic English. (http://en.wikipedia.org/wiki/Basic_English)
4. The ABC of Basic English (1932). (http://www2.educ.fukushima-u.ac.jp/~ryota/word-list.html)
5. Kaufmann, A.: Introduction a la Thèorie Des Sous-Ensembles Flous. Masson et C. Editeurs, Paris. V.1-4, 1975/1977
6. Rebecca, S.: Wills Google's Page Rank: The math behind the search engine. http://www.cems.uvm.edu/~tlakoba/AppliedUGMath/other_Google/Wills.pdf

Interaction Using Qualitative Data

Vadim L. Stefanuk

Abstract To overcome some problems with deep understanding of fuzzy values, certain learning finite automaton was put into a fuzzy environment. Previously, such a device has been studied in the probabilistic environment, where the classic technique of standard Markov chains was applicable. The new study became possible due to several previous results by the present author, namely the axiomatic of fuzzy evidence accumulation and the theory of generalized Markov chains. The mathematical results, obtained in the paper, prove that the learning automaton has the property of asymptotic optimality. We propose to use this property for measuring membership functions in case of values analogous to singletons or point functions. It is claimed that the obtained results might lead to a fuzzy value measurement procedure resembling statistics developed in probability area.

Keywords Fuzzy environment · Probabilistic environment · Finite automata with learning · Asymptotic optimality · Generalized Markov chain · Fuzzy singletons

1 Introduction

The exchange with quantitative information plays an important role in the interaction among people and technical devices. The temperature at home or in the street, the cost of goods or services—without those precise figures our life would be meaningless. In the theory of man–machine systems [1], the transmission of the quantitative information in the form of dial or menu readings for various crisp values does not create any problems.

V.L. Stefanuk (✉)
Institute for Information Transmission Problems, Russian Academy of Sciences,
Bolshoi Karetny per., 19, 101447 Moscow, Russian Federation
e-mail: stefanuk@iitp.ru

V.L. Stefanuk
Peoples' Friendship University of Russia, Mikluho-Maklaya str. 6, 117198 Moscow,
Russian Federation

© Springer International Publishing Switzerland 2016   279
L.A. Zadeh et al. (eds.), *Recent Developments and New Direction
in Soft-Computing Foundations and Applications*, Studies in Fuzziness
and Soft Computing 342, DOI 10.1007/978-3-319-32229-2_20

However, people commonly use for communication the *qualitative information,* speaking on the temperature in the room in terms of *high, low, comfortable, suitable,* or reasoning on the cost of goods and services as *expensive, cheap, and reasonably priced.* To support such qualitative information, various schemes have been proposed, including fuzzy sets [1], gray numbers, and probability theory [2].

These schemes are commonly used in the flexible interfaces intended to establish a man–machine contact, which satisfies "both sides." Among various schemes, the axiomatic of fuzzy set theory is the most developed and the most popular in applications. Yet, the formal analysis shows that many problems are still not completely resolved.

For instance, in fuzzy set theory by Zadeh [1], which is extensively used in the area of expert systems [3, 4], the meaning of the fuzzy membership does not reach the level of transparency comparable to that of probability schemes. The reason is that the fuzzy set theory does not provide anything similar to the statistics in probabilistic approach, which actually brings meaning to the probability and makes it suitable for practical applications.

Another problem is the lack of well-understood mechanisms for defining fuzzy membership values as well as the mechanisms for direct understanding of obtained fuzzy values by a person. The use of natural language (NL) description does not resolve the problem due to the lack of mathematical apparatus which allows going from one NL expression to another using some formal tools. It is this problem that probably led Prof. L.A. Zadeh to formulation of the well-established axiomatic theory for fuzzy sets [1].

It is important to stress that though the probability theory has obtained certain advantages due to the statistical support, this general problem is not removed completely as statistic does not allow expressing such a vague consideration as *"probable," very probable,* end, etc. People do use the phrases *with high probability,* or *it is improbable,* the meaning of which is quite clear for the people. Yet, to convey the meaning to technical systems, or to teach a technical system to use such unclear expressions is still an open problem.

Our paper presents an attempt to find some solution to above-mentioned problems, using the mathematical model of a learning finite automaton, provided that it is being put into the fuzzy environment [5].

Our formal analysis is based on the two previous results. The first is our axiomatic for the problem of evidence accumulation [6].

This axiomatic has been used in our system SEISMO for prognosis of seismic phenomena. In Fig. 1, we demonstrate the results obtained with SEISMO in case of middle-term prediction of an earthquake.

The second important result is our generalization of the concept of Markov chains [7]. Classic Markov chain describes probabilistic events that changed in accordance with the Markov chain logic [8].

Fig. 1 SEISMO: middle-term forecast (the week of the event)

18.12.91-25.12.91
25.12.91-01.01.92
01.01.92-08.01.92
08.01.92-15.01.92
15.01.92-22.01.92
22.01.92-29.01.92
29.01.92-05.02.92
05.02.92-12.02.92
12.02.92-19.02.92
19.02.92-26.02.92
26.02.92-04.03.92
04.03.92-11.03.92

There are a number of various generalizations in the literature for the classic Markov chains; however, the presence of some probabilistic phenomena is always assumed. Yet, our generalization is suitable for any chains with the Markov property, and it does not necessary to assume the probabilistic character of the values involved. To avoid misunderstanding in the paper, we will refer to our generalization as the *Markov-Stefanuk chain* [7].

Our paper is organized in the following way.

Following the approach described in [6] in the first Chapter, an expression is derived for summary fuzzy effect of two fuzzy values: current automaton state and the penalty/reward obtained by the automaton. The mathematical expression has the form different from the one proposed by Zadeh [1], but in fact agrees with his logic. This expression simplifies the equations, describing the behavior of fuzzy system in a fuzzy environment.[1]

Second Chapter considers the behavior linear tactic automaton, which was introduced and studied by Tsetlin in a probabilistic environment, which penalize or rewarding the automaton with fixed probabilities [9]. However in the second chapter, it is being put into a fuzzy environment, which issues the rewards and penalties with some fuzzy membership function. We demonstrate how from assumed ergodicity of the generalized Markov chain [7] it becomes possible to obtain formulas, relating fuzzy values of inner states of automaton with it actions in the external fuzzy media.

In the third Chapter, the obtained formulas are used for inference some final assertions on behavior of the linear tactic automaton (LTA). It is shown that in accordance with our expectation, the character of LTA behavior does not essentially changes due to transfer of LTA from probabilistic environment to the fuzzy one.

In the fourth Chapter, we present some discussion of membership function used and why there is a possibility to work with *fuzzy singletons*. The fuzzy singletons differ from the common singletons known from literature [11] as they open some new possibility for defining values of unknown membership functions in a certain class.

---

[1]We obtained one particular version of T-norms that is well known in fuzzy set theory.

In the fifth Chapter, it is shown that the obtained property of asymptotic optimality opens, in principal, the possibility for measuring fuzzy singletons, i.e., singletons with arbitrary membership values.

In conclusion, it is discussed that the experimental verification of obtained expressions meets some serious difficulty as unlike probability theory, where statistics give certain grounds and the fuzzy set theory does not have theoretically confirmed experimental base. For this reason, the results obtained and the steps used in this process are today the only evidences of correctness of our analysis. From the other side, the obtained results provide us with the hope for building some analog of statistics, but oriented to the area of fuzzy systems.

For Markov-Stefanuk chain, it is possible to consider the ergodicity property and to find final values in case this chain has one ergodicity class [7].

2 Collecting Evidences with Axiomatic Tools

Following the approach described in [7], an expression is derived for total fuzzy effect of two fuzzy values: current automaton state and the penalty/reward obtained by the automaton. The mathematical expression has the more "smooth" form, which is different from the maximum expression proposed by Zadeh, but in fact follows his theory logic. Our expression simplifies the equations, describing the behavior of fuzzy system in a fuzzy environment.

Let us consider first very simple learning machine shown in Fig. 2. It is a finite automaton, which has only two inner *states* (1, 2) and two external *actions* (1, 2). This automaton is being put into some environment giving to it a *feedback*, which consists from *rewards* or *penalties*, depending on the actions performed by the automaton.

The upper part of this figure shows transition between its states, when it obtained a penalty when it was in the state 1 (left state in this figure), performing action **1**; or obtained the penalty when it was in the state 1 (right in the figure), performing the action **2**.

The bottom graph shows transition between its states, when it obtained a reward when it was in the state 1, performing action **1**; or obtained the reward, when it was in the state 1, performing the action **2**.

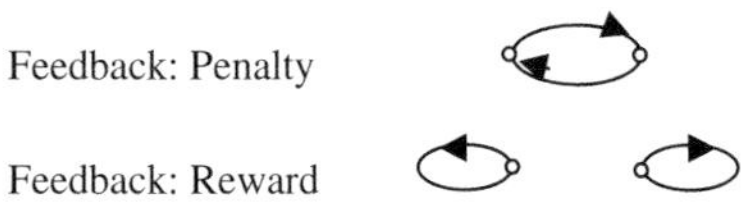

Fig. 2 The simplest learning automaton with two actions

Feedback: Penalty

Feedback: Reward

2.1 Fuzzy Environment

When this automaton is in a fuzzy environment, it comes to its inner states in accordance with some fuzzy scheme, when $\mu_1^{(1)}(t)$ denotes the membership function to the state 1, corresponding to the action **1** in the moment $t = 1, 2, \dots$. The same way one may define $\mu_1^{(2)}(t)$—the membership function to the state 1 that corresponds to the action **1** in the moments $t = 1, 2, \dots$.

Let $\lambda^{(1)}$ is a fuzzy singleton meaning the penalty for the action **1** performed in the state 1, and $\lambda^{(2)}$ be a fuzzy singleton meaning the penalty for the action **2** performed in the state 2.

As the automaton does not change its state, when the feedback is neutral, it is natural to consider that the singleton value $(1 - \lambda^{(1)})$ corresponds to the reward for the action **1** performed in the state 1, and $(1 - \lambda^{(2)})$ is the reward for the action **2**.

As the penalty/reward at the moment t defines *the next time* fuzzy states $\mu_1^{(i)}(t+1)$, $i = 1, 2,$, one may be sure that $\mu_1^{(1)}(t)$ and $\mu_1^{(2)}(t)$ *do not depend* on the feedback at the moment t. Hence, following Prof. L.A. Zadeh "the total fuzzy value" is defined with the following classic expression:

$$f(x, y) = \min\{x, \ y\}, \text{ where } x = \mu_1^{(i)}(t); \ y = \lambda^{(i)}, i = 1, 2. \tag{1}$$

However, this expression is difficult for analytic study of our automaton behavior. Hence taking the axiomatic of evidence summarization from our publication [6], we add two new axioms, concerning the total result, namely:

$$f(0, 0) = f(1, 0) = f(0, 1) = 0 \tag{2a}$$

$$f(1, \alpha) = f(\alpha, 1) = \alpha \tag{2b}$$

It is easy to see that these axioms are true for the above formula of L.A. Zadeh. However, it is important for us that they are true also for the next expression for the total sum:

$$f(x, y) = x \times y, \tag{3}$$

where $x = \mu_1^{(i)}(t); \ y = \lambda^{(i)}, i = 1, 2$. Fig. 3 shows graphically the difference between (1) and (3).

2.2 Ergodicity

It easy to see that the behavior of our simplest automaton is described with an ergodic Markov-Stefanuk chain. The parameters of this generalized Markov chain

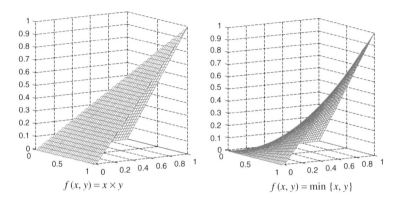

$f(x, y) = x \times y$ $f(x, y) = \min\{x, y\}$

Fig. 3 The comparison of expressions (3) and (1)

are not known yet. Its ergodicity will be established later after the equality of flows of the chain values will be demonstrated [7].

Due to the expected ergodicity, the *final* singleton meaning transition from the left state to right state should be equal to the final singleton meaning transition from right state of automaton to the left its state:

$$\mu_1^{(1)} \lambda^{(1)} = \mu_1^{(2)} \lambda^{(2)} \tag{4}$$

Hence, if $\lambda^{(1)} \geq \lambda^{(2)}$, then $\mu_1^{(1)} \leq \mu_2^{(2)}$. It means that our simplest automaton is able to learn to reduce the number of punishments!

Using same technique, one may study the automaton with linear tactics, proposed by Tsetlin [9]. The linear tactics automaton is an extension of our simplest automaton considered above.

3 Behavior of Linear Tactics Automaton

In all the states belonging to the left-hand side shown in Fig. 1, the automaton performs the first action (**1**) and obtains the penalty (in the upper graph) and the reward (in the bottom graph). Similar situation is valid for the action (**2**).

The linear tactic automaton was studied in [9] in a *random environment*, which issues the penalties and rewards with some fixed probabilities.

In the present paper, the situation is different from that studied by M.L. Tsetlin as now this automaton is put into a fuzzy environment and performs in correspondence with the fuzzy membership functions similar to the simplest automaton described above.

Fig. 4 State transitions for the LTA

Again, it may be shown that the performance of the linear automaton is controlled with a generalized Markov chain [7], i.e., Markov-Stefanuk chain. It turned out possible to obtain the final expressions, describing the behavior of linear automata, taking into account also some our results from [6].

Let memory depth of the automaton shown in Fig. 4 is equal to n that means that the linear tactic automaton has exactly n inner states for each of its actions (**1** and **2**). Assuming ergodicity property, one may obtain the following equations. (The states are numbered in the way that the "deepest state" shown in Fig. 4 has number n.)

$$\lambda^{(1)}\mu_i^{(1)} = (1 - \lambda^{(1)})\mu_{i-1}^{(1)}, \quad i = 1,\ldots,n, \tag{5}$$

where $(1 - \lambda^{(1)})-$ is the reward for the action **1** and $\mu_i^{(1)}-$ is the membership function for the state i on the left line of Fig. 4, corresponding to the first action, i.e., action **1**.

Let us temporally fix values $\mu_1^{(1)}$ and $\mu_1^{(2)}$. Then from (4) and (5), one has the following:

$$\mu_k^{(1)} = \mu_1^{(1)}\left(\frac{1-\lambda^{(1)}}{\lambda^{(1)}}\right)^{k-1} \text{ и } \mu_k^{(2)} = \mu_1^{(2)}\left(\frac{1-\lambda^{(2)}}{\lambda^{(2)}}\right)^{k-1}, k = 1,\ldots,n \tag{6}$$

As the $\mu_1^{(1)}$ and $\mu_1^{(2)}$ are related with (4), we have:

$$\mu_k^{(1)} = \mu_1^{(1)}\left(\frac{1-\lambda^{(1)}}{\lambda^{(1)}}\right)^{k-1} \text{ и } \mu_k^{(2)} = \mu_1^{(1)}\frac{\lambda^{(1)}}{\lambda^{(2)}}\left(\frac{1-\lambda^{(2)}}{\lambda^{(2)}}\right)^{k-1}, k = 1,\ldots,n \tag{7}$$

If $M^{(1)}$ is the membership function (singleton) for the first action and $M^{(2)}$ is the membership function for the second action, then it may be shown [5] that the following expressions are valid for the final values (singletons):

$$M^{(1)} = 1 - \prod_{k=1}^{n}(1 - \mu_k^{(1)}) = 1 - \prod_{k=1}^{n}\left(1 - \mu_1^{(1)}\left(\frac{1-\lambda^{(1)}}{\lambda^{(1)}}\right)^{k-1}\right) \tag{8a}$$

$$M^{(2)} = 1 - \prod_{k=1}^{n}(1 - \mu_k^{(2)}) = 1 - \prod_{k=1}^{n}\left(1 - \mu_1^{(1)}\frac{\lambda^{(1)}}{\lambda^{(2)}}\left(\frac{1-\lambda^{(2)}}{\lambda^{(2)}}\right)^{k-1}\right) \tag{8b}$$

3.1 Asymptotic Optimality

From the last two expressions, one has:

$$\frac{M^{(1)}}{M^{(2)}} = \frac{1 - \left\{ \prod\limits_{k=1}^{n} \left(1 - \mu_1^{(1)} \left(\frac{1-\lambda^{(1)}}{\lambda^{(1)}}\right)^{k-1}\right) \right\}}{1 - \left\{ \prod\limits_{k=1}^{n} \left(1 - \mu_1^{(1)} \frac{\lambda^{(2)}}{\lambda^{(2)}} \left(\frac{1-\lambda^{(2)}}{\lambda^{(2)}}\right)^{k-1}\right) \right\}} \tag{9}$$

where $\mu_1^{(1)}$ is some positive value, satisfying $\mu_1^{(1)} \leq 1$ (details are in [5]). As in the case of simplest automaton, one may observe the learning property of this linear tactics automaton. Indeed, if the reward for the first action is greater than the reward for the second action, i.e., $\lambda^{(1)} > \lambda^{(2)}$, then one obtains $M^{(2)} > M^{(2)}$.

Moreover, if in addition the following inequality holds

$$\frac{1 - \lambda^{(2)}}{\lambda^{(2)}} > 1$$

then from (8) under $n \to \infty$, the following is valid:

$$M^{(1)} \to 0, \ M^{(2)} \to 1,$$

That is, Tsetlin's automaton with linear tactics being put into a fuzzy environment has the property of *asymptotic optimality as* shown previously in [9] for the probabilistic environments. From the expression (9), it is obvious that the asymptotical optimality is impossible, if both inequalities are true

$$\frac{1 - \lambda^{(i)}}{\lambda^{(i)}} \leq 1, i = 1, 2. \tag{10}$$

4 Comments on Singletons

The membership functions used in this paper usually were referred to as *fuzzy points*, i.e., single element fuzzy sets. They may be found in many publications as *singletons* [10]. However, usually for singletons, it is assumed that membership function in this point is equal to 1.

Thus, the paper [10] says that in realizations, where a fuzzy system is being built from some set of elementary functional elements, non-singleton fuzzification usually creates problems. The author of [10] shows that application of defuzzification of the type DCOG may be modeled with the use of singleton architecture.

In the present paper, we use a bit different concept of a singleton. We referred to it as fuzzy singleton. The fuzzy singleton may take any value from the interval [0–1]. Similar values have been used in expert system MYCIN, where a simple heuristic rule was applied to combine evidences in favor of some phenomenon [3].

The use of fuzzy singletons corresponds to the membership functions from general fuzzy set theory of Prof. Zadeh, presenting a convenient tool for numerical calculations. Indeed, let us denote s the set of two elements, where 0 corresponds to penalty for an automata and 1 corresponds to the reward.

Then, the value of *feedback* represents the traditional fuzzy value λ, with the help of couple of "delta functions λ_1 and λ_o, provided that $\lambda_o + \lambda_1 = 1$ is true. (The last assumes that there no neutral feedback.)

Thus, from the fuzzy property, it follows that by push button "feedback" on the learning machine, the latter may obtain penalty (0) or reward (1). That is why in our real learning machine made in Moscow State University [11], there were two separate push buttons: one for punishments and the other for rewards.

As it is assumed that then in our mathematical analysis, it was natural to restrict with a single value, namely *fuzzy singleton for punishment* λ_0, and to use *fuzzy singleton* $(1 - \lambda_0)$ *for reward*. In the result, we use the fuzzy singleton $\lambda_0^{(1)}$ for punishment feedback for the action **1** and the fuzzy singleton $\lambda_0^{(2)}$ for punishment feedback for the action **2**, or simply $\lambda^{(1)}$ and $\lambda^{(2)}$.

5 The Possibility of Measuring with Precision

It was the use of fuzzy singletons that let us obtain the above expressions, describing final behavior of the linear tactic automaton designed by M.L.Tsetlin [9] when it was put into a fuzzy environment [5]. It follows that this finite automaton in a fuzzy environment has the important property of *asymptotic optimality* [9]. In other words, this automaton lets one to establish, which of the following relations is true $\lambda^{(1)} < \lambda^{(2)}$ or $\lambda^{(1)} \geq \lambda^{(2)}$.

First of all, it is important to stress that the values, which define the penalties and rewards obtained by the finite automaton, *are not observable values.*

If the penalties or rewards are issued by a person, then these values are defined with a fuzzy considerations in his/her brain. In technical systems and for theoretical analysis, these values may be logically deduced from some other non-observable factors.

From the other side, the values $\lambda^{(1)}$ and $\lambda^{(2)}$ are not observable in the same sense as $p(A)$, the probability of a certain event A, is not observable either. Yet, in the probability theory, we have an indirect way of approximate calculation of $p(A)$ by collecting statistics of events.

Is there some possibility to make a fuzzy value known with some precision?

A positive answer to this question follows from important property of asymptotic optimality. The latter means this automaton allows to learn is it true that $\lambda_1 > \lambda_2$ or we have an opposite $\lambda_1 \leq \lambda_2$ under $n \to \infty$.

Hence, the learning automaton described above lets one collect some "statistics." Indeed, such automaton lets one to define the position of unknown λ within the set of ordered values, $\left(\lambda^{(1)}, \lambda^{(2)}, \ldots, \lambda^{(s)} \right)$. The ordering may be established by the several applications of the linear tactic automaton. The reliability of the ordering increases with the value of n in accordance with the property of its asymptotic optimality.

This ordering allows establishing the membership value with a prescribed precision. For instance, for this purpose, the following ruler of "reference membership values" may be used such as (0.1, 0.2, 0.3, 0.4, 0.5, 0.6, 0.7, 0.8, and 0.9).

Obviously, this ruler and the described automata allow measuring some fuzzy value with 10 % precision. It was mentioned in our paper [12], where some other procedure was discussed as well ("Boston device").

Actually, the automaton shown in Fig. 3 due to the restriction (10) forces us to restrict with the following ruler of functions (0.5, 0.6, 0.7, 0.8, and 0.9). We will discuss this problem elsewhere.

Presently, it is not clear how to physically generate the penalties if the value λ is given. The procedures of fuzzification and defuzzification were aimed to it. However, this problem is removed if the penalties and rewards are created by a person, who is operating with the fuzzy values in a way he/she understands it [10]. The same person in exactly same way should then define fuzzy values which are sent to the input of machine. It is important as otherwise the result obtained from some technical system may be incorrectly interpreted by a person.

Please note that in [10], a number of different ways of *understanding fuzziness* are demonstrated. Some problems may be avoided if there will be organized a preliminary tutoring of a group of users, involved in man–machine interaction [13]. In the process of the tutoring, they will develop common understanding of what is the fuzzy membership function.

Thus, we understand that a person may formulate a fuzzy value for some fact and say something like "I believe that $\lambda^{(1)} = 0.4$" to reward the machine for its action **1**. Yet, we do not understand presently how the person comes to this decision.

The difficulty is the lack of the clear understanding, how the value $\lambda^{(1)} = 0.4$ is used for feedback.

6 Conclusion

The expressions obtained in the paper shows that the behavior of Tsetlin's automaton is asymptotically optimal. It means that the character of the behavior does not change very much due to transition from probabilistic environment to the fuzzy one. The theoretical results correspond to expectation and hence are rather reliable.

Yet, the experimental verification of the result may lead to a difficulty, which is almost of philosophical kind.

In the theoretical analysis, these values are formally deduced from some other non-observable factors. Unlike to probability theory which finds serious support in statistics, the fuzzy theory does not have yet experimental base, that is, theoretically justified. That is, why the obtained mathematical results and correctness of all our approach presently are the most convincing evidence correctness of the analysis demonstrated in this paper and correctness of our results.

From the other side, it should be mentioned that in a real situation of exchange with technical systems, the probabilistic scheme is not 100 % justified either. When we pushed the buttons on our learning machine, where we observed a collective behavior, the person acted using some intelligent considerations. Yet in our practice, these considerations were closer to some fuzzy considerations, not probabilistic ones. Indeed, it would be difficult to imagine that the person has in his head a precise idea of certain probability values, such as say 0.4.

Fuzzy set theory has many questions that should be answered. And we hope that the theoretical developments of the present paper might make the fuzzy theory handier, more justified and more suitable for real-life applications. Very important hope present author has with respect to the possibility of designing some analogy for the procedure of collecting statistics, but this time for the area of fuzzy systems.

Acknowledgments This work was partially supported by the Russian Fund for Basic Research (RFBR), Grant #12-07-00209a, and by the Presidium of Russian Academy of Science, Programs П15 and 1.5П. Current publication is an extended version of our plenary presentation made during 4th World Conference on Soft Computing, (abstract of plenary talk, pp. 43–44), Berkeley, USA, 2014 (http://www.wconsc-2014-berkeley.com/keynote.html).

References

1. Zadeh, L.A.: Fuzzy sets. Inf. Control. 8, 338–348 (1965)
2. Kolmogorov, A.N.: Zur Theorie der Markoffschen Ketten. Math. Ann. **101**, 126–136 (1929)
3. Shortliffe, E.H.: Computer-Based Medical Consultations: MYCIN. Elsevier/North Holland, New York (1976)
4. Stefanuk, V.L.: Dynamic expert systems. KYBERNETES Int. J. Syst. Cybern. **29**(5/6), 702–709 (2000)

5. Stefanuk, V.L.: Behavior of Tsetlin's learning automata in a fuzzy environment. In: Second World Conference on Soft Computing (WConSC).), pp. 511–513, Letterpress, Azerbaijan, Baku (2012)

6. Stefanuk, V.L.: Should one trust evidences? In: Proceedings of the All-country AI Conference, vol. 1, pp. 406–410, Moscow (1988)

7. Stefanuk, V.L.: Deterministic Markovian chains. Inf. Process. **11**(4), 702–709 (2011)

8. Romanovskii V.I.: Discrete Markov chains. Гостехиздат Moscow: Gostechizdat, pp. 436 (1949)

9. Tsetlin, M.L.: Some problems of finite automata behaviour. Doklady USSR Acad. Sci. **139**(4), (1961)

10. Munakata, T.: Fundamentals of the New Artificial Intelligence. Neural, Evolutionary, Fuzzy and More. Springer, USA (2008)

11. Stefanuk, V.L.: An example of collective behaviour of two automata. Autom. Remote Control. **24**(6), 781–784 (1963)

12. Stefanuk, V.L.: Discovery of values of membership functions. In: VII International Science and Practice Conference Integral Models and Soft Computing in Artificial Intelligence, Kolomna: Fizmatlit, T.3, c.1338–1343, Russia (2013)

13. Stefanuk, V.L.: On man-machine interaction with qualitative data. In: Proceedings of 12th IFAC/IFIP/IFORS/IEA Symposium on Analysis, Design, and Evaluation of Human-Machine Systems, 11–15 August 2013, Las Vegas, USA

Part VII
Optimization and Differential Equations

Analysis of Chaotic and Stochastic Causes Started in Solutions to Deterministic Nonlinear Differential Equations

T.Q. Rzayev

Abstract The report attempts to make a comparative analysis of ChP- and SP-based information approach and to identify the factors that cause the occurrence of ChP solutions in deterministic equations.

1 Introduction

Since 70s of the last century, a new direction of work on the identification and analysis of chaotic processes (ChPs) arising in deterministic solutions of nonlinear differential equations (NLDE) both open system and closed system (CS). The main tool for identifying ChP is computer modeling and simulation of systems of equations of their parameters.

Papers devoted to this area repeatedly noted that ChP looks no different from the stochastic processes (SPs). Therefore, attempts were made to identify the signs typical of ChP, on these grounds to distinguish ChP from SP. However, many of the proposed features also typical for SP. Besides concepts, deterministic equations and ChP are compatible, and therefore, the occurrence of ChP in deterministic systems is difficult to perceive. And the emergence of ChP in such systems is manifested as a fact, but the causes of this fact are not paying attention.

The report attempts to make a comparative analysis of ChP- and SP-based information approach and to identify the factors that cause the occurrence of ChP solutions in deterministic equations.

T.Q. Rzayev (✉)
The Azerbaijan Technical University, Baku, Azerbaijan
e-mail: kerimovarn.1963@gmail.com

© Springer International Publishing Switzerland 2016
L.A. Zadeh et al. (eds.), *Recent Developments and New Direction in Soft-Computing Foundations and Applications*, Studies in Fuzziness and Soft Computing 342, DOI 10.1007/978-3-319-32229-2_21

2 Statement of the Problem

As is-well known physical systems (PhSs) control in real conditions in general are exposed to controlled and uncontrolled disturbing influences. In this case, the equation can be represented CS in the following form:

$$\dot{x} = F(x, u, z, a), \quad x(0) = x_0, \tag{1}$$

where x, u is the vector of phase variables and control actions; z is the vector controlled disturbances, a is the vector coefficients of the equation for a given structure of the display operator F. This factor accumulates uncontrolled disturbance—$\xi = (\xi_1, \xi_2, \ldots)$ taking place in the PhS; x_0—the initial value of the vector of phase variables x.

When deterministic operator F and clearly defined function $z(t)$ and values a, x_0 Eq. (1) is considered to be deterministic. This equation is highly abstracted representation of the PhS and the special case of the stochastic equation.

Constructing the Eq. (1) takes a number of prerequisites. Often suggested that PhS stationary functions $z(t)$ and $\xi(t)$ are the ergodic-stationary processes.

At the known forecast of function $z(t)$, $t_0 \le t \le T$ or its absence from the Eq. (1), the following turns out:

$$\dot{x} = F(x, u, a), \quad x(0) = x_0. \tag{2}$$

Here, operator F has other structure, and coefficient has other value.

If we synthesize control in the first equation by certain criterion, it is possible to receive the law of control of a kind:

$$u(t) = \rho(x(t), z(t - \tau)), \quad t \ge \tau \le T, \tag{3}$$

where for stationary $z(t)$ $\tau = 0$, and in the second equation we will receive the law of control a kind:

$$u(t) = \rho(x(t)) \tag{4}$$

Taking into account control laws accordingly in (1) and (2), the equation of open system turns out:

$$\dot{x} = F(x, a) \tag{5}$$

The works devoted to research ChP in closed systems, i.e. in the CS, basically use the determined nonlinear equation of a kind (2) is third order of certain structure and in open systems use equation of a kind (5) third order with certain structure or their discrete variants.

The operator of image F in the Eq. (1) expresses a surface in space x, u, z in the Eq. (2) expresses a surface in space x, u, and in the Eq. (5) expresses a surface in space x. Often characteristics PhS cover wide area in space of corresponding variables and have difficult enough form. Attempt to approximate all surface of such characteristics PhS one equations has not crowned success. It is connected by that; the equations received thus have very difficult structure and consequently small applied the importance. Proceeding from told, the surface of general characteristic ChP breaks into small areas, number, and which numbers we will designate v, N accordingly, and these areas are approximated by the equations concerning simple structure. Capacity of each v area is defined, proceeding from desirable structure of the equation and degree of complexity of characteristic PhS in this area.

It is necessary to notice that for everyone PhS, proceeding from technical regulations, the limited number of operating conditions is defined. The vicinity of each such mode makes corresponding working area. And working areas in overwhelming majority of cases do not cover a general characteristic considered PhS, including the CS. Therefore for the system analysis, its equation is made only for the working areas which number is much less than total working areas on a surface of a general characteristic of system can located in the neighborhood or is isolated (Fig. 1).

The system during each moment of time is only in one of working areas and with change of its parameters (x_0, u, and therefore a vector of factors) there is quantitative and probably qualitative change in its decision in the given area. Thus, quantitative change in the decision occurs regularly, and qualitative in steps during the moments when there is a qualitative change in roots of the corresponding characteristic equation. Such approach underlies the qualitative analysis of the equations of systems [1, 2]. As a result of the qualitative analysis of the decision of the equation of system by its computer simulation and imitation of parameters, such values of the last at which ChP arises are defined. It does not pay sufficient attention to the causes of ChP in decisions NLDE.

Fig. 1 Location workspaces
on the overall surface of F

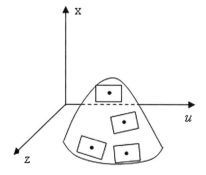

3   Solution of the Problem

Considering that known signs are insolvent for qualitative definition of randomness or stochastic of the strange processes arising in decisions of the equations of systems, for this purpose more effective approach based on the important characteristic indicator of-entropy of the theory of the information of K. Shennon is offered [3]. Thus, proceeding from the substantial analysis of chaotic and stochastic processes, it is shown that these processes have the identical (stochastic) nature, but different degrees of uncertainty. Here fairly following.

Theorem 1 *Degree of uncertainty of the StP always is less than in ChP.*
   It is easily possible to prove it proceeding from that that the StP has the established static indicators (probabilities, the moments, distribution functions, correlation functions, etc.) which carry certain information on these processes, promote their ordering and carrying out of certain operations over them. And the StP theory has old history and the fulfilled powerful mathematical apparatus. Therefore, it is not casual that for the analysis of many known in physicist ChP, for example, in macrosystems at molecular and electronic levels, they are transferred to area of the ChP and are investigated with the use of a mathematical apparatus of the last. It is possible to carry Fermi's, Boze's, Einstein's, Maxwell's, Boltzmann's, etc. works, which have established the laws of distribution. However, there is a fair question: If it were ChP, it is adequately possible to describe StP means why it not to name StP.
   Thus, ChP it is possible to carry to the SP category, the found which statistics contain big uncertainty so statistical hypotheses about them partially or completely are not carried out. In [3], HP refers to the joint venture with a distribution close to uniform. Based on this ratio and the entropy of normal $H^n(x)$, evenly $H^p(x)$, and other $H^i(x)$, $i = 1, 2, \ldots$, distributions.

$$H^n(x) \leq H^i(x) \leq H^p(x)$$

   Use of a mathematical apparatus of the StP (methods of probability theory and the mathematical statistics) in each concrete case, for example, certain demands make to a kind of function of distribution of random variables, ergodic-stationary, the StP, etc.
   It is known that the StP displays in itself random variables (RV) and casual events since the StP represents time function RV, and RV is the set of casual events. Therefore, it is possible to present the StP as system RV. As such system also, it is possible to present RV.
   Another important issue in the identification and analysis of HP solutions in deterministic NLDE is to determine the causes of such processes. As noted above, ChP to identify solutions of equations by computer modeling and simulation of their parameters as a fact, but do not explain the causes of such processes in deterministic systems, which causes misunderstanding. Here, the following assertions:

Theorem 2 *The exact solution of deterministic NLDE under certain parameter values in their decisions the last occurrence of ChP or StP impossible.*

Consequence. ChP (StP) without appropriate abstraction cannot be represented by a deterministic equation.

Theorem 3 *For occurrence of chaotic or stochastic process in the decision influence on the NLDE, decision of corresponding character is necessary.*

To prove Theorem 3, *it suffices to prove the existence of the relevant impacts on the solution of equations. In fact, such effects exist. For their determination must meaningfully analyze the principles and technique of constructing solutions of equations and move past. On the basis of this determination, we note the following causes corresponding strange processes—ChP or StP in decisions NLDE:*

1. *Neglect coherences of separate degrees of freedom x_i in the equations at their structural and parametrical identification. In existing practice, identification for each degree of freedom is carried out independently with use of the scalar metrics. For this reason, the received equations are badly joined, i.e., in joints of the corresponding equations are formed artificial backlashes (tolerance zones), that breaks decision regularity. And with increase in degrees of freedom, the number люфтов increases and even faster strangeness degree grows in character of the decision.*
2. *The errors of identification stimulating bad compatibility of the equations of degrees of freedom.*
3. *Measurement errors, coding, transmission, signal processing and solving equations problems. Such errors are usually classified as statistical. Obviously, these can cause a strange process in solutions of equations.*
4. *Imbalance between the equations of the system for an arbitrary variation of their coefficients.*

Given that the number of impacts on the solution of a deterministic NLDE and varied greatly and can be represented as a generalized effect with a wide range of frequencies. It is clear that the frequency spectra are available with different intensity in different directions and act on the solution and thus cause complicated random (stochastic) processes there in.

Theorem 3 *implies the following:*

Theorem 4 *To hold the system in an unstable mode in closed systems necessary to control the impact was enough to compensate for the existing weak disturbing effect.*

Generalized disturbance taking place at the decision NLDE usually rather weak.

As is known, nonlinear system, unlike linear, has established several modes (states). Figure 2 *presented six steady-state regimes: unstable node (1), a stable cycle (2), seat (3.6), and an unstable cycle (5).*

If we assume that some system contains all the modes shown in Fig. 2 *and it is in one of them, then changing the parameters of the equation can make the transition to other modes. Also coming in the works devoted to the analysis of solutions NLDE. However, it should be noted that a change in the parameters of the equation*

Fig. 2 Steady nonlinear systems

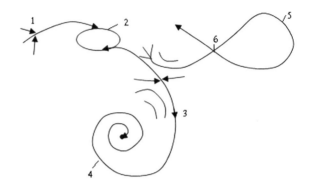

from the center of his field of approximation, in any direction glidants steady-state current in the appropriate direction. There exist values of parameters of the equation, in which the bifurcation occurs (qualitative change) steady state (NLDE solutions).

Different modes have different sensitivity to weak disturbing influences (noise) chaotic or stochastic nature. Most sensitive to weak disturbing influences are unstable, and the least sensitive—stable in the large—is stable regimes. Stable states with varying degrees of stability, less than absolutely stable, which occupy an intermediate position. Moreover, with the increasing stability of the system in a stable steady state increases its damping, and hence filtering ability. Thus, the degree of randomness of the system is inversely proportional to its degree of stability.

In connection with the foregoing, weak disturbances in open systems in general remove them from the unstable steady state. However, in closed systems unstable regimes are held in their states by using control action. Here, we have the following assertion:

Theorem 5 *ChP (StP) may occur in the solution of a linear differential equation and unstable linear system.*

Perhaps also forced excitation ChP (StP) in deterministic decision NLDE. It is often possible to randomization control action.

Above under weak stability is understood to mean a stable state from which it can come under the influence of weak influence of the type of errors that occur when solving NLDE.

It should be noted that the control action, unstable restraint system in steady-state operation, acquires the properties of the perturbing effects to the accuracy of its filtration system or other means.

In support of the theorem in Fig. 3 are four examples. The first example shows a balloon disposed on the top of the paraboloid surface; the second—the ball located in the well at a low plane; and the third—the antenna; and the fourth pole, located in the vertical steady state. It is obvious that the slightest disturbances they will pull them out and possibly unstable weak stability.

Fig. 3 Examples of steady and unstable weak stability regimes

For retention of these systems in the steady state unstable need small effort (control in the appropriate direction). Such an effort in four areas in the figures shown in the form of rods and cables with arrows. Perhaps retention system in a fragile state with a weak control the three symmetric directions.

References

1. Andrievskiy, B.P., Fradko, A.L.: Control of chaos. Methods and Applications. Methods I, II applications (surveys). Autom Remote Control **65**(4), 505–533 (2004)
2. Neymark, Y.I., Landa, T.S.: Stochastic and Chaotic Oscillations. M: Nauka (1987)
3. Moon, F.: Chaotic Oscillations. M: Mir (1990)
4. Rzayev, T.: Information approach analyses of chaotic and stochastic processes in nonlinear control systems. In: IV International conference "Problems of Cybernetics and Information", vol. 2, pp. 77–79. Baku, Sept 2012

Soft Computing Approaches for Two-Dimensional Beamforming

Rama Kiran, Pradip Sircar and Nishchal K. Verma

Abstract Last decade has seen constant growth in wireless technology. Still there is requirement for higher data rates. Current technologies have nearly maximized the use of temporal and spectral techniques to improve capacity and data transfer speeds. But additional spatial dimension is not yet exploited. We can improve capacity of cellular systems by canceling interfering signals using directional arrays. This process is known as beamforming. There are numerous studies available for beamforming mostly using uniform linear arrays but little work has been done on other array configurations. Constrained beamforming techniques with planar array configurations are to be developed for capacity improvement of wireless systems in 3D space. In this work, we have employed the bacterial foraging optimization algorithm (BFOA) and genetic algorithm (GA) for constrained beamforming using uniform planar array and uniform circular arrays.

1 Introduction

In the last decade, wireless technology has seen rapid growth and is playing an increasing role in the lives of people throughout the world. Larger numbers of people are relying on the technology directly or indirectly. Last decade has seen introduction of many complex cellular standards in order to achieve higher data rates. The current standards use temporal and spectral techniques to improve capacity and data transfer speeds. But spatial dimension is not fully exploited yet. This can be done by using directional antenna arrays. In wireless communication, situation arises where we have

R. Kiran (✉) · P. Sircar · N.K. Verma
Department of Electrical Engineering, Indian Institute of Technology Kanpur, Kanpur 208016, Uttar Pradesh, India
e-mail: ramakiranab@gmail.com

P. Sircar
e-mail: sircar@iitk.ac.in

N.K. Verma
e-mail: nishchal@iitk.ac.in

© Springer International Publishing Switzerland 2016
L.A. Zadeh et al. (eds.), *Recent Developments and New Direction in Soft-Computing Foundations and Applications*, Studies in Fuzziness and Soft Computing 342, DOI 10.1007/978-3-319-32229-2_22

to receive the signal transmitted by a particular user undistorted and reject all signals from other users which cause interference. This can be accomplished by placing beam along the desired direction and placing nulls along other interfering directions. This process is known as constrained beamforming. In [1], an algorithm has been proposed which forms beam along with a particular user direction and cancels all other interfering signals by optimizing directivity of beam pattern. When there are more number of constraints than the number of array elements, or when there are more number of interfering signals along a particular plane, the method fails to operate. This problem is solved by modified constrained beamforming in [2]. But modified constrained beamforming is not optimized with respect to sidelobe levels. In literature, there are several works which employ soft computing algorithms for beamforming [3–6]. But most of these works employ linear arrays and use only magnitude of array weights. In this work, we consider complex array weights so that both the phase and the magnitude of array weights are chosen. In the first part of this paper, we propose the bacterial foraging optimization algorithm (BFOA) and genetic algorithm (GA) for null formation and sidelobe reduction for beamforming using 2D array configurations of uniform rectangular array (URA) and uniform circular array (UCA). In the second part, we propose beamforming by maximizing the signal to interference plus noise ratio (SINR) for URA and UCA.

2 Two-Dimensional Array Configurations and Array Factors

We consider 2D array configurations, namely the URA and UCA, for constrained beamforming.

2.1 Uniform Rectangular Array

In URA, the array elements are arranged in a plane.

Consider an array with M elements along X-direction and N elements along Y-direction as shown in Fig. 1. Let the distance between two consecutive array element along X-direction be d_x and that of along Y-direction be d_y. Let us consider a point source located in the far-field region of the array at a direction (θ, ϕ). Since the source is in far-field, we can assume that the waves arriving the array are planar. The array factor for the URA is given by,

$$AF(\theta, \phi) = \sum_{m=1}^{M} \sum_{n=1}^{N} w(m, n) e^{j(m-1)Kd_x \cos(\theta)\cos(\phi)} \times e^{j(n-1)Kd_y \cos(\theta)\sin(\phi)} \quad (1)$$

where $K = 2\pi/\lambda$, λ is the wavelength of incident wave.

Fig. 1 Uniform planar array
(UPA) with $M \times N$ elements
in XY plane

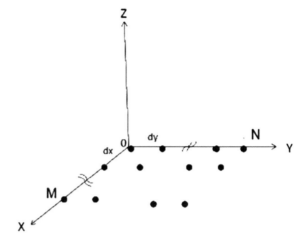

2.2 Uniform Circular Array

In UCA, the array elements are arranged on the perimeter of a circle uniformly. Consider an N element circular array as shown in Fig. 2. Let a be the radius of the circle and the *nth* element is located on the perimeter of the circle at an angle ϕ_n. Assuming far-field condition, the array factor is given by,

$$AF(\theta, \phi) = \sum_{n=1}^{N} w(n) e^{j(Ka)\sin(\theta)\cos(\phi-\phi_n)} \tag{2}$$

Fig. 2 Uniform circular
array (UCA)

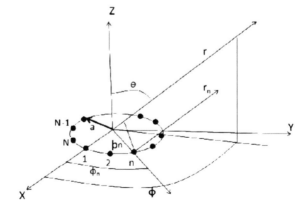

3 Soft Computing Optimization

We consider the bacterial foraging optimization algorithm (BFOA) [7] and genetic algorithm (GA) [8] for constrained beamforming with sidelobe level reduction and beamforming by maximizing the signal to interference plus noise ratio (SINR).

3.1 Bacterial Foraging Optimization Algorithm

The BFOA optimizes the given fitness function by mimicking foraging behavior of *E. Coli* bacteria. During foraging process of bacteria, movement is achieved by tensile flagella. Flagella help bacteria to tumble or swim. Bacteria swim along straight line on nutrient-rich surface and tumble frequently at noxious places to find a nutrient gradient.

Suppose that we want to find the minimum of fitness function $J(\theta)$ where $\theta \in \Re^p$. The BFOA mimics the four principal mechanisms observed in a bacteria foraging process, namely chemotaxis, swarming, reproduction, and elimination-dispersal to solve non-gradient optimization problem. Let us define chemotaxis step to be tumble followed by tumble or tumble followed by run. Let j be the index for the chemotactic step, k be the index for the reproduction step, and l be the index for the elimination-dispersal step. Also, let p be the dimension of search space, S be the total number of bacteria in the population, N_c be the number of chemotactic steps, N_s be the swimming length, N_{re} be the number of reproduction steps, P_{ed} be the elimination-dispersal probability of particular bacteria, and $C(i)$ be the size of the step taken in the random direction specified by tumble.

Let $P(j, k, l) = \left\{ \theta^i(j, k, l) | i = 1, 2, 3, \ldots, S \right\}$ represent the position of each member in the population of the S bacteria at *jth* chemotactic step, *kth* reproduction step, and *lth* elimination-dispersal event. Let $J(i, j, k, l)$ denote the cost at the location of the *ith* bacterium $\theta^i(j, k, l) \in \Re^p$. Below, we briefly describe the four prime steps in BFOA.

Step 1 *Chemotaxis*: This step simulates the movement of *E. Coli bacteria* through swimming and tumbling. The *E. Coli* bacterium can move in two different ways. It can swim for a period of time in the same direction or it may tumble, and alternate between these two modes of operation for the life time. Suppose $\theta^i(j, k, l)$ represents the *ith* bacterium at *jth* chemotactic, *kth* reproductive, and *lth* elimination-dispersal step. Then, the chemotaxis movement of the bacterium may be represented by

$$\theta^i(j+1, k, l) = \theta^i(j, k, l) + C(i) \frac{\Delta(i)}{\sqrt{\Delta(i)^{\mathrm{T}} \Delta(i)}} \qquad (3)$$

where Δ indicates a vector in the random direction.

Step 2 *Swarming*: In chemotaxis phase, it is observed that bacteria move in groups. They form spatiotemporal patterns in nutrient-rich surfaces. The group formation is done by releasing attractants. This group behavior may be represented by the following function in BFOA,

$$J_{cc}(\theta, P(j, k, l)) = \sum_{i=1}^{S} J_{cc}(\theta, \theta^i(j, k, l))$$

$$= \sum_{i=1}^{S} \left\{ -d_{\text{attractant}} e^{\left[-w_{\text{attractant}} \sum_{m=1}^{P} (\theta_m - \theta)^2 \right]} + h_{\text{repellant}} e^{\left[-w_{\text{repellant}} \sum_{m=1}^{P} (\theta_m - \theta)^2 \right]} \right\}$$

(4)

where $J_{cc}(\theta, \theta^i(j, k, l))$ is the objective function value to be added to actual fitness function. The coefficients $d_{\text{attractant}}$, $w_{\text{attractant}}$, $h_{\text{repellant}}$, and $w_{\text{repellant}}$ are to be chosen properly.

Step 3 *Reproduction*: The least healthy bacteria die in this step and healthy bacteria split into two and placed in same location.

Step 4 *Elimination and Dispersal*: Gradual or sudden changes in environment where bacteria lives may occur due to various reasons. The bacteria living in that region may die or disperse due to these environmental changes. This is simulated by killing bacteria with probability P_{ed} and placing new bacteria at random place.

3.2 Genetic Algorithm

The genetic algorithm (GA) is a heuristic search algorithm which is inspired by natural selection of genes. It optimizes given fitness function in iterations involving selection, crossover, and mutations of chromosomes. The genetic algorithm begins with randomly assigning values in the range of expected solution regions. It evaluates the fitness function which is to be optimized for these values of chromosomes and allocates reproductive opportunities to chromosomes. The chromosomes which represent better solution are given better reproduction chances than poor chromosomes which represent poorer solutions.

Selection of chromosomes: The fitness function is evaluated at all the chromosomes, and the number of copies that are passed to the next generation is proportional to their fitness function values.

Crossover: The selected chromosomes are represented as binary-coded strings. Each binary-coded chromosome is paired with another binary-coded chromosome. Then, each pair is crossed over at randomly selected positions.

Mutation: The crossed-over chromosomes are flipped bitwise with certain probability.

Termination: The genetic algorithm (GA) is terminated when the solution satisfies the minimum criteria, and fixed number of iterations is reached or allocated number of computations is completed.

4 Beamforming by Sidelobe Reduction and by Optimizing the SINR of Desired Signal

Our objective is to reduce the side lobes in constrained beamforming. This can be accomplished by reducing the integral of beam pattern in desired side lobe range. The fitness function which achieves this is given by,

$$\sum_i \frac{1}{\Delta\phi_i \times \Delta\theta_i} \int_{\phi_{i1}}^{\phi_{i2}} \int_{\theta_{i1}}^{\theta_{i2}} |AF(\theta, \phi, w_R + jw_I)|^2 d\phi d\theta \tag{5}$$

The values of beam pattern at null direction can be reduced and the main beam can be formed using minimization of the fitness function

$$\sum_k |AF(\theta_k, \phi_k, w_R + jw_I)|^2 + |AF(\theta_0, \phi_0, w_R + jw_I)|^2 \tag{6}$$

Therefore, the total fitness function to be minimized is given by

$$\begin{aligned} \text{Fitness}(w_R + jw_I) = &\sum_i \frac{1}{\Delta\phi_i \times \Delta\theta_i} \int_{\phi_{i1}}^{\phi_{i2}} \int_{\theta_{i1}}^{\theta_{i2}} |AF(\theta, \phi, w_R + jw_I)|^2 d\phi d\theta \\ &+ \sum_k |AF(\theta_k, \phi_k, w_R + jw_I)|^2 + |AF(\theta_0, \phi_0, w_R + jw_I)|^2 \end{aligned} \tag{7}$$

where $w_R + jw_I$ is the weight vector of the sensors in the array. The difference angles $\Delta\phi_i$ and $\Delta\theta_i$ are the lengths of *ith* interval. ϕ_k and θ_k correspond to *kth* null.

In wireless communication, the signal to interference plus noise ratio (SINR) is an important factor which measures the effect of noise and interference on the desired signal. It is desirable to have the SINR as high as possible. Here, we propose beamforming techniques which maximize the SINR of the desired signal using soft computing methods, namely the BFOA and GA.

Let us assume that the direction of desired signal is (θ_d, ϕ_d) and (θ_k, ϕ_k) be the interference directions for different k. Let P be the power of the desired signal and the power of the interfering signals at the antenna array. Let η be the noise power at the antenna array. The total power of the desired signal at array output is given as

$$P_{\text{desired}} = |AF(\theta_d, \phi_d)|^2 P \tag{8}$$

The total noise plus interference power is given as

$$P_{\text{undesired}} = \sum_k |AF(\theta_k, \phi_k)|^2 P + \sum_{i=1}^{MN} |w_i|^2 \eta \qquad (9)$$

Therefore, the SINR is given by

$$\text{SINR} = \frac{P_{\text{desired}}}{P_{\text{undesired}}} = \frac{|AF(\theta_d, \phi_d)|^2 P}{\sum_k |AF(\theta_k, \phi_k)|^2 P + \sum_{i=1}^{MN} |w_i|^2 \eta} \qquad (10)$$

Writing above SINR as function of array weight

$$\text{SINR}(w_R + jw_I) = \frac{|AF(\theta_d, \phi_d, w_R + jw_I)|^2 P}{\sum_k |AF(\theta_k, \phi_k, w_R + jw_I)|^2 P + \sum_{i=1}^{MN} |w_i|^2 \eta} \qquad (11)$$

5 Simulation and Results

The arrays considered for simulation are 3×3 uniform rectangular array (URA) with spacing $d_x = 0.5\lambda$ and $d_y = 0.5\lambda$ and 9 element uniform circular array (UCA) with radius $a = 0.7\lambda$. The following are the BFOA parameters used: $N_c = 10$, $N_s = 10$, $N_{\text{re}} = 10$, $N_{\text{ed}} = 3$, $P_{\text{ed}} = 0.2$, $C(i) = 0.001$, $P = 9$ and $S = 30$.

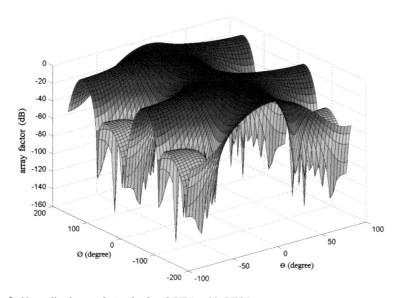

Fig. 3 Normalized array factor for 3×3 URA with BFOA

Fig. 4 Normalized array
factor for 3 × 3 URA with
GA

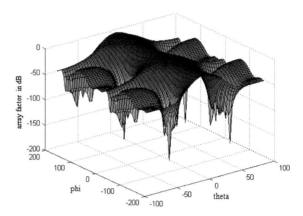

For the genetic algorithm, we used 30 genes over 2000 generations and chromosomes of length 32 bits. We have considered array weights as complex vector and
we have applied soft computing algorithms for computing real and imaginary parts
of the array vector. We have considered the entire range of (θ, ϕ) for sidelobe
reduction. The results are shown in Figs. 3, 4, 5, and 6 and Tables 1, 2, 3, 4.

The arrays considered for the SINR optimization beamforming are 3 × 3 URA
with spacing $d_x = 0.5\lambda$ and $d_y = 0.5\lambda$ and 9 element UCA with radius $a = 0.7\lambda$.
The following are the BFOA parameters used: $N_c = 10$, $N_s = 10$, $N_{re} = 10$,
$N_{ed} = 10$; $P_{ed} = 0{:}2$, $C(i) = 0{:}005$, $P = 9$, and $S = 30$. For the genetic algorithm,

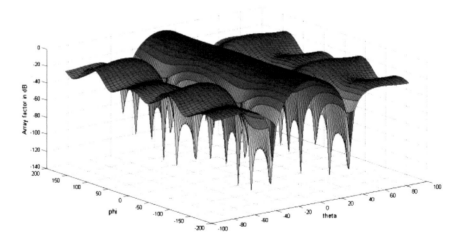

Fig. 5 Normalized array pattern using BFOA for 9 element UCA

Fig. 6 Normalized array pattern using GA for 9 element UCA

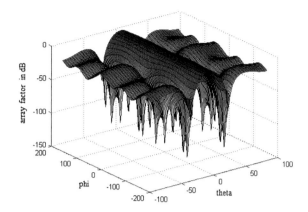

Table 1 Comparison between different beamforming techniques for 3 × 3 URA

	BFOA		Genetic		Constrained	Conventional
Null direction (dB)	Sidelobe level (θ, ϕ)	Values at null (dB)	Sidelobe level (dB)	Values at null (dB)	Sidelobe level (dB)	Sidelobe level (dB)
(90, 45)	−44.20	−52.69	−41.76	−43.13	−7.89	−29.63
(62.5, 60)	−44.47	−45.88	−38.33	−32.42	−7.89	−29.63
(60, 210)	−44.48	−52.09	−22.34	−53.10	−7.89	−29.74
(45, 300)	−44.94	−9.7	−22.34	−45.41	−8.47	−32.03

Table 2 Comparison between different beamforming techniques for 9 element UCA

	BFOA		Genetic			Conventional	Constrained
Null direction (θ, ϕ)	Values at null (dB)	Sidelobe level (dB)	Values at null (dB)	Sidelobe level (dB)		Values at null (dB)	Sidelobe level (dB)
(90, 45)	−40.78	−20.23	−17.60	−20.95		−1.05	−18.31
(62.5, 60)	−32.16	−20.23	−19.12	−20.95		−2.30	−18.31
(60, 210)	−30.74	−35.56	−10.84	−11.10		−2.83	−18.31
(45, 300)	−41.56	−35.56	−31.53	−10.67		−2.85	−18.31

we used 30 genes over 2000 generations and chromosome length of 32 bit. We have considered array weights as complex vector and we have applied soft computing algorithms for computing real and imaginary parts of the array vector. The results are shown in Figs. 7, 8, 9, 10, 11, 12, 13, 14.

Table 3 Performance of various beamforming algorithms for 3 × 3 URA

Values at nulls (in dB)					
Nulls	BFOA	Genetic	Conventional	Constrained	Modified constrained
1	−70.87	−80.86	−26.29	−13.84	−15.65
2	−96.32	−71.99	−23.69	−12.46	−17.18
3	−105.65	−81.59	−23.69	−11.07	−17.92
4	−91.46	−86.45	−26.29	−14.08	−15.92
5	−89.91	−81.47	−58.52	−18.06	−21.07
6	−92.89	−86.45	−16.71	−10.98	−34.37
7	−94.95	−81.59	−1.64	−24.08	−32.18
8	−90.50	−71.99	−1.64	−24.08	−47.74
9	−84.08	−80.86	−16.71	−24.08	−31.93
10	−79.23	−59.20	−58.52	−21.07	−47.74

Table 4 Performance of various beamforming algorithms for 9 element UCA

Values at nulls (in dB)					
Nulls	BFOA	Genetic	Conventional	Constrained	Modified constrained
1	-80.24	−66.32	−22.74	−27.56	−35.55
2	−85.61	−53.68	−22.43	−24.89	−38.78
3	−90.82	−46.44	−22.13	−27.53	−34.65
4	−102.98	−42.22	−21.85	−28.83	−39.78
5	−94.77	−39.61	−21.58	−39.82	−48.67
6	−98.15	−37.95	−21.32	−29.01	−42.48
7	−84.57	−36.90	−21.08	−39.26	−40.38
8	−93.87	−36.24	−20.86	−28.97	−28.56
9	−92.11	−35.84	−20.64	−29.03	−30.32
10	−81.22	−35.60	−20.44	−29.14	−31.32

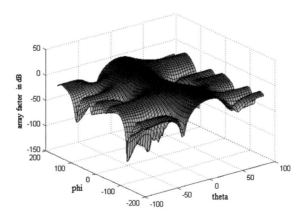

Fig. 7 Normalized array factor for 3 × 3 URA with GA

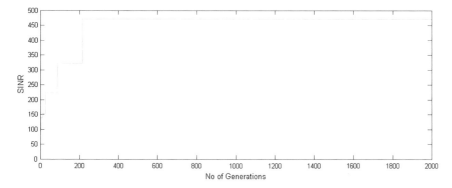

Fig. 8 SINR improvement after number of iterations for 3 × 3 URA with GA

Fig. 9 Normalized array factor for 3 × 3 URA with BFOA

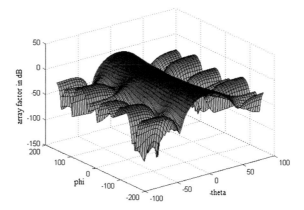

Fig. 10 SINR improvement after number of iterations for 3 × 3 URA with BFOA

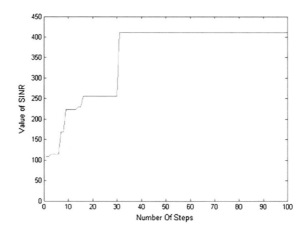

Fig. 11 Normalized array
factor for 9 element UCA
with BFOA

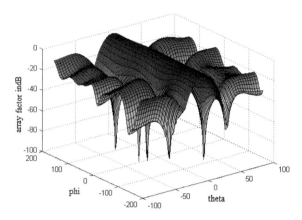

Fig. 12 SINR improvement
after number of iterations for
9 element UCA with BFOA

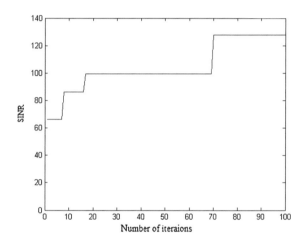

Fig. 13 Normalized array
factor for 9 element UCA
with GA

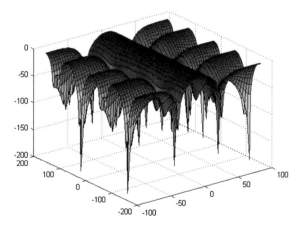

Fig. 14 SINR improvement after number of iterations for 9 element UCA with GA

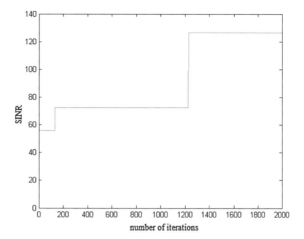

number of iterations

6 Conclusion

In this paper, we have proposed sidelobe reduction for null steering beamforming using soft computing algorithms, namely the genetic algorithm (GA) and bacterial foraging optimization algorithm (BFOA). We have compared the results with the conventional, constrained, and modified constrained beamforming. In the later part of the paper, we have performed beamforming by optimizing the SINR of the desired signal using soft computing algorithms and compared results with other methods.

References

1. Kuhwald, T., Boche, H.: A constrained beam forming algorithm for 2D planar antenna arrays. In: Vehicular Technology Conference, 1999. VTC. 1999-Fall. IEEE VTS 50th, vol. 1, pp. 1–5 (1999)
2. Myneni, H., Sircar, P.: Two-dimensional beam forming and interference reduction using different arrays. In: Recent Advances in Information Technology (RAIT), 2012 1st International Conference on, pp. 794–799 (2012)
3. Abu-Al-Nadi, D., Ismail, T., Mismar, M.J.: Synthesis of linear array and null steering with minimized side-lobe level using particle swarm optimization. In: Antennas and Propagation (EuCAP), 2010 Proceedings of the Fourth European Conference on, pp. 1–4 (2010)
4. Khan, M., Tuzlukov, V.: Null steering beamforming for wireless communication system using genetic algorithm. In: Microwave Technology Computational Electromagnetics (ICMTCE), 2011 IEEE International Conference on, pp. 289–292 (2011)
5. Mandal, D., Das, S., Bhattacharjee, S., Bhattacharjee, A., Ghoshal, S.: Linear antenna array synthesis using novel particle swarm optimization. In: Industrial Electronics Applications (ISIEA), 2010 IEEE Symposium on, pp. 311–316 (2010)
6. Petrella, N., Khodier, M., Antonini, M., Ruggieri, M., Barbin, S., Christodoulou, C.: Planar array synthesis with minimum sidelobe level and null control using particle swarm

optimization. In: Microwaves, Radar Wireless Communications, 2006. MIKON 2006. International Conference on, pp. 1087–1090 (2006)

7. Passino, K.: Biomimicry of bacterial foraging for distributed optimization and control. Control Syst. IEEE **22**(3), 52–67 (2002)

8. Whitley, D.: A genetic algorithm tutorial. Stat. Comput. **4**(2), 65–85 (1994). [Online]. Available: http://dx.doi.org/10.1007/BF00175354

Part VIII
Evolutionary Methods in Applications

Design of Ensemble Neural Networks for Predicting the US Dollar/MX Time Series with Particle Swarm Optimization

Martha Pulido, Patricia Melin and Oscar Castillo

Abstract This paper shows the use of particle swarm optimization (PSO) in the design of a neural network ensemble with type-1 and type-2 fuzzy integration of responses for time series prediction. The considered time series in this paper for testing the hybrid approach is the US/Dollar MX time series. Simulation results show that the hybrid ensemble approach, combining neural networks and fuzzy logic, produces good prediction of the dollar time series.

Keywords Ensemble neural networks · Optimization · Particle optimization swarm · Optimization · Time series prediction

1 Introduction

Time series is a collection of data recorded over a period of time. Time series are used in statistics, signal processing, pattern recognition, econometrics, mathematical finance, weather forecasting, electroencephalography, control engineering, astronomy, communications engineering, and earthquake prediction. An analysis of the history of a time series can be used by management to make current decisions and plans based on long-term forecasting. We usually assume that past patterns will continue into the future. Long-term forecasts extend more than 1 year into the future; 5-, 10-, 15-, and 20-year projections are common. Long-range predictions are essential to allow sufficient time for the procurement, manufacturing, sales, finance, and other departments of a company to develop plans for possible new plants, financing, development of new products, and new methods of assembling.

M. Pulido (✉) · P. Melin · O. Castillo
Tijuana Institute of Technology, Tijuana, B.C., Mexico
e-mail: marthapulido_84@hotmail.com

P. Melin
e-mail: pmelin@tectijuana.mx

O. Castillo
e-mail: ocastillo@tectijuana.mx

© Springer International Publishing Switzerland 2016
L.A. Zadeh et al. (eds.), *Recent Developments and New Direction in Soft-Computing Foundations and Applications*, Studies in Fuzziness and Soft Computing 342, DOI 10.1007/978-3-319-32229-2_23

317

Forecasting the level of sales, both short-term and long-term, is practically dictated by the very nature of business organizations. Competition for the consumer's dollar, stress on earning a profit for the stockholders, a desire to procure a larger share of the market, and the ambitions of executives are some of the prime motivating forces in business. Thus, a forecast (a statement of the goals of management) is necessary to have the raw materials, production facilities, and staff available to meet the projected demand [1].

The main contribution of the paper is the proposed model of a neural network ensemble that is optimized with the particle swarm optimization (PSO) method. The optimization method determines the number of modules of the neural network ensemble, number of layers, and number of neurons per layer, and thus obtains the best architecture of the ensemble neural network. After obtaining this architecture, the results are aggregated with type-1 and type-2 fuzzy systems, and the inputs to the fuzzy system are the responses according to the number of network modules and these are the number of inputs of the fuzzy system. In this case, the maximum number of inputs that are being considered 5 inputs and one output with two Gaussian membership functions and these will be granulated in two linguistic variables that are low and high forecast, and the forecast output will also be low high and thereby obtain the forecast error for this series of the US/Dollar MX time. The proposed hybrid ensemble of neural networks with fuzzy response aggregation and its optimization with PSO is the main contribution of the paper, as this hybrid approach has not been proposed previously in the literature for this kind of time series prediction problems. This paper is used for this series of the US/Dollar MX time series because the problem is quite complex and the time series is chaotic and for this reason an ensemble model is justified, but also we had previously work with other series.

The rest of the paper is organized as follows: Sect. 2 describes the concepts of time series, Sect. 3 describes the concepts of ensemble neural networks, Sect. 4 describes the concepts of fuzzy system, Sect. 5 describes the concepts of PSO, Sect. 6 describes the problem and the proposed method of solution, Sect. 7 describes the simulation results of the proposed method, and Sect. 8 has the conclusion part.

2 Time Series

It is called time series to a set of observations on values that a variable (quantitative) at different times. The time series are widely used as various organizations today require knowledge of the future behavior of certain phenomena in order to plan, prevent, etc. That is, time series is used to predict what will happen with a variable in the future from the behavior of that variable in the past.

Data can behave in different ways over time, it may be a trend, which is the long-term component that represents the growth or decline in the time series on a high this period. We can also have a cycle, which is the movement waveform that occurs around the trend; or it may not have a defined or random, there (annual, biannual, etc.). Seasonal variations, which are defined as patterns that are repeated year after year in a fixed period of time [2–9], are also important.

3 Ensemble Neural Networks

Neural networks are conceived as abstractions of the neurobiological structures (brain) found in nature and have the characteristic of being able to help disordered systems, i.e., information systems. The way they work is fundamentally different from that used by conventional computers. Microscopic brain processors (neurons) operating in parallel and qualitatively exhibit more noise than the elements that make computers.

A neural network is a system of parallel processors connected as a directed graph. Schematically, each processing element (neuron) of the network is represented as a node. These connections establish a hierarchical structure that is trying to emulate the physiology of the brain as it looks for new ways of processing to solve real-world problems. What is important in developing the techniques of NN is if its useful to learn behavior, recognize and apply relationships between objects and plots of real-world objects themselves. In this sense, artificial neural networks have been applied to many problems of considerable complexity. Its most important advantage is in solving problems that are too complex for conventional technologies, problems that have no solution, and/or that the algorithm of the solution is very difficult to find [10, 11].

A neural network ensemble is a learning paradigm where a collection of a finite number of neural networks is trained for the same task, which shows that the generalization ability of a neural network system can be significantly improved through assembling a number of neural networks, i.e., training many neural networks and then combining their predictions [12–14].

4 Fuzzy Systems

Fuzzy logic is an area of soft computing that enables a computer system to reason with uncertainty [15]. A fuzzy inference system consists of a set of if-then rules defined over fuzzy sets. Fuzzy sets generalize the concept of a traditional set by allowing the membership degree to be any value between 0 and 1 [16]. This corresponds, in the real world, to many situations where it is difficult to decide in an unambiguous manner if something belongs or not to a specific class. The basic structure of a fuzzy inference system consists of three conceptual components: a rule base, which contains a selection of fuzzy rules; a database (or dictionary), which defines the membership functions used in the rules; and a reasoning mechanism that performs the inference procedure [17].

The concept of a type-2 fuzzy set was introduced by Zadeh (1975) as an extension of the concept of an ordinary fuzzy set (henceforth called a "type-1 fuzzy set"). A type-2 fuzzy set is characterized by a fuzzy membership function, i.e., the membership grade for each element of this set is a fuzzy set in [0,1], unlike a type-1 set where the membership grade is a crisp number in [0,1]. Such sets can be used in situations where there is uncertainty about the membership grades themselves, e.g., uncertainty in the shape of the membership function or in some of its parameters. Consider the transition from ordinary sets to fuzzy sets. When we cannot determine the membership of an element in a set as 0 or 1, we use fuzzy sets of type-1. Similarly, when the situation is so fuzzy that we have trouble determining the membership grade even as a crisp number in [0,1], we use fuzzy sets of type-2 [18].

5   Particle Swarm Optimization

Since its introduction in 1995 [19–21], PSO has seen many improvements and applications. Most of the basic PSO modifications are aimed at improving the convergence of PSO and increasing the diversity of the swarm. Before an in-depth discussion of these changes can occur, it is necessary to discuss the original PSO algorithms, since these were presented in 1995.

A PSO algorithm maintains a swarm of particles, where each particle represents a potential solution. In analogy with the paradigms of evolutionary computation, a swarm is similar to a population, while a particle is similar to an individual. In simple terms, the particles are "flown" through a multidimensional search space, where the position of each particle is adjusted according to its own experience and that of its neighbors. Let x_i denotes the position i in the search space at time step t; unless otherwise stated, t denotes discrete time steps. The position of the particle is changed by adding a velocity, $v_i(t)$, to the current position, i.e.,

$$x_i(t+1) = x_i(t) + v_i(t+1) \tag{1}$$

with $x_i(0) \sim U(X_{\min}, X_{\max})$.

Is the velocity vectors the one that drives the optimization process and reflects both the experimental knowledge of the particles and the information exchanged in the vicinity of particles?

For gbest PSO, the particle velocity is calculated as:

$$v_{ij}(t+1) = v_{ij}(t) c_1 r_1 \left[y_{ij}(t) - x_{ij}(t) \right], + c_2 r_2(t) \left[\hat{y}_j(t) - x_{ij}(t) \right] \tag{2}$$

where $v_{ij}(t)$ is the velocity of the particle i in dimension j at time step t, c_1 y c_2 are positive acceleration constants used to scale the contribution of cognitive and social skills, respectively; y $r_{1j}(t)$, y $r_{2j}(t) \sim U(0,1)$ are the random values in the range [0,1].

The best personal position in the next time step $t + 1$ is calculated as:

$$y_i(t+1) = \begin{cases} y_i(t) & \text{if } f(x_i(x_i(t+1)) \geq f \ y_i(t)) \\ x_i(t+1) & \text{if } f(x_i(x_i(t+1)) > f \ y_i(t)) \end{cases} \tag{3}$$

where $f : \mathbb{R}^{nx} \to \mathbb{R}$ is the fitness function, as with EAs, measuring fitness function closely corresponding to the optimal solution; for example, the objective function quantifies the performance or the quality of a particle (or solution).

The overall best position, $\hat{y}$ (t) at time step t, is defined as:

$$\hat{y}(t)\varepsilon\{y_o(t), \ldots, y_{n_s}(t)\}f(y(t)) = \min \{f(y_o(t)), \ldots, f(y_{n_s}(t)),\} \tag{4}$$

where n_s is the total number of particles in the swarm. Importantly, the above equation defining and establishing $\hat{y}$ the best position is uncovered by either of the particles so far as this is usually calculated from the best position or best personal. The overall best position may be selected from the actual swarm particles, in the following case:

$$\hat{y}(t) = \min\{f(x_o(t)), \ldots, f(x_{n_s}(t)),\} \tag{5}$$

6   Problem Statement and Proposed Method

The goal of this work was to implement a particle swarm algorithm to optimize the ensemble neural network architectures, for each of the modules, and thus to find a neural network architecture that yields optimum results in each of the time series to be considered. In Fig. 1, we have the historical data of each time series prediction, then the data are provided to the modules that will be optimized with particle swarm algorithm for the ensemble network, and then the results of the modules are integrated with integration based on the type-1 and type-2 fuzzy system.

Data of the US Dollar/MX Peso time series: The data consist of 800 points that correspond to a period from 07/04/08 to 09/05/11 (as shown in Fig. 2). In this case, 70 % of the data are used for the ensemble neural network trainings and 30 % to test the network [22]. This time series is labeled number 1.

For the optimization of the structure of the ensemble neural network, a genetic algorithm is used. A brief description of the genetic algorithm details is presented below.

The objective function is defined to minimize the prediction error:

$$\text{EM} = \left(\sum_{i=1}^{D} |a_i - x_i|\right)/D \tag{6}$$

Fig. 1 The general
architecture of the ensemble
neural network

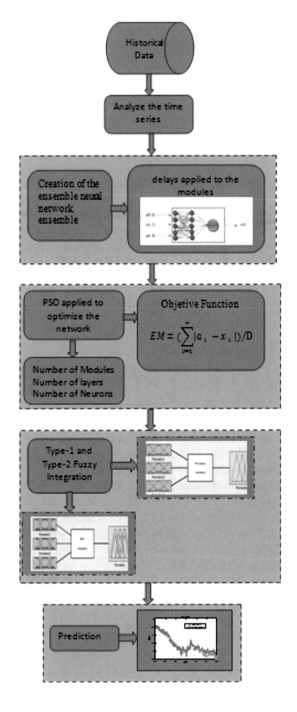

Fig. 2 Exchange rate time series

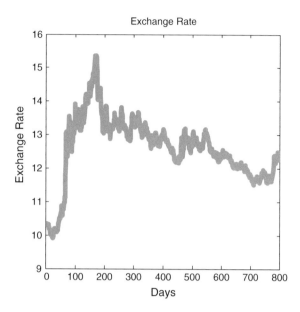

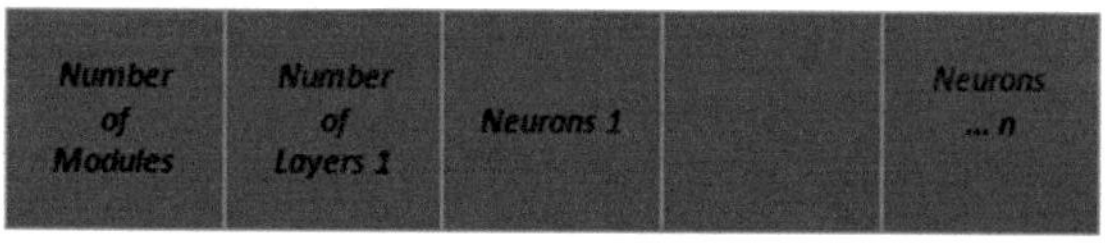

Fig. 3 Particle structure to optimize the ensemble neural network

where a corresponds to the predicted data depending on the outputs of the network modules, X represents real data, D is the number of data, and EM is the total prediction error.

The corresponding chromosome structure is shown in Fig. 3.

Figure 3 illustrates the particle structure to optimize the ensemble neural network, where the parameters that are optimized are the number of de modules, number of layers, and number of neurons.

The parameters for PSO, C_1 and C_2, have a value of 2, and the maximum speed is 1. Constriction coefficient of linear increase $(C) = (0–0.9)$ and inertia weight with linear decrease $(W) = (0.9–0)$; these parameters were used as they perform a manual testing, and the search space is related to the number of modules, which are number of 1–5 modules, number of layers of 1–3, and neurons number of 1–30. The network parameters used to train the neural network ensemble for each of the ensembles are 100 epochs, learning rate of = 0.01, the error goal of 0.01, and training method of Levenberg–Marquardt (LM) (*trainlm*); these parameters were used in this work because previous tests were performed and managed to obtain a good prediction error. The parameters for the PSO algorithm are 100 particles and 100 iterations, and these parameters were used because an optimization with a genetic algorithm was considered previously and to compare it with this method, we used similar parameters.

Fig. 4 Fuzzy inference
system for integration of the
ensemble neural network

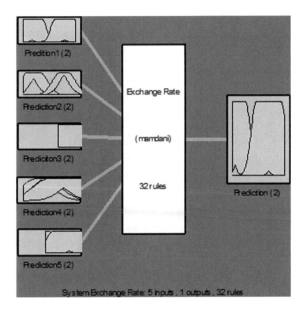

Figure 4 shows a type-2 fuzzy system consisting of 5 inputs depending on the number of modules of the neural network ensemble and one output. Each input and output linguistic variable of the fuzzy system uses 2 Gaussian membership functions. The performance of the type-2 fuzzy integrators is analyzed under different levels of uncertainty to find out the best design of the membership functions for the 32 rules of the fuzzy system.

Figure 5 represents the possible 32 rules of the fuzzy system; we have 5 inputs in our fuzzy system, with 2 membership functions and outputs with 2 membership functions. These fuzzy rules are used for the type-1 and type-2 fuzzy systems.

7   Simulations Results

This section presents the simulation and test results obtained by applying the proposed prediction method to the time series exchange rate.

Using a particle swarm algorithms to optimize the structure of the ensemble neural network and having 2 modules at the most, the best architecture achieved is shown in Fig. 6.

In this architecture, there are two layers in each module. In module 1, in the first layer there are 20 neurons and 26 neurons in the second layer. In module 2, there are 3 neurons in the first layer and 23 neurons in the second layer. The LM training method was used; we also applied 3 delays to the network.

Fig. 5 Type-2 fuzzy system for the US/Dollar MX time series

When using a particle swarm algorithm to optimize the structure of the ensemble neural network and considering 5 modules at the most, the best achieved architecture for time series is shown in Table 1, in row number four.

Tables 1 and 2 show experiments are the integration type-1 and type-2 of the results of the neural network ensemble.

Fuzzy integration is performed initially by implementing a type-1 fuzzy system in which the best result was in the experiment of row number 8 of Table 2 with an error of 0.4251.

Fuzzy integration is performed by implementing a type-1 fuzzy system in which the achieved results were as follows: For the best evolution with a degree of uncertainty of 0.3, a forecast error of 0.4040 was obtained, and with a degree of uncertainty of 0.4, a forecast error of 0.4088, and with a degree of uncertainty of 0.5, a forecast error of 0.4039 was obtained, as shown in Table 3.

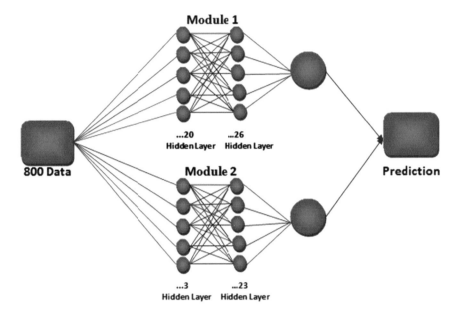

Fig. 6 Ensemble neural network architecture for time series number 2

Table 1 Particle swarm results for the ensemble neural network

No.	Iterations	Particles	Number of modules	Number of layers	Number of neurons	Duration	Prediction error
1	100	100	2	3	27, 13, 15 24, 15, 25	02:01:52	0.0028523
2	100	100	2	2	16, 27 14, 24	01:04:08	0.0027317
3	100	100	2	3	7, 27, 11 19, 13, 15	01:27:07	0.0018746
4	100	100	2	2	20, 26 3, 32	01:54:33	0.0015855
5	100	100	2	2	19, 12 18, 14	01:40:33	0.0019656
6	100	100	3	1	11 13 5	01:17:07	0.0038019
7	100	100	2	2	16, 19 3, 12	01:28:19	0.0042207
8	100	100	3	2	24, 18 15, 23 15, 27	02:20:03	0.0031941
9	100	100	2	3	7, 12, 19 21, 18, 22	01:45:00	0.0029591
10	100	100	2	3	20, 23, 23 17, 10, 11	02:05:09	0.0027134

Table 2 Results of type-1 fuzzy integration

Experiment	Prediction error with fuzzy integration type-1
Experiment 1	0.4860
Experiment 2	0.1724
Experiment 3	0.7310
Experiment 4	1.8360
Experiment 5	0.9306
Experiment 6	1.7777
Experiment 7	0.5433
Experiment 8	0.4251
Experiment 9	2.3293
Experiment 10	0.4653

Table 3 Results of type-2 fuzzy integration

Experiment	Prediction error 0.3 uncertainty	Prediction error 0.4 uncertainty	Prediction error 0.5 uncertainty
Experiment 1	1.2346	0.7402	0.9029
Experiment 2	0.4627	0.4756	0.4157
Experiment 3	0.5313	0.6192	0.6717
Experiment 4	1.3986	1.3056	1.2210
Experiment 5	0.5394	0.4194	0.4269
Experiment 6	0.7996	1.2783	1.2804
Experiment 7	0.5866	0.5346	0.4039
Experiment 8	0.4040	0.4088	0.4073
Experiment 9	4.1251	5.4394	5.5476
Experiment 10	1.8488	1.8799	1.9070

7.1 Comparison of Results

A comparison of results for this time series of the Exchange Rate "Optimization of type-2 fuzzy integration in ensemble neural networks for predicting the US Dollar/MX pesos time series" [23] shows that the best result was using an ensemble neural network architecture with 1 layer using 3 delays. The error obtained by the average integration was 0.0021415 and the architecture of the ensemble neural network was with 5 modules. In this paper, the best result when applying the particle swarm algorithm to optimize the ensemble neural network was 0.0015855, (as shown in Table 1) and the architecture of the ensemble neural network was with 2 modules. This shows that our hybrid ensemble neural approach produces better results for this time series.

Good results were obtained with type-2 fuzzy system because are unable to directly handle rule uncertainties, are very useful in circumstances where it is difficult to determine an exact, and measurement uncertainties.

8 Conclusion

In this paper, we design with PSO a neural network ensemble to find the best architecture of this network. Applying optimization techniques, similar results regarding the prediction error were obtained; applying the particle swarm, an architecture of 2 modules is obtained. In this architecture, we have two layers in each module. In module 1, in the first layer we have 20 neurons and 26 neurons in the second layer. In module 2, we used 3 neurons in the first layer and 23 neurons in the second layer, and the LM training method was used; 3 delays for the network were considered. We can conclude that optimization using particle swarm and after testing type-1 and type-2 fuzzy integration, the best result of type-1 fuzzy integration was 0.4251 and type-2 fuzzy integration, for the best evolution with a degree of uncertainty of 0.3 a forecast error of 0.4040 was obtained, and with a degree of uncertainty of 0.4 a forecast error of 0.4088 and with a degree of uncertainty of 0.5 a forecast error of 0.4039 was obtained. The PSO algorithm has potential to be the best to reduce the architecture of ensemble neural network. This algorithm faces the problems that there may be a better solution and particles move through the solution space and are evaluated according to some fitness criterion after each time step. The main advantage of this approach over other minimization strategies is the globalization of the search process.

Acknowledgments We would like to express our gratitude to CONACYT, and Tijuana Institute of Technology for the facilities and resources granted for the development of this research.

References

1. Brockwell, P.T., Davis, R.A.: Introduction to Time Series and Forecasting, pp. 1–219. Springer, New York (2002)
2. Cowpertwait, P., Metcalfe, A., Time Series.: Introductory Time Series with R, pp. 2–5. Springer, Dordrecht (2009)
3. Davey, N., Hunt, S., Frank, R.: Time Series Prediction and Neural Networks. University of Hertfordshire, Hatfield (1999)
4. Castillo, O., Melin, P.: Hybrid intelligent systems for time series prediction using neural networks, fuzzy logic, and fractal theory. IEEE Trans. Neural Netw. **13**(6), 1395–1408 (2002)
5. Castillo, O., Melin, P.: Simulation and forecasting complex economic time series using neural networks and fuzzy logic. In: Proceedings of the International Neural Networks Conference, vol. 3, pp. 1805–1810 (2001)

6. Castillo, O., Melin, P.: Simulation and forecasting complex financial time series using neural networks and fuzzy logic. In: Proceedings the IEEE the International Conference on Systems, Man and Cybernetics, vol. 4, pp. 2664–2669 (2001)
7. Pulido, M., Melin, P.: Genetic optimization of ensemble neural networks for complex time series prediction. In: IJCNN, pp. 202–206 (2011)
8. Pulido, M., Melin, P.: Optimization of type-2 fuzzy integration in ensemble neural networks for predicting the dow jones time series. Napfis (2012)
9. Pulido, M., Melin, P.: A new method for type-2 fuzzy integration in ensemble neural networks based on genetic algorithms. Studies in Computational Intelligence. Recent Adv. Hybrid Intell. Syst. **451**, 173–182
10. Hansen, L.K., Salomon, P.: Neural network ensembles. IEEE Trans. Pattern Anal. Mach. Intell. **12**(10), 993–1001 (1990)
11. Multaba, I.M., Hussain, M.A.: Application of Neural Networks and Other Learning. Technologies in Process Engineering. Imperial Collage Press, London (2001)
12. Jang, J.S.R., Sun, C.T., Mizutani, E.: Neuro-Fuzzy and Soft Computing. Prentice Hall, Englewood Cliffs (1996)
13. Sharkey, A.: One combining Artificial of Neural Nets. Department of Computer Science University of Sheffield, U.K. (1996)
14. Sharkey, A.: Combining Artificial Neural Nets: Ensemble and Modular Multi-net Systems. Springer, London (1999)
15. Zadeh, L.A.: Fuzzy Sets and Applications: Selected Papers. In: Yager, R.R., et al. Wiley, New York (1987)
16. Castillo, O., Melin, P.: Type-2 Fuzzy Logic Theory and Applications, pp. 29–43. Springer, Berlin (2008)
17. Cheng, C.H., Chen, T.L., Theo, H.J., Chiang, C.H.: Fuzzy time-series based on adaptive expectation model for TAIEX forecasting. Expert Syst. Appl. **34**(2), 1126–1132 (2008)
18. Karnik, N., Mendel, J.M.: Introduction to type-2 fuzzy logic systems. IEEE Trans. Signal Process. **2**, 915–920 (1998)
19. Kennedy, J., Eberhart, R.C.: Particle swarm optimization. In: Proceedings Intelligent Symposium, pp. 80–87 (2003)
20. Eberhart, R.C., Kennedy, J.: A new optimizer particle swarm theory. In: Proceedings of the Sixth Symposium on Micromachine an Human Science, pp. 39–43 (1995)
21. Eberhart, R.C.: Fundamentals of Computational Swarm Intelligence, pp. 93–129. Wiley, New York (2005)
22. Mexico Bank Database: http://www.banxico.org.mx (30 Aug 2011)
23. Pulido, M., Melin, P., Ocastillo.: Optimization of type-2 fuzzy integration in ensemble neural networks for predicting the US Dollar/MX pesos time series. In: IFSA World Congress and NAFIPS Annual Meeting (IFSA/NAFIPS), pp. 1508–1512 (2013 Joint)

Genetic Optimization of Type-1 and Interval Type-2 Fuzzy Integrators in Ensembles of ANFIS Models for Time Series Prediction

Jesus Soto, Patricia Melin and Oscar Castillo

Abstract This paper describes the Mackey-Glass time series prediction using genetic optimization of type-1 and interval type-2 fuzzy integrators in Ensembles of adaptive neuro-fuzzy inferences systems (ANFIS) models, with emphasis on its application to the prediction of chaotic time series. The considered chaotic problem is the Mackey-Glass time series that is generated from the differential equations, so this benchmark time series is used to the test of performance of the proposed Ensemble architecture. We used the interval type-2 and type-1 fuzzy systems to integrate the outputs (forecasts) of each of the ANFIS models in the Ensemble. Genetic algorithms (GAs) were used for the optimization of memberships function (with linguistic labels "Small, Middle, and Large") parameters of the fuzzy integrators. In the experiments, the GAs optimized the Gaussians, generalized bell and triangular membership functions for each of the fuzzy integrators, thereby increasing the complexity of the training. Simulation results show the effectiveness of the proposed approach.

1 Introduction

The analysis of the time series consists of a (usually mathematical) description of the movements that compose it and then builds models using movements to explain the structure and predict the evolution of a variable over time [1]. The fundamental procedure for the analysis of the time series is described below:

J. Soto (✉) · P. Melin · O. Castillo
Division of Graduates Studies and Research, Tijuana Institute of Technology,
Tijuana, Mexico
e-mail: jesvega83@gmail.com

P. Melin
e-mail: pmelin@tectijuana.mx

O. Castillo
e-mail: ocastillo@tectijuana.mx

© Springer International Publishing Switzerland 2016
L.A. Zadeh et al. (eds.), *Recent Developments and New Direction in Soft-Computing Foundations and Applications*, Studies in Fuzziness and Soft Computing 342, DOI 10.1007/978-3-319-32229-2_24

1. Collecting data time series, trying to ensure that these data are reliable.
2. Representing the time series qualitatively noting the presence of long-term trends, cyclical variations, and seasonal variations.
3. Plot a graph or trend line length and obtain the appropriate trend values using either method of least squares.
4. When are present seasonal variations, it may be appropriate to remove seasonality before modeling of the time series (seasonally adjusted series) since the emphasis is usually on nonseasonal fluctuations of the time series.
5. Adjust the seasonally adjusted trend.
6. Represent cyclical variations obtained in step 5.
7. Combine the results of steps 1–6 and any other useful information to make a prediction (if desired) and if possible discuss the sources of error and their magnitude.

Therefore, the abovementioned ideas can assist in the important problem of prediction in time series. Along with common sense, experience, skill, and judgment of the researcher, such mathematical analysis can, however, be of value for predicting the short, medium, and long term.

Genetic algorithms (GAs) are adaptive methods which may be used to solve search and optimization problems. They are based on the genetic process of living organisms. Over generations, the populations evolve in line with the nature of the principles of natural selection and survival of the fittest, postulated by Darwin, in imitation of this process; GAs are capable of creating solutions to real-world problems. The evolution of these solutions to optimal values of the problem depends largely on proper coding them. The basic principles of GAs were established by Holland [2, 3] and are well described in the texts of Goldberg [4, 5] and Lawrence [6]. The evolutionary modeling of fuzzy logic systems (FLS) can be considered as an optimization process where the part or all a fuzzy system parameters constitute a search spaces of model operational (our case), cognitive, and structural.

This paper reports the results of the simulations, in which the Mackey-Glass time series [7, 8] prediction uses genetic optimization of type-1 and interval type-2 fuzzy integrators in Ensembles of adaptive neuro-fuzzy inferences systems (ANFIS) models. The results for each ANFIS were evaluated by the method of the root mean square error (RMSE). For the integration of the results of each modular in the Ensemble of ANFIS, we used the following integration methods: type-1 fuzzy system, interval type-2 fuzzy systems, and integration by genetically optimized interval type-2 FIS.

The selection of the time series for the simulations was based on the fact that these time series are widely quoted in the literature by different researchers [1, 8–13], which allows to compare the results with other approaches such as neural networks and linear regression.

Fig. 1 The Mackey-Glass
time series

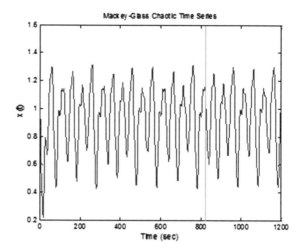

2   Mackey-Glass Time Series

The problem of predicting future values of a time series has been a point of
reference for many researchers. The aim is to use the values of the time series
known at a point $x = t$ to predict the value of the series at some future point
$x = t + P$. The standard method for this type of prediction is to create a mapping
from D points of a Δ spaced time series, $(x(t - (D - 1)\Delta) \ldots x(t - \Delta), x(t))$, to a
predicted future value $x(t + P)$. To allow a comparison with the previous results in
this work [11, 13–16], the values $D = 4$ and $\Delta = P = 6$ were used.

One of the chaotic time series data used in many studies is defined by the
Mackey-Glass [7, 8] time series, whose mathematical model is given by:

$$x(t) = \frac{0.2x(t - \tau)}{1 + x^{10}(t - \tau)} - 0.1x(t - \tau) \tag{1}$$

For obtaining the values of the time series at each point, we applied the
Runge-Kutta method for the solution of Eq. (1). The integration step was set at 0.1,
with initial condition $x(0) = 1.2$, $\tau = 17$, and $x(t)$ is then obtained for $0 \leq t \leq 1200$
(Fig. 1) (we assume $x(t) = 0$ for $t < 0$ in the integration).

3   Adaptive Neuro-fuzzy Inference System "ANFIS"

3.1   ANFIS Models

There have been proposed systems that have fully achieved the combination of
fuzzy systems with neural networks, and one of the most intelligent hybrid systems

Fig. 2 Represented the
ANFIS architecture

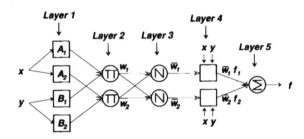

is the ANFIS (adaptive neuro-fuzzy inference system method) as referred by R. Jang [10, 17] (Fig. 2), which is a method for creating the rule base of a fuzzy system, using the algorithm of backpropagation training from the data collection process. Its architecture is functionally equivalent to a fuzzy inference system of Takagi–Sugeno–Kang [18].

The basic learning rule of ANFIS is the gradient descent backpropagation [17], which calculates the error rates (defined as the derivative of the squared error for each output node) recursively from the output to the input nodes.

As a result, we have a hybrid learning algorithm [10], which combines the gradient descent and least squares estimation. More specifically in the forward step of the hybrid learning algorithm, functional signals (output nodes) are processed toward layer 4 and the parameters of consequence are identified by least squares. In the backward step, the premise parameters are updated by gradient descent.

4 Ensemble Learning

The Ensemble consists of a learning paradigm where multiple component learners are trained for a same task, and the predictions of the component learners are combined for dealing with future instances [19, 20]. Since an Ensemble is often more accurate than its component learners, such a paradigm has become a hot topic in recent years and has already been successfully applied to optical character recognition, face recognition, scientific image analysis, medical diagnosis, and time series prediction [21].

5 Interval Type-2 Fuzzy Logic Systems

Type-2 fuzzy sets are used to model uncertainty and imprecision; originally, they were proposed by Zadeh [22, 23] and they are essentially "fuzzy–fuzzy" sets in which the membership degrees are type-1 fuzzy sets [14, 15, 24, 25] (Fig. 3).

The basic structure of a type-2 fuzzy system implements a nonlinear mapping of input to output space. This mapping is achieved through a set of type-2 if-then fuzzy rules, each of which describes the local behavior of the mapping.

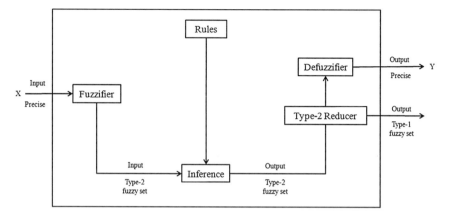

Fig. 3 Basic structure of the interval type-2 fuzzy logic system

Fig. 4 Interval type-2 membership function

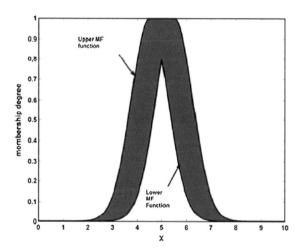

The uncertainty is represented by a region called footprint of uncertainty (FOU). When $\mu_{\tilde{A}}(x, u) = 1, \forall u \in l_x \subseteq [0, 1]$, we have an interval type-2 membership function (Fig. 4).

The uniform shading for the FOU represents the entire interval type-2 fuzzy set, and it can be described in terms of an upper membership function $\bar{\mu}_{\tilde{A}}(x)$ and a lower membership function $\underline{\mu}_{\tilde{A}}(x)$.

A FLS described using at least one type-2 fuzzy set is called a type-2 FLS. Type-1 FLSs are unable to directly handle rule uncertainties, because they use type-1 fuzzy sets that are certain [26–28]. On the other hand, type-2 FLSs are very useful in circumstances where it is difficult to determine an exact certainty value, and there are measurement uncertainties.

6   Genetic Algorithms

GAs are adaptive heuristic search algorithms based on the evolutionary ideas of natural selection and the genetic process [29]. The basic principles of GAs were first proposed by John Holland in 1975, inspired by the mechanism of natural selection, where stronger individuals are likely the winners in a competing environment [5, 6, 30–35]. GAs assumes that the potential solution of any problems of an individual can be represented by a set of parameters [36]. These parameters are added as the genes (individuals) of a chromosome and can be structured by string of values in binary or real form. A positive value, generally known as a fitness value, is used to reflect the degree of "goodness" of the chromosome for the problem which would be highly related with its objective value. The pseudocode of a GAs is as follows:

1. *Start with a randomly generated population of n individuals (candidate a solutions to problem).*
2. *Calculate the fitness of each individual in the problem.*
3. *Repeat the following steps until an offspring have been created:*

   a. *Select a pair of parent individual from the concurrent population, the probability of selection being an increasing function of fitness. Selection is done with replacement, meaning that the same individual can be selected more than once t that the same individuals can be selected more than once to become a parent.*
   b. *With the probability (crossover rate), perform crossover to the pair at a randomly chosen point to form two offspring.*
   c. *Mutate the two offspring at each locus with probability (mutate rate) and place the resulting individual in the new population.*

4. *Replace the current population with the new population.*
5. *Go to step 2.*

The simple procedure just described above is the basic one for most applications of GAs found in the literature.

7   General Architecture of the Proposed Method

The proposed method combines the Ensemble of ANFIS models and the use of interval type-2 and type-1 fuzzy systems as response integrators (Fig. 5).

This architecture is divided into 5 sections, where the first phase represents the database to simulate in the Ensemble of ANFIS, which in our case is the dataset of the Mackey-Glass [15, 17] time series. From the Mackey-Glass time series, we extracted the first 800 pairs of data points (Fig. 1) [10, 12, 16], similar to [11].

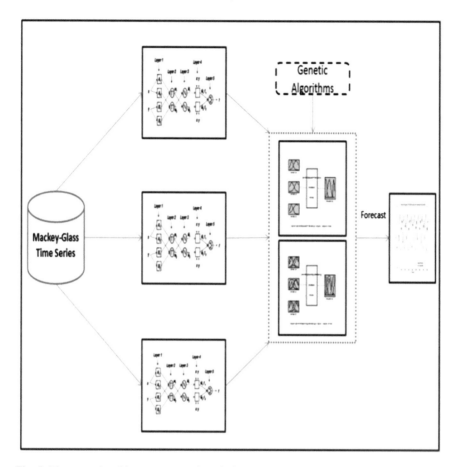

Fig. 5 The general architecture proposed method

We predict x (t) from three past (delays) values of the time series, that is, x $(t - 18)$, x $(t - 12)$, and x $(t - 6)$. Therefore, the format of the training data is as follows:

$$[x(t - 18), x(t - 12), x(t - 6); x(t)] \tag{2}$$

where $t = 19$–818 and $x(t)$ is the desired prediction of the time series, $x(t - 18)$ represent the first input variable, $x(t - 12)$ represent the second input variable, and x $(t - 6)$ represent the third input variable.

In the second phase, training (the first 400 pairs of data are used to train the ANFIS) and validation (the second 400 pairs of data are used to validate the ANFIS models) are performed sequentially in each ANFIS model, where the number of ANFIS to be used can be from 1 to n depending on what the user wants to test, but in our case, we are dealing with a set of 3 ANFIS in the Ensemble.

In the third phase, we have to generate the results of each ANFIS used in the previous section, and in the fourth phase, we integrate the overall results of each ANFIS; such integration done by type-1 and interval type-2 fuzzy integrators are of Mamdani kind, but each fuzzy integrators is optimized (GAs) in the membership functions [14, 15]. Finally, the outcome or the final prediction of the Ensemble ANFIS learning is obtained.

8 Simulations Results

This section presents the results obtained through experiments on the architecture of genetic optimization of type-1 and interval type-2 fuzzy integrators in Ensembles of ANFIS models for the time series prediction, which show the performance that was obtained from each experiment to simulate the Mackey-Glass time series.

8.1 Design of the Type-1 Fuzzy Inference Systems Integrator

The design of the type-1 fuzzy inference systems integrator is of Mamdani type and has 3 inputs (ANFIS1, ANFIS2, and ANFIS3 predictions) and 1 output (forecast), so each input will be assigned three MFs with linguistic labels "Small, Middle, and Large" and the output will be assigned 3 MFs "OutANFIS1, OutANFIS2, and OutANFIS" (Fig. 6) and have if-then 9 rules.

The MFs of the type-1 fuzzy integrator will be changing to different membership functions MFs (Gaussians, Generalized Bell, and Triangular) to observe the behavior of each of them and determine which one provides better forecast of the time series.

8.2 Design of the Interval Type-2 Fuzzy Inference Systems Integrator

The design of the interval type-2 fuzzy inference systems integrator is of Mamdani type and has 3 inputs (ANFIS1, ANFIS2, and ANFIS3 predictions) and 1 output (forecast), so each input will be assigned three MFs "Small and Large" and the output will be assigned 3 MFs "OutANFIS1, OutANFIS2, and OutANFIS3" (Fig. 7) and have 8 if-then rules.

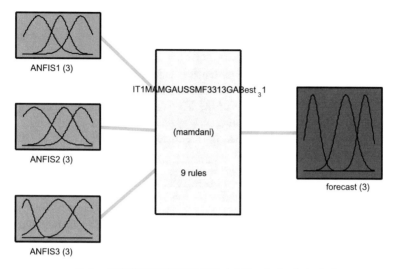

System IT1MAMGAUSSMF3313GABest$_3$1: 3 inputs, 1 outputs, 9 rules

Fig. 6 Structure of the type-1 fuzzy inference system integrator

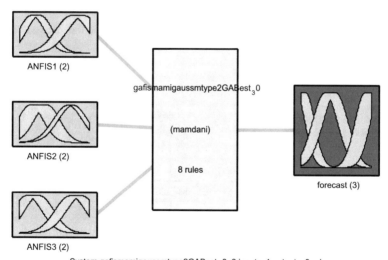

System gafismamigaussmtype2GABest$_3$0: 3 inputs, 1 outputs, 8 rules

Fig. 7 Structure of the interval type-2 fuzzy inference system integrator

The MFs of each interval type-2 fuzzy integrator will be changing to different membership functions MFs (Gaussians, Generalized Bell, and Triangular) to observe the behavior of each of them and determine which one provides better forecast of the time series.

8.3 Design the Representation or Structure of Genetic Algorithms

The GAs are used to optimize the parameters values of the MFs in each type-1 and interval type-2 fuzzy integrators. The representation of GAs is of real values, and the chromosome size will depend of the MFs that are used in each design of the type-1 and interval type-2 fuzzy inference system integrators.

The objective function is defined to minimize the prediction error as follows:

$$f_{(t)} = \sqrt{\frac{\sum_{t=1}^{n} (a_t - p_t)^2}{n}} \tag{3}$$

where a corresponds to the real data of the time series, p corresponds to the output of each fuzzy integrator, t is the sequence time series, and n is the numbers of data points of time series.

The general representation of the chromosome (individuals) represents the utilized fuzzy membership functions (Figs. 8 and 9). In these figures, the first section represents *search space* "solutions" to explore the GAs, the second section represents the input–output variables of the fuzzy integrators, and the third section is represented as follows: The first row represents the input "name" (ANFIS1, ANFIS2, and ANFIS3) and output "name" (FORECAST) variables of the fuzzy integrators; the second row represents the linguistic label of the MFs containing each input variable (MFs1 "Small," MFs2 "Middle", and MFs3 "Large") and output

Fig. 8 Representation of the chromosome for the optimization of the type-1 fuzzy Gaussian membership functions

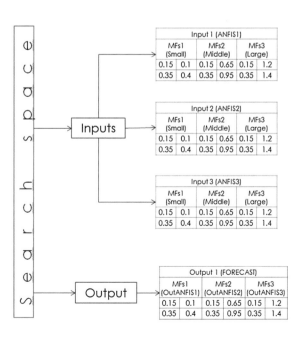

Fig. 9 Representation of the
chromosome for the
optimization of the interval
type-2 fuzzy Gaussian
membership functions

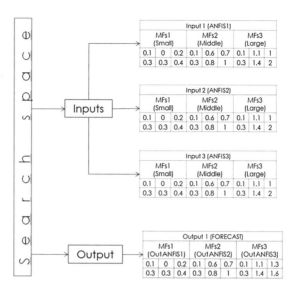

Fig. 9 Representation of the chromosome for the optimization of the interval type-2 fuzzy Gaussian membership functions

variable (MFs1 "OutANFIS1," MFs2 "OutANFIS2" and MFs3 "OutANFIS1") of the fuzzy integrators; the third row represents the MFs parameter "PL = Lower Parameter" where $PL_1, \ldots, PL_N$ (0.15,...,1.2) are the size parameters of the MFs; and the fourth row represents the MFs parameter "PU = Upper Parameter" where $PU_1, \ldots, PU_N$ (0.35,...,1.4) are the size parameters of the MFs that correspond to each input and output. The number of parameters varies according to the kind of membership function of the type-1 fuzzy system (e.g., two parameter are needed to represent a Gaussian MF's are "sigma and mean") is illustrated in Fig. 8 and that of interval type-2 fuzzy system (e.g., three parameters are needed to represent "igaussmtype2"; MFs are "sigma, mean1, and mean2") is illustrated in Fig. 9.

Therefore, the number of parameters that each fuzzy integrator has depends on the kind of membership functions assigned to each input and output variable.

The GAs used have the following parameters: The selection method is the stochastic universal sampling, the percentage of crossover or recombine is 0.8, the mutation is 0.1, and the size population is 100 individuals; we used 100 generations for the evolutions that replace the new population and 31 iterations (test "running" the GAs for the obtain the sample result statistical). There are fundamentals parameters to test the performances of the GAs.

8.4 Genetic Optimization of Type-1 Fuzzy Integration Using Gaussian MFs

In the design of the type-1 fuzzy integrator, we consider three input variables and one output variable, so the input/output variables have three MFs with the linguistic

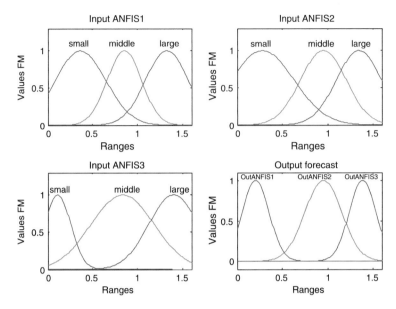

Fig. 10 Optimization of the Gauss MFs (input and output) parameters

label (MFs1 "Small," MFs2 "Middle," and MFs3 "Large"). Therefore, the number of parameters that are used in the representation of the chromosome is 24, because Gauss MFs used two parameters (sigma and mean) for their representation in the type-1 fuzzy systems integrator. The results obtained for the optimization of the Gauss MFs with GAs are the following: the parameters obtained with the GAs for the type-1 fuzzy Gauss MFs (Fig. 10). The forecast data (Fig. 11) are generated by the optimization of the type-1 fuzzy integrators. Therefore, the obtained evolution error with the GAs for this integration is 0.013616.

Fig. 11 Forecast generated by the genetic optimization of type-1 fuzzy integrators with Gauss MFs

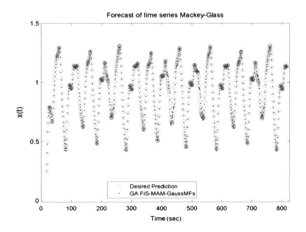

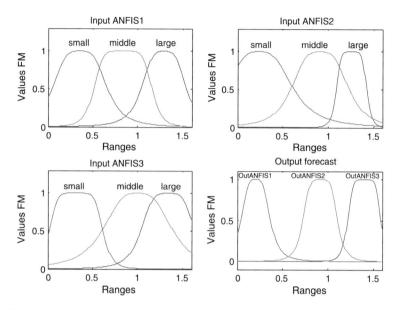

Fig. 12 Optimization of the GBell MFs (input and output) parameters

8.5 Genetic Optimization of Type-1 Fuzzy Integration Using Generalized Bell MFs

In the design of the type-1 fuzzy integrator, we consider three input variables and one output variable, so the input/output variables have three MFs with the linguistic label (MFs1 "Small," MFs2 "Middle," and MFs3 "Large"). Therefore, the number of parameters that are used in the representation of the chromosome is 36, because generalized MFs used three parameters (sigma, $b+$ and mean) for their representation in the type-1 fuzzy systems integrator. The results obtained for the optimization of the Generalized Bell MFs with GAs are the following: the parameters obtained with the GAs for the type-1 fuzzy Generalized Bell MFs (Fig. 12). The forecast data (Fig. 13) are generated by optimization of the type-1 fuzzy integrators. Therefore, the obtained evolution error with the GAs for this integration is 0.013434.

8.6 Genetic Optimization of Type-1 Fuzzy Integration Using Triangular MFs

In the design of the type-1 fuzzy integrator, we consider three input variables and one output variable, so the input/output variables have three MFs with the linguistic label (MFs1 "Small," MFs2 "Middle," and MFs3 "Large"). Therefore, the number

Fig. 13 Forecast generated
by the genetic optimization of
type-1 fuzzy integrators with
GBell MFs

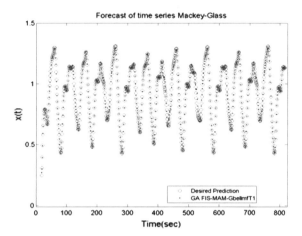

of parameters that are used in the representation of the chromosome is 36, because
Triangular MFs used three parameters (a, b, and c) for their representation in the
type-1 fuzzy systems integrator. The results obtained for the optimization of the
Triangular MFs with GAs are the following: the parameters obtained with the GAs
for the type-1 fuzzy Triangular MFs (Fig. 14). The forecast data (Fig. 15) are
generated by optimization of the type-1 fuzzy integrators. Therefore, the obtained
evolution error with the GAs for this integration is 0.033860.

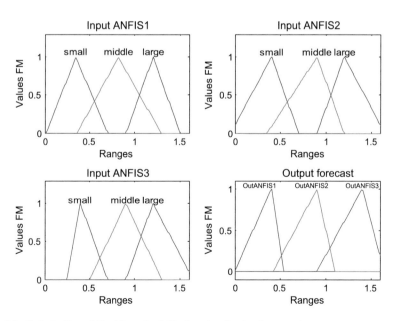

Fig. 14 Optimization of the triangular MFs (input and output) parameters

Fig. 15 Forecast generated by the genetic optimization of type-1 fuzzy integrators with triangular MFs

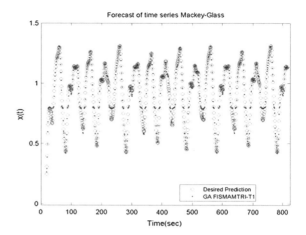

8.7 Genetic Optimization of the Interval Type-2 Fuzzy Integration Using Gauss "igaussmtype2" MFs

In the design of the interval type-2 fuzzy integrator, we consider three input variables and one output variable, so the input/output variables have three MFs with the linguistic label (MFs1 "Small," MFs2 "Middle," and MFs3 "Large").

Therefore, the number of parameters that were used in the representation of chromosome is 36, because "igaussmtype2" MFs used three parameters (Sigma, Mean1, and Mean2) for their representation in the interval type-2 fuzzy systems.

The results obtained to the optimization of the "igaussmtype2" MFs with GAs are the following: the parameters obtained with the GAs for the type-1 fuzzy Gauss MFs (Fig. 16).

The forecast data (Fig. 17) are generated by optimization of the interval type-2 fuzzy integrators. Therefore, the obtained evolution error with the GAs for this integration is 0.011248.

8.8 Genetic Optimization of the Interval Type-2 Fuzzy Integration Using Generalized Bell "igbelltype2" MFs

In the design of the interval type-2 fuzzy integrator, we consider three input variables and one output variable, so the input/output variables have three MFs with the linguistic label (MFs1 "Small," MFs2 "Middle," and MFs3 "Large").

Therefore, the number of parameters of the MFs that were used in the representation of chromosome is 72, because "igbelltype2" MFs used six parameters (a1, b1, c1, a2, b2, and c2) for their representation in the interval type-2 fuzzy systems.

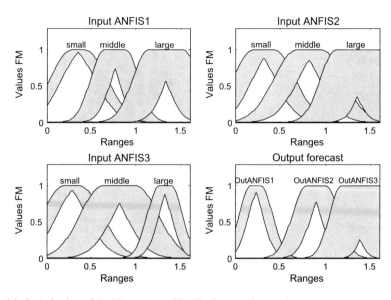

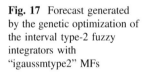

Fig. 16 Optimization of the "igaussmtype2" MFs (input and output) parameters

Fig. 17 Forecast generated
by the genetic optimization of
the interval type-2 fuzzy
integrators with
"igaussmtype2" MFs

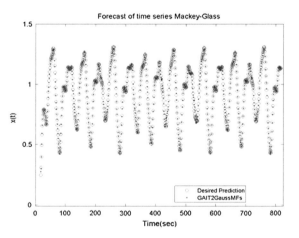

The results obtained to the optimization of the "igbelltype2" MFs with GAs are the following: the parameters obtained with the GAs for the interval type-2 fuzzy "igbelltype2" MFs (Fig. 18).

The forecast data (Fig. 19) are generated by optimization of the interval type-2 fuzzy integrators. Therefore, the obtained evolution error with the GAs for this integration is 0.016540.

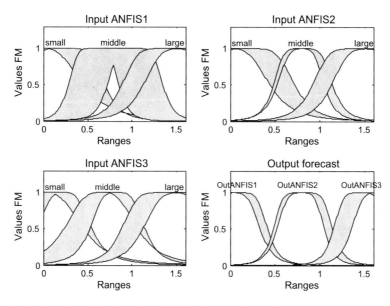

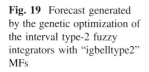

Fig. 18 Optimization of the "igbelltype2" MFs (input and output) parameters

Fig. 19 Forecast generated by the genetic optimization of the interval type-2 fuzzy integrators with "igbelltype2" MFs

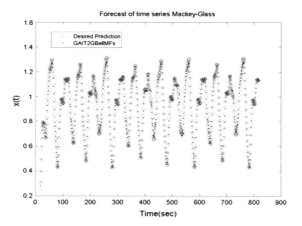

8.9 Genetic Optimization of the Interval Type-2 Fuzzy Integration Using Triangular "itritype2" MFs

In the design of the interval type-2 fuzzy integrator, we consider three input variables and one output variable, so the input/output variables have three MFs with the linguistic label (MFs1 "Small," MFs2 "Middle," and MFs3 "Large"). Therefore, the number of parameters that were used in the representation of chromosome is 72, because "itritype2" MFs used six parameters (a1, b1, c1, a2, b2, and c2) for their

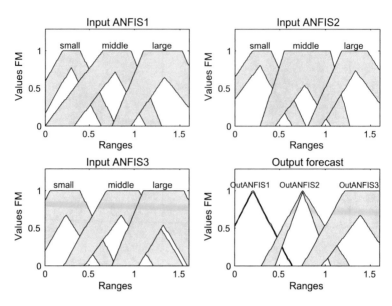

Fig. 20 Optimization of the "itritype2" MFs (input and output) parameters

Fig. 21 Forecast generated
by the genetic optimization of
the interval type-2 fuzzy
integrators with "itritype2"
MFs

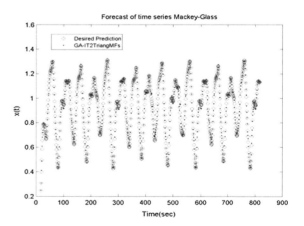

representation in the interval type-2 fuzzy systems. The results obtained to the
optimization of the "itritype2" MFs with GAs are the following: the parameters
obtained with the GAs for the interval type-2 fuzzy "itritype2" MFs (Fig. 20). The
forecast data (Fig. 21) are generated by optimization of the interval type-2 fuzzy
integrators. Therefore, the obtained evolution error with the GAs for this integration
is 0.017381.

Table 1 Best and average results of the prediction error of Mackey-Glass

Genetic optimization of type-1 and interval type-2 fuzz integrators						
Prediction error	Type-1 fuzzy integrator			Interval type-2 fuzzy integrator		
	Gauss	GBell	Triangular	igaussmtype2	igbelltype2	hritype2
Best (RMSE)	0.013616	0.013434	0.033860	0.011248	0.016540	0.017381
Average (RMSE)	0.013888	0.013838	0.033931	0.011595	0.027234	0.017494

8.10 Comparisons Results of the Fuzzy Integrators

Table 1 shows the results of the experiments that were obtained with the proposed method (Fig. 5). This table shows the best results and average results obtained for the optimization of the fuzzy integrators with GAs (using 31). The best and average results for the type-1 fuzzy integrator are using Generalized Bell MFs, which obtained a best prediction error of 0.013434 and average prediction error of 0.014217. The best and average results for the interval Type-2 fuzzy integrator are using Gaussian (igaussmtype2) MFs, which obtained a best prediction error of 0.011248 and average prediction error of 0.011888.

9 Conclusions

In conclusion, we can say that the results obtained with the proposed architecture of the Mackey-Glass time series prediction using genetic optimization of type-1 and interval type-2 fuzzy integrators in Ensembles of ANFIS models have been good and positive in predicting time series.

The interval type-2 fuzzy integrator is better than the type-1 fuzzy integrator, because in most of the experiments that were performed with the proposed architecture of Ensembles of ANFIS, the best and average results of the interval type-2 fuzzy are better than the results of the type-1 fuzzy integrator (show Table 1). Therefore, the proposal offers efficient results in the prediction of such time series, which can help us make decisions and avoid unexpected events in the future.

References

1. Brocklebank, J.C., Dickey, D.A.: SAS for Forecasting Series, pp. 6–140. SAS Institute Inc., Cary (2003)
2. Holland, J.H.: Outline for a logical theory of adaptive systems. J. Assoc. Comput. Mach. **3**, 297–314 (1962)
3. Holland, J.H.: Adaptation in natural and artificial systems. University of Michigan Press, Ann Arbor (1975)

4. Goldberg, D.E., Kalyanmoy, D.: A comparative analysis of selection schemes used in genetic algorithms. In: G.J.E. Rawlins (eds.) Foundations of Genetic Algorithms, pp. 69–93. Morgan Kaufmann Publishers, San Mateo (1991)
5. Goldberg, D.E., Korb, B., Kalyanmoy, D.: Messy genetic algorithms: motivation, analysis, and first results. Complex Syst. **3**, 493–530 (1989)
6. Lawrence, D.: Handbook of Genetic Algorithms. Van Nostrand Reinhold, New York (1991)
7. Mackey, M.C., Glass, L.: Oscillation and chaos in physiological control systems. Science **197**, 287–289 (1997)
8. Mackey, M.C.: "Mackey-Glass". McGill University, Canada, http://www.sholarpedia.org/-article/ Mackey-Glass_equation, 5 Sept 2009
9. Brockwell, P.D., Richard, A.D.: Introduction to Time Series and Forecasting. Springer, New York, pp 1–219 (2002)
10. Jang, J.S.R.: ANFIS: Adaptive-network-based fuzzy inference systems. IEEE Trans. Syst Man Cybern. **23**, 665–685 (1992)
11. Melin, P., Soto, J., Castillo, O., Soria, J.: A new approach for time series prediction using ensembles of ANFIS models. Experts Syst Appl, El-Sevier **39**(3), 3494–3506 (2012)
12. Wang, C., Zhang, J.P.: Time series prediction based on ensemble ANFIS. In: Proceedings of the Fourth International Conference on Machine Learning and Cybernetics, Guangzhou, 18–21 Aug 2005
13. Werbos, P.: Beyond regression: new tools for prediction and analysis in the behavioral sciences. Ph.D. thesis, Harvard University (1974)
14. Castro, J.R., Castillo, O., Melin, P., Rodríguez, A.: Hybrid learning algorithm for interval type-2 fuzzy neural networks. GrC, pp. 157–162 (2007)
15. Castro, J.R., Castillo, O., Melin, P., Rodriguez, A.: A hybrid learning algorithm for interval type-2 fuzzy neural networks: the case of time series prediction, vol. 15a, pp. 363–386. Springer, Berlin (2008)
16. Pulido, M., Mancilla, A., Melin, P.: Redes ensemble con integración Difusa para Pronosticar Series de Tiempo Complejas. Tijuana Institute of Technology, Mexico, 21 Sept 2009
17. Jang, J.S.R.: Rule extraction using generalized neural networks. En Proceedings of te 4th IFSA World Congress, pp. 82–86 (1991)
18. Takagi, T., Sugeno, M.: Derivation of fuzzy control rules from human operation control actions. In: Proceedings of the IFAC Symposium on Fuzzy Information, Knowledge Representation and Decision Analysis, pp. 55–60 (1983)
19. Rumelhart, D.E., Hinton, G.E., Williams, R.J.: Learning internal representations by error propagation. In: Parallel Distributed Processing: Explorations in the Microstructure of Cognition, vol. 1, pp. 318–362. MIT Press, Cambridge (1986)
20. Takagi, T., Sugeno, M.: Derivation of fuzzy control rules from human operation control actions. In: Proceedings of the IFAC Symposium on Fuzzy Information, Knowledge Representation and Decision Analysis, pp. 55-60 (1983)
21. Zhou, Z.H., Wu, J., Tang, W.: Ensembling neural networks: many could be better than all. Artif. Intell. **137**(1–2), 239–263 (2002)
22. Zadeh, L.A.: Fuzzy logic. Computer **1**(4), 83–93 (1988)
23. Zadeh, L.A.: Fuzzy logic = computing with words. IEEE Trans. Fuzzy Syst. **4**(2), 103 (1996)
24. Jang, J.S.R.: Fuzzy modeling using generalized neural networks and Kalman fliter algorithm. In: Proceedings of the Ninth National Conference on Artificial Intelligence (AAAI-91), pp. 762–767 (1991)
25. Melin, P., Mendoza, O., Castillo, O.: An improved method for edge detection based on interval type-2 fuzzy logic. Expert Syst Appl. **37**(12), 8527–8535 (2010)
26. Mendel, J.M.: Why we need type-2 fuzzy logic systems. Article is provided courtesy of Prentice Hall, By Jerry Mendel, 11 May 2001
27. Mendel, J.M.: Uncertain rule-based fuzzy logic systems: introduction and new, directions, pp. 25–200. Prentice Hall, USA (2000) (Ed)
28. Mendel, J.M., Mouzouris, G.C.: Type-2 fuzzy logic systems. IEEE Trans. Fuzzy Syst. **7**, 643–658 (1999)

29. Goldberg, D.E.: Genetic Algorithms in Search, Optimization, and Machine Learning. Addison-Wesley Publishing Company, Boston (1989)
30. Castillo, O., Melin, P.: Optimization of type-2 fuzzy systems based on bio-inspired methods: a concise review. Appl. Soft Comput. **12**(4), 1267–1278 (2012)
31. Castro, J.R., Castillo, O., Martínez, L.G.: Interval type-2 fuzzy logic toolbox. Eng. Lett. **15**(1), 89–98 (2007)
32. Chua, T.W., Tan, W.W.: Genetically evolved fuzzy rule-based classifiers and application to automotive classification. Lect. Notes Comput. Sci. **5361**, 101–110 (2008)
33. Cordon, O., Gomide, F., Herrera, F., Hoffmann, F., Magdalena, L.: Ten years of genetic fuzzy systems: current framework and new trends. Fuzzy Sets Syst. **141**, 5–31 (2004)
34. Cordon, O., Herrera, F., Hoffmann, F., Magdalena, L.: "Genetic Fuzzy Systems: Evolutionary Tuning and Learning of Fuzzy", Knowledge Bases. World Scientific, Singapore (2001)
35. Eiben, A.E., Smith, J.E.: Introduction to Evolutionary Computation, pp. 37–69. Springer, Berlin (2003)
36. Cordon, O., Herrera, F., Villar, P.: Analysis and guidelines to obtain a good uniform fuzzy partition granularity for fuzzy rule-based systems using simulated annealing. Int. J. Approximate Reasoning **25**, 187–215 (2000)

Sustainable Supplier Selection: A New Differential Evolution Strategy with Automotive Industry Application

S.K. Jauhar and M. Pant

Abstract Modern times, highly competitive, global operating environment, sustainability plays a very vital role. Changing climatic conditions and environmental deterioration has multi dimensional impact on every sphere of life forms and life form driving processes. Public hold on corporations responsible for ecological misconduct in their supply chains getting more firm with time. To counter the threat it's a high time industries started taking initiatives for sustainability in their supply chains. In spite of that, suppliers often are unsuccessful to appropriately contribute in these initiatives. Hence, present paper justifies the supplier involvement in sustainable initiatives in supply chain management (SCM) with using differential evolution (DE) to select the efficient sustainable suppliers providing the maximum fulfillment for the sustainable criteria determined. Finally, two illustrative cases on automotive industry validate the application of the present approach.

1 Introduction

In recent years, an increasing environmental awareness has favoured the emergence of the new sustainable supply chain management (SSCM) paradigm; thus, also in the supplier selection problem, sustainable criteria have been incorporated. Sustainable supplier selection (SSS) is a vital segment of SSCM. In today's world, in view of the growing awareness concerning sustainability in business firm, the SSS would be the vital element in the process of managing a sustainable supply chain.

Developing an efficient and robust sustainable supply chain is a crucial task for the success of a business firm. One of the most significant factors that help in building a strong sustainable supply chain is the SSS process. Traditionally, only

S.K. Jauhar (✉) · M. Pant
Indian Institute of Technology, Roorkee 447667, UK, India
e-mail: suniljauhar.iitr@gmail.com

M. Pant
e-mail: millidm@gmail.com

© Springer International Publishing Switzerland 2016
L.A. Zadeh et al. (eds.), *Recent Developments and New Direction in Soft-Computing Foundations and Applications*, Studies in Fuzziness and Soft Computing 342, DOI 10.1007/978-3-319-32229-2_25

353

the management aspects such as lead time, quality, price, late deliveries, rate of rejected parts, and service quality of the supply chain were considered for selecting a potential supplier. However, with the growing environmental issues researchers are also paying attention to factors, such as greenhouse effect, reusability, and carbon dioxide (CO_2) emission, jointly known as carbon foot printing. The resulting problem is called "*sustainable supplier selection*," where a balance is maintained between the management and environment concerns.

The literature survey has highlighted that the penetration of sustainable issues within the Supplier Selection Problem literature is still quite limited, as confirmed by the relatively low number of papers published. Few SSS methodologies have been suggested by the practitioners in earlier, to find a solution to the SSS problem. A SSS problem fundamentally is a multi-criteria practice. It is a judgment of tactical significance to enterprises.

It is natural that for a particular product, huge amounts of suppliers/vendors are available in the market that fulfills some preliminary criteria. However, the real task is to determine the most suitable set of suppliers (or key suppliers) subject to management as well as environmental aspects. Now, it is the duty of the purchasing managers to identify the most suitable clusters of sustainable suppliers for their product.

Evaluation and above selection of sustainable suppliers is a complex process and depend on a large number of qualitative as well as quantitative factors to ensure a cost effective model without compromising with the quality. The nature of this decision usually is difficult and unstructured. Optimization practices might be useful tools for these types of decision-making problems. During last few years, differential evolution (DE) has arisen as a dominating tool used for solving a variety of problems arising in numerous fields. DE algorithm proposed by Stron and Price in 1995 [1]. It is a population set-based evolutionary algorithm that has been applied successfully to a wide-ranging issues [2–6].

In this research, we present an approach to solve the multiple-criteria SSS problem with the application of DE, for data envelopment analysis (DEA)-based mathematical model, and validated this approach with the help of two cases on automotive industry. In the current study, we present first case used DE to solve the traditional supplier selection problem of automobile industry without considering environmental issues and then second case used DE for solving the supplier selection problem of automobile industry with considering environmental issues (called as SSS problem).

This research article is divided into eight sections. Subsequent to the introduction in Sect. 1, the SSCM in automotive industry, SSS, and methodology are briefed in Sects. 2, 3, and 4, respectively. Section 5 describes the mathematical model formulation with DEA used in this article. Section 6 describes the DE algorithm for SSS and two cases on automotive industry discussed in Sect. 7. Finally, conclusion with summary drawn from the current research is given in Sect. 8.

2 SSCM in Automotive Industry

Literature analysis seem to reveal that there is a growing attention towards the topic. The sustainable development has turn out to be a buzzword that acknowledged a lot of attentions in numerous fields such as manufacturing [7], business development [8], tourism [9], and agriculture [10]. Also, in SCM both academics and general practitioner contemplate the sustainable concerns in their workings.

Sustainable SCM is the managing of resources, information's and capital flows, as well as collaboration between firms alongside the supply chain, meanwhile taking into account the objectives from all three dimensions, such as financial, ecological and societal, of sustainable growth derived from client and investor wants [11]. Business organizations are gradually identifying that the effective management of sustainable supply chains is a most important driver of value creation as well as environmental performance.

The significance of SSCM has emerged as a result of the present-day commercial conditions of worldwide competition, globalization of supply chains, small-product life cycles, immediate modifications in technologies, the need to deliver greater levels of consumer service, and the continual force to decrease prices, increase asset use in addition to take care of environment concerns. Reference [12] discusses that SSCM is a critical corporate matter in these industries that offers incredible potential for achieving better environmental performance, consumer fulfillment, pull down operating expenditures, reducing inventory investments in addition to achieving better fixed asset usage. Figure 1 presents a flow diagram of sustainable supply chain with stages and relationships field.

The manufacturing of the worldwide automotive vehicles is taking on by not great in size of companies with their operational activities expands across the world. These companies must contain their role as environmental overseers if the automotive vehicles are not turn out to be the leading contributor to degrading worldwide environmental condition. The Worldwatch Institute states that the worldwide automotive vehicles has become greater in size as of 50 million units in 1950 to in excess of 550 million automobiles in 2004, with approximations of above 5 billion automobiles by the year 2050 [14]. The environmental consequences of the present

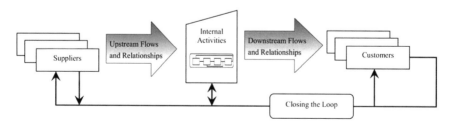

Fig. 1 Sustainable supply chain with stages and relationships field. *Source* [13]

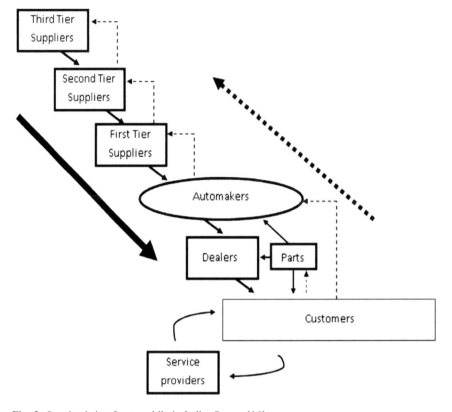

Fig. 2 Supply chain of automobile in India. *Source* [16]

as well as forthcoming development of automotive vehicles have been seen as a major research area [15]. The Fig. 2 presents an easy illustration of the automobile industry supply chain organization in India with its three-tier supplier's involvement.

3 Sustainable Supplier Selection

An efficient supplier selection process needs to be in place and of paramount importance for successful SCM, supplier selection and evaluation forms an integral part of a supply chain. A wrong choice or decision may lead to unpleasant circumstances and in worst case may even lead to the deterioration of the entire supply chain's financial and operational position.

Supplier selection comprises numerous criteria including price, lead time, quality, speed, delivery performance, reliability, etc., and often encompasses the choice of one meanwhile give up the other. The supplier selection practices are comprehensively studied in the literature with multi-criteria decision-making models. These models comprise such practices, as the DEA, analytic hierarchy process (AHP), analytic network process (ANP), case-based reasoning (CBR), fuzzy set theory, genetic algorithm (GA), mathematical programming, simple multi-attribute rating technique (SMART), and their hybrid variants. Different researchers have studied the works in the past relating the supplier evaluation and selection problem [17–31].

For successful SCM, it is important to pay attention to environmental requirements in supply chain process; as a result, similarly in the suppliers selection issues, sustainable criteria have been integrated. The SSS problem can be well-defined as "a traditional supplier selection problem in which, among the others, as well environmental friendly aspects are taken into account in order to select and evaluate suppliers performances" [32]. Automotive vehicles are very complicated products that have need of a great amount of outsourcing to suppliers for their assemblage [33]. Bought resources, parts and additional external input account for a great portion of overall expenses [34]. At present, the portion is between 60 and 80 % of the overall manufacturing spending, interpreting the automobile segment one of the most supplier-relying industries [35, 36].

Its turn out to be gradually more obvious that the image and above common man perception of an automobile SCM with respect to sustainability is not only rely on its individual environmental performance, but as well on the environmental performance of its supply chain associates, and in specific the "suppliers" primary member of the supply chain [37], so the SSS is primarily held accountable for the sustainable automotive SCM. Reference [38] states that the participation of suppliers in the development of environmental friendly automotive paint to be vital. Further in recent times [39] examined the role of suppliers in plant-level ecological enrichments in the Canadian printing industry in addition to described the significance of cooperation for supplier investments in environmental friendly technologies.

To select the potential suppliers, two focuses comprising significance: One is the degree of the selection criteria and second one is the suppliers' sustainable performance; these two focuses need to be verified with the appropriate decision makers [40]. Toward accomplishment a sustainable supply chain, entire associates in the chain from raw material suppliers to topmost administrators must have natural liking in relation to sustainability. Even now, comprehensive SCM study is yet to be accomplished on how corporations can contain suppliers in sustainable management practices in addition to involve them into collaboration in sustainable activities. Therefore, to manage this problem and cope with the imprecision that is be found in the SSS problem, use of DE algorithm is explored in this study.

4 Methodology

To measure and analyze the relative efficiency of automotive suppliers, we follow a four-step methodology on cases:

1. Design a criteria containing input and output criteria,
2. Select a problem,
3. Formulate the mathematical model of the SSS problem with the help of DEA, and
4. Apply DE on mathematical model.

   The present model can be carrying out for any quantity of suppliers and there is no limitation, by using this model, the company can obtain a recommended combination of efficient suppliers.

5 Mathematical Model Formulation with DEA

DEA-based method is used for determining the efficiencies of decision-making units (DMUs) on the basis of multiple inputs and outputs [41]. DMU can comprise of business firms, divisions of huge groups such as institution of higher education, schools, hospitals, power plants, police stations, tax offices, prisons, and a set of organizations [42–45]. The DMU well-describe in this research work using input as well as output criteria are as follows:

The performance of DMU is estimated in DEA by the concept of efficiency or productivity, which the proportion of weights sum of outputs (o/p) to the weights sum of (i/p) inputs [46]:

$$\text{Efficiency} = \frac{\text{Weighted sum of o/p}}{\text{Weighted sum of i/p}} \tag{1}$$

The two basic DEA models are the Charnes, Cooper and Rhodes (CCR) model [47] and the Banker, Charnes and Cooper (BCC) model [48]; these two models distinguish on the returns-to-scale assumed. The former assumes constant returns-to-scale whereas the latter assumes variable returns-to-scale [41]. In the current study, we use CCR model which is well-defined further down: Suppose that there are N DMUs and each unit have I input and O outputs then the efficiency of mth unit is achieved by resolving the below model which is presented by Charnes et al. [47].

$$\text{Max} \quad E_m = \frac{\sum_{k=1}^{O} w_k \text{Output}_{k,m}}{\sum_{l=1}^{i} z_l \text{Input}_{l,m}}$$

$$0 \leq \frac{\sum_{k=1}^{O} w_k \text{Output}_{k,n}}{\sum_{l=1}^{i} z_l \text{Input}_{l,n}} \leq 1; \quad n = 1, 2, \ldots, m \ldots N \tag{2}$$

$$w_{k,z_l} \geq 0; \quad \forall k, l$$

where

E_m is the efficiency of the mth DMU, $k = 1$ to O, $l = 1$ to I and $n = 1$ to N.
$Output_{k,m}$ is the kth output of the mth DMU and w_k is weight of output $Output_{k,m}$.
$Input_{l,m}$ is the lth input of mth DMU and z_l is the weight of $Input_{l,m}$.
$Output_{k,n}$ and $input_{l,n}$ are the kth output and lth input, respectively, of the nth DMU where $n = 1, 2, \ldots, m \ldots N$.

The fractional program shown in "(2)" can be converted in a linear program which is shown in "(3)":

$$\text{Max } E_m \sum_{k=1}^{O} w_k \text{Output}_{k,m}$$

s.t.

$$\sum_{l=1}^{I} z_l \text{Input}_{l,m} = 1. \tag{3}$$

$$\sum_{k=1}^{O} w_k \text{Output}_{k,n} - \sum_{l=1}^{I} z_l \text{Input}_{l,n} \leq 0, \quad \forall n$$

$$w_k, z_l \geq 0; \quad \forall k, l$$

To calculate the efficiency score for each DMU, we run the above program run N times. A DMU is considered efficient if the efficiency score is 1; otherwise, it is considered as inefficient.

6 DE Algorithm

DE algorithm was proposed by Stron and Price in 1995 [1]. It is a type of evolutionary algorithm, used to most effective use of (optimize) the functions. It is a population set-based evolutionary algorithm for global optimization. In the current study, we have used DE/rand/1/bin scheme and DE algorithm from reference [1].

6.1 Pseudocode for the DE Algorithm

1	Begin
2	Generate uniformly distribution random population $P=\{X_{1,G}, X_{2,G},...,$ $X_{NP,G}\}$. $X_{i,G} = X_{lower} + (X_{upper} - X_{lower})*rand(0,1)$, where $i = 1, 2,..,NP$
3	Evaluate $f(X_{i,G})$
4	While (Termination criteria is met)
5	{
6	For $i=1:NP$
7	{
8	Select three random vector $X_{r1,G}, X_{r2,G}, X_{r3,G}$ where $i \neq r_1 \neq r_2 \neq r_3$
9	Perform mutation operation
10	Perform crossover operation
11	Evaluate $f(U_{i,G})$
12	Select fittest vector from $X_{i,G}$ and $U_{i,G}$ to the population of next generation
13	}
14	Generate new population $Q=\{X_{1,G+1}, X_{2,G+1},..., X_{NP,G+1}\}$
15	} /* end while loop*/
16	END

6.2 Constraints Handling

For the constraint problems, various methods have been suggested in literature. A survey of different methods for constraint handling can be found in [49] and [50]. In this paper, Pareto ranking method is used for handling the constraints [51].

6.3 Parameter Setting for the DE Algorithm

In present research article, we have applied DE to solve the DEA-based mathematical model. The parameter settings for DE are given in Table 1.

The program is implemented is DEV C++ and all the uniform random number is generated using the inbuilt function *rand ()* in DEV C++. The fitness value is taken as the average fitness value in 30 runs and the program is terminate when reach to max iteration.

Table 1 Parameter setting for DE	Pop size (NP)	100
	Scale factor (F)	0.5
	Crossover rate (Cr)	0.9
	Max iteration	3000

A buyer (decision maker) can effect an assessment (supplier evaluation) with the ability to choose of weight system. For this purpose with the help of program which is implemented is DEV C++, we intended to generate all the uniform random number (in between 0 to 1) using the inbuilt function *rand ()* in DEV C++, to assist the selection of the weights for input as well as output criteria in a manner to permit the control of the result for the sustainable supplier evaluation and assessment practice.

7 Cases on Automotive Industry

7.1 Case 1: Traditional Supplier Selection Problem

This first hypothetical case of an Indian automobile part manufacturing company at northern part of India is presented here to illustrate the present approach. In the present study, we have considered a case of selecting best suppliers out of 12 potential suppliers. To efficient suppliers selection, companies need to keep comprehensive supplier information files which should include a list of items available from each supplier, such as the as price, late deliveries, rate of rejected parts, and service quality. The criteria considered for selection as given below.

7.1.1 Designing a Criteria

In this supplier selection case, two set of criteria were formulated: input (traditional purchasing criteria) and output criteria. The criteria considered for selection as given below:

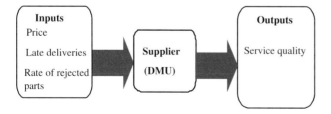

1. Input criteria: which include price, late deliveries, rate of rejected parts.
2. Output criteria: including service quality of the product and services.

In this study, a supplier is considered efficient if the efficiency score is 1; otherwise, it is considered as inefficient. Data for service quality of items is taken from concept of service quality dimension based on 12 questionnaires including 27 questions given by [52].

7.1.2 Selection of a Problem

This case presented here to illustrate the present approach; the underlying data are shown in Table 2 (Source 52) with the supplier's database covering input (traditional purchasing criteria) such as price, late deliveries, rate of rejected parts and output criteria is service quality of an item provided in the shipment of a company.

7.1.3 Formulation of Mathematical Model

On the basis of the above data, the DEA model of Kth DMU with the help of "(3)" will be as follows:

$$\text{Max} \quad SQ_m$$
$$\text{s.t.}$$
$$z_1 PR_m + z_2 LD_m + z_3 RR_m = 1 \tag{4}$$
$$w SQ_n - (z_1 PR_n + z_2 LD_n + z_3 RR_n) \leq 0$$
$$\forall n = 1, \ldots m, \ldots 12$$

Table 2 Data of 12 different suppliers

Criteria	Inputs			Output
Suppliers	Price (PR)	Late deliveries (LD) %	Rate of rejected (RR) %	Service quality (SQ)
1	290	7	3	95
2	240	3	5	98
3	300	4	6	12
4	255	5	3	100
5	295	10	8	65
6	250	3	3	110
7	245	7	4	92
8	285	6	4	73
9	270	6	6	75
10	270	12	4	81
11	285	3	5	112
12	275	5	8	85

7.1.4 Applying DE on Mathematical Model

After applying DE on traditional supplier selection problem, in Table 3 average efficiency and weights results of all DMUs are given. From this Table, we can see that for suppliers 2, 6, and 11, the efficiency score is 1, so these suppliers are assumed to be 100 % efficient while efficiency score for all other suppliers are less than 1. So these suppliers are not as efficient and among these, supplier no. 3 is probably the most inefficient in comparison with all other suppliers.

In Table 4, results of all DMUs are given, and Fig. 3 shows the histogram of all suppliers with their efficiency score.

7.1.5 Discussion

The research of efficient traditional supplier selection problem for automotive industry can obtain a recommended combination of efficient suppliers 2, 6, and 11 using DE algorithm gives the better solution.

For the current research conducted in 12 suppliers, the results are as follows:

Table 3 Average efficiency and weights in 30 runs

Supplier	Value of input and output weights				Efficiency
	Z_1	Z_2	Z_3	W	
1	0	0	0.33337	0.00909282	0.863818
2	0.00352436	0.0225639	0.0172946	0.0102042	1
3	0.00333337	0	0	0.00757585	0.0909102
4	0	0	0.333337	0.00909102	0.909102
5	0.00338987	0	0.42974e017	0.00830173	0.539612
6	0.00023935	0.158028	0.156064	0.00909102	1
7	0	0	0.250003	0.00681827	0.62728
8	0.00350881	0	0.0241e−017	0.00797458	0.582144
9	0.00370374	0.62202e0	0	0.00841761	0.631321
10	0.00370374	0	0.15911e−018	0.00841761	0.681826
11	0.00013141	0.317989	0.0038805	0.00892868	1
12	0	0.191415	0.00536679	0.00536679	0.456178

Table 4 Suppliers efficiency score

Suppliers	Efficiency	Suppliers	Efficiency
1	0.863818	7	0.62728
2	1	8	0.582144
3	0.0909102	9	0.631321
4	0.909102	10	0.681826
5	0.539612	11	1
6	1	12	0.456178

Fig. 3 Histogram of all suppliers with their efficiency score

1. Suppliers 2, 6, and 11, the efficiency score is 1, so these suppliers are assumed to be 100 % efficient.
2. Supplier 3 is probably the most inefficient in comparison with all other suppliers.
3. Suppliers 2, 6, and 11 would be the most suitable set of suppliers (or key suppliers).
4. By using this DE, the company can obtain a recommended combination of efficient suppliers.
5. Combination of suppliers 2, 6, and 11 would be the recommended supplier set while the company needing single-item suppliers.
6. The results of the case indicate that the DE algorithm can solve the problem effectively.

7.2 Case 2: Sustainable Supplier Selection Problem

The second case study presented in this paper stands a hypothetical automobile industry in India (X Company). After verifying a group of criteria in a view point of sustainable merits, some criteria including lead time, quality, price, service quality, reusability, and CO_2 emissions of the delivered products are derived for SSS problem.

7.2.1 Designing a Criteria

In the current study, we split the criteria in two manners: the input and output criteria. The input criteria are the traditional supplier selection criteria, such as lead time, price, and quality of the delivered goods. The output criteria are the reusability and CO_2 emission of the product and services. We assume that the reusability and CO_2 emission are the output of the examined model.

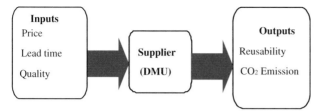

1. Management criteria: Lead time, quality, and price; and
2. Environmental (green) criteria: Reusability and CO_2 emission.

Reusability concept is taken from [53] and for CO_2 emissions, *LOCOG* Guidelines on Carbon Emissions of Products and Services—Version 1 [54], is considered.

7.2.2 Selection of a Problem

The data are shown in Table 5 with the supplier's database covering management (input) as well as environmental (output) criteria of an item provided in the shipment of automobile company.

Table 5 Data for numerical example

Criteria	Management criteria (inputs)			Environmental criteria (outputs)	
Suppliers	Lead time (L) (Day)	Quality (Q) (%)	Price (P) (Rs.)	Reusability (R) (%)	CO_2 emissions (CE) (g)
1	2	80	107	70	30
2	1	70	161	50	10
3	3	90	269	60	15
4	4	65	270	30	12
5	2	55	260	40	18
6	5	70	201	50	20
7	3	85	111	66	14
8	2	95	300	35	28
9	1	67	197	60	16
10	4	72	157	44	28
11	5	51	170	41	14
12	3	58	106	49	10
13	2	72	255	32	25
14	4	60	117	40	29
15	5	63	245	22	5
16	3	90	299	10	9
17	1	87	101	42	15
18	2	82	206	70	18

7.2.3 Formulation of Mathematical Model

Based on the basis of the above data, the DEA model of Kth DMU with the help of "(3)" will be as follows:

$$\text{Max} \quad SQ_m + CE_n$$
$$\text{s.t.}$$
$$z_1 L_m + z_2 Q_m + z_3 P_m = 1 \tag{5}$$
$$w_1 SQ_n + w_2 CE_n - (z_1 L_m + z_2 Q_m + z_3 P_m) \leq 0$$
$$\forall n = 1, \ldots m, \ldots 18$$

7.2.4 Applying DE on Mathematical Model

After applying DE on SSS problem, in Table 6 average efficiency and weight results of all DMUs are given. From this Table, we can see that for suppliers 1, 9, and 14, the efficiency score is 1, so these suppliers are assumed to be 100 % efficient while efficiency score for all other supplier are less than 1. So these suppliers are not as efficient and among these, supplier no. 16 is probably the most inefficient in comparison with all other suppliers.

In Table 7, results of all DMUs are given, and Fig. 4 shows the histogram of all suppliers with their efficiency score.

7.2.5 Discussion

The research of efficient SSS practice can acquire a desirable cluster of competent sustainable suppliers 1, 9, and 14 using DE algorithm gives the better solution.

For the current research conducted in 18 suppliers, the results are as follows:

1. Suppliers 1, 9, and 14, the efficiency score is 1 so these suppliers are assumed to be 100 % sustainable efficient.
2. Supplier 16 is probably the most inefficient in comparison with all other suppliers.
3. Suppliers 1, 9, and 14 would be the most suitable set of suppliers (or key suppliers).
4. By using this DE algorithm, the business firms can acquire desirable clusters of competent sustainable suppliers.
5. Combination of suppliers 1, 9, and 14 would be the desirable clusters of competent sustainable suppliers set; meanwhile, the business firms requiring single-item sustainable suppliers.

Table 6 Average efficiency and weights in 30 runs

Suppliers	Value of input and output weight					Efficiency
	Z_1	Z_2	Z_3	W_1	W_2	
1	0.100129	0.00204427	0.00594674	0.0124772	0.0042648	1
2	0.9899	0	0	0.0166683	0.08283e−017	0.833417
3	0	0.0111122	0.18753e−016	0.0124103	0.33556e−01	0.744619
4	0	0.0153862	0	0.0165195	0.00248747	0.525436
5	0	0.0181836	0	0.011392	0.0219116	0.850091
6	0	0.0142871	0.19188e−020	0.0153397	0.00230969	0.81318
7	0	0.71429e−018	0.00900991	0.0137737	0.58488e−01	0.909066
8	0.442549	0.00121055	0	0	0.0327347	0.916572
9	0.989	0	0	0.00337086	0.0498655	1
10	0.0794687	0.00947534	0.71282e−009	0.42912e−008	0.0305688	0.855927
11	0	0.0196098	0.0887e−019	0.0210538	0.0031707	0.907594
12	0	0.0167855	0.000250394	0.0195676	0.29418e−018	0.958812
13	0.0944849	0.0112657	0	0.36807e−018	0.0363442	0.908605
14	0.0069216	0.0121538	0.00207852	0.00215503	0.0315172	1
15	0	0.0158746	0.68761e−019	0.0177283	0.14697e−017	0.390023
16	0.0728352	0.00868438	0.75703e−019	0	0.0280174	0.252156
17	0.963	0	0	0	0.0625062	0.937594
18	0	0.0117559	0.000175325	0.0137047	0.01723e−014	0.959931

Table 7 Suppliers efficiency

Suppliers	Efficiency	Suppliers	Efficiency
1	1	10	0.855927
2	0.833417	11	0.907594
3	0.744619	12	0.958812
4	0.525436	13	0.908605
5	0.850091	14	1
6	0.81318	15	0.390023
7	0.909066	16	0.252156
8	0.916572	17	0.937594
9	1	18	0.959331

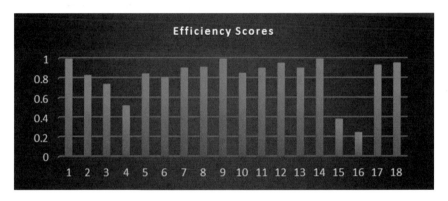

Fig. 4 Histogram of all suppliers with their efficiency score

8 Conclusion and Summary

SSS is a challenging task among thousands of potential suppliers. The present study shows DE algorithm as a tool for selecting the optimal sustainable suppliers. In this research work, we present a novel SSS approach for automotive industry. The first step is to build a criterion set that comprising both input as well as output factors, which is appropriate for real-world applications.

We then present an approach to solve the multiple-criteria SSS problem with the application of DE, for DEA-based mathematical model. By using this tactics, the business firms can acquire a desirable cluster of competent sustainable suppliers. Numerical results validate the efficiency of DE for dealing with such problems.

The main motivation of this research was to gain an understanding of the mechanics of DE algorithm and to determine the accuracy of DE in generating the optimum solutions for the DEA-based mathematical model, which is the underlying optimization problem for the aforementioned purchasing system. To our best information, this is the first time that the DE algorithm is applied to the SSS.

This research article presents a DEA model for SSS practice in SCM. The key offerings of this research are précised as below:

1. SSS in SCM: to date, there are a small number of researches seeing sustainable concern in the supplier selection practice.
2. The selection criteria on the basis of sustainable concern are collected by means of the literature after that these are put into the mathematical model for the SSS practice.
3. The present model can be carrying out for any quantity of suppliers and criteria's in the great size business firms.
4. In spite of the fact that lots of efforts have been made for the supplier selection, taking into consideration sustainable concern for this problem remains a demanding task.
5. In this study, the goal was the application of DE algorithm to the efficient SSS in the SCM.
6. Two cases on automotive industry validate the application of the present approach.
7. Future research may explore the practice of the DE algorithm to find a solution to more difficult problem such as multi-objective SSS.

References

1. Storn, R., Price, K.: Differential evolution—a simple and efficient adaptive scheme for global optimization over continuous spaces. Technical Report TR-95-012, Berkeley, CA (1995)
2. Plagianakos, V., Tasoulis, D., Vrahatis, M.: A review of major application areas of differential evolution. In: Advances in Differential Evolution, vol. 143, pp. 197–238. Springer, Berlin (2008)
3. Wang, F., Jang, H.: Parameter estimation of a bio reaction model by hybrid differential evolution. In: Proceedings of IEEE Congress on Evolutionary Computation (CEC-2000), pp. 410–417 (2000)
4. Joshi R., Sanderson, A.: Minimal representation multi sensor fusion using differential evolution. IEEE Trans. Syst. Man Cybern. Part A Syst. Hum. 29(1), 63–76 (1999)
5. Ilonen, J., Kamarainen, J., Lampine, J.: Differential evolution training algorithm for feed-forward neural networks. Neural Process. Lett. 17(1), 93–105 (2003)
6. Ali, M., Siarry, P., Pant, M.: An efficient differential evolution based algorithm for solving multi-objective optimization problems. Eur. J. Oper. Res. 217(2), 404–416 (2012)
7. Jayal, A., Badurdeen, F., Dillon Jr, O., Jawahir, I.: Sustainable manufacturing: modeling and optimization challenges at the product, process and system levels. CIRP J. Manufact. Sci. Technol. 2, 144–152 (2010)
8. Floridi, M., Pagni, S., Falorni, S., Luzzati, T.: An exercise in composite indicators construction: assessing the sustainability of Italian regions. Ecol. Econ. 70, 1440 (2011)
9. Luthe, T., Schuckert, M.: Socially responsible investing—implications for leveraging sustainable development. In: Trends and Issues in Global Tourism, pp. 315–321 (2011)
10. Paoletti, M., Gomiero, T., Pimentel, D.: Introduction to the special issue: towards a more sustainable agriculture. Crit. Rev. Plant Sci. 30(1), 2–5 (2011)
11. Buyukozkan, G., Çifçi, G.: A novel fuzzy multi-criteria decision framework for sustainable supplier selection with incomplete information. Comput. Ind. 124, 252 (2010)

12. McLean, S.: Finding strategic advantage through SCM. Pulp and Paper International **41**(10), 28–31 (1999)
13. Sarkis, J.: "A boundaries and flows perspective of green supply chain management. Supply Chain Manage. Int. J. **17**(2), 202–216 (2012)
14. Parker, P.: Environmental initiatives among Japanese automakers: new technology, EMS, recycling and life cycle approaches. Environments **29**(3), 91–113 (2001)
15. Nieuwenhuis, P., Wells, P.: The Automotive Industry and the Environment: A Technical, Business and Social Future. Woodhead, Cambridge (2003)
16. Supply Chain of Automobile Industry India.png (source http://commons.wikimedia.org/wiki/File:Supply_Chain_of_Automobile_Industry_India.png)
17. Agarwal, P., Sahai, M., Mishra, V., Bag, M., Singh, V.: A review of multi-criteria techniques for supplier evaluation and selection. Int. J. Ind. Eng. Comput. **2**. (2011) doi:10.5267/jijiec/2011/06/004
18. Weber, C.A., Current, J.R., Benton, W.C.: Vendor selection criteria and methods. Eur. J. Oper. Res. **50**(1), 2–18 (1991)
19. Degraeve, Z., Labro, E., Roodhooft, F.: An evaluation of vendor selection models from a total cost of ownership perspective. Eur. J. Oper. Res. **125**(1), 34–58 (1991)
20. Boer, L., Labro, E., Morlacchi, P.: A review of methods supporting supplier selection. Eur. J. Purchasing Supply Manage. **7**(2), 75–89 (2000)
21. Holt, G.D.: Which contractor selection methodology? Int. J. Project Manage. **16**(3), 153–164 (1998)
22. Aamer, A.M., Sawhney, R.: Review of suppliers selection from a production perspective. In: Proceedings of IIE Conference, pp. 2135–2140 (2004)
23. Ho, W., Xu, X., Dey, P.K.: Multi-criteria decision making approaches for supplier evaluation and selection: a literature review. Eur. J. Oper. Res. **202**, 16–24 (2010)
24. Tahriri, F., Osman, M.R., Ali, A., Yusuff, R.M.: A review of supplier selection methods in manufacturing industries. Suranaree J. Sci. Technol. **15**(3), 201–208 (2008)
25. Cheraghi, S.H., Dadashzadeh, M., Subramanian, M.: Critical success factors for supplier selection: an update. J. Appl. Bus. Res. **20**(2), 91–108 (2011)
26. Jauhar, S.K., Pant, M.: Recent trends in supply chain management: a soft computing approach. In: Proceedings of Seventh International Conference on Bio-Inspired Computing: Theories and Applications (BIC-TA 2012). Springer, India (2013)
27. Jauhar, S.K., Pant, M., Deep, A.: An approach to solve multi-criteria supplier selection while considering environmental aspects using differential evolution. In: Swarm, Evolutionary, and Memetic Computing, pp. 199–208. Springer International Publishing (2013)
28. Jalhar, S.K., Pant, M., Nagar, M.C.: Differential evolution for sustainable supplier selection in pulp and paper industry: a DEA based approach. Comput. Methods Mater. Sci. **15** (2015)
29. Jauhar, S., Pant, M., Deep, A.: Differential evolution for supplier selection problem: a DEA based approach. In: Proceedings of the Third International Conference on Soft Computing for Problem Solving, pp. 343–353. Springer, India (2014)
30. Jauhar, S.K., Pant, M.: Genetic algorithms, a nature-inspired tool: review of applications in supply chain management. In: Proceedings of Fourth International Conference on Soft Computing for Problem Solving, pp. 71–86. Springer, India (2015)
31. Jauhar, S.K., Pant, M., Abraham, A.: A novel approach for sustainable supplier selection using differential evolution: a case on pulp and paper industry. In: Intelligent Data analysis and its Applications, vol. II, pp. 105–117. Springer International Publishing (2014)
32. Genovese, A., Koh, S.C.L., Bruno, G., Bruno, P.: Green supplier selection: a literature review and a critical perspective. In: Paper Presented at IEEE 8th International Conference on Supply Chain Management and Information Systems (SCMIS), Hong Kong (2010)
33. Simpson, D.F., Power, D.J.: Use the supply relationship to develop lean and green suppliers. Supply Chain Manage. Int. J. **19**(1), 60–68 (2005)
34. Lee, S., Klassen, R.D.: Drivers and enablers that foster environmental management capabilities in small and medium-sized suppliers in supply chains. Prod. Oper. Manage. **17**(6), 573–586 (2008)

35. Van Weele, A.: Purchasing and Supply Chain Management. Cengage Learning EMEA, Andover (2010)
36. Scannell, T., Vickery, S., Droge, C.: Upstream supply chain management and competitive performance in the automobile supply industry. J. Bus. Logistics **21**(3), 23–48 (2000)
37. Awaysheh, A., Klassen, R.D.: The impact of supply chain structure on the use of supplier social responsible practices. Int. J. Oper. Prod. Manage. **30**(12), 1246–1268 (2010)
38. Geffen, C., Rothenberg, S.: Suppliers and environmental innovation. Int. J. Oper. Prod. Manage. **20**(2), 166–186 (2000)
39. Klassen, R., Vachon, S.: Collaboration and evaluation in the supply chain. Prod. Oper. Manage. **12**(3), 336–352 (2003)
40. Büyüközkan, G., Çifçi, G.: A novel fuzzy multi-criteria decision framework for sustainable supplier selection with incomplete information. Comput. Ind. **62**(2), 164–174 (2011)
41. Dimitris, K.S., Lamprini, V.S., Yiannis, G.S.: Data envelopment analysis with nonlinear virtual inputs and outputs. Eur. J. Oper. Res. **202**, 604–613 (2009)
42. Ramanathan, R.: An Introduction to Data Envelopment Analysis: A Tool for Performance Measurement. Sage Publication Ltd., New Delhi (2003)
43. Wen, U.P., Chi, J.M.: Developing green supplier selection procedure: a DEA approach. In: 2010 IEEE 17th International Conference onIndustrial Engineering and Engineering Management (IE&EM), pp. 70, 74. 29–31 Oct 2010. doi:10.1109/ICIEEM.2010.5646615
44. Dobos, I., Vörösmarty, G.: Supplier selection and evaluation decision considering environmental aspects 149. sz. Mőhelytanulmány, HU ISSN 1786-3031 (2012)
45. Jauhar, S. K., Pant, M.: Sustainable Supplier's Management Using Differential Evolution. Problem Solving and Uncertainty Modeling through Optimization and Soft Computing Applications, 239 (2016)
46. Data envelopment analysis: models and extensions. Production/Operation Management Decision Line, pp. 8–11 (2000)
47. Charnes, A., Cooper, W.W., Rhodes, E.: Measuring the efficiency of decision making units. Eur. J. Oper. Res. **2**(6), 429–444 (1978)
48. Banker, R.D., Charnes, A., Cooper, W.W.: Some models for estimating technical and scale inefficiencies in data envelopment analysis. Manage. Sci. **30**, 1078–1092 (1984)
49. Jouni, L.: A constraint handling approach for differential evolution algorithm. In: Proceeding of IEEE Congress on Evolutionary Computation (CEC 2002), pp. 1468–1473 (2002)
50. Coello, C.A.C.: Theoretical and numerical constraint handling techniques used with evolutionary algorithms: a survey of the state of the art. Comput. Methods Appl. Mech. Eng. **191**(11–12), 1245–1287 (2002) (Differential Evolution for Data Envelopment Analysis 319)
51. Ray, T., Kang, T., Chye, S.K.: An evolutionary algorithm for constraint optimization. In: Whitley, D., Goldberg, D., Cantu-Paz, E., Spector, L., Parmee, I., Beyer, H.G. (eds.) Proceeding of the Genetic and Evolutionary Computation Conference (GECCO 2000), pp. 771–777 (2000)
52. Shirouyehzad, H., Lotfi, F.H., Dabestani, R.: A data envelopment analysis approach based on the service quality concept for vendor selection. In: International Conference on Computers & Industrial Engineering, 2009. CIE 2009, pp. 426, 430, 6–9 July 2009. doi: 10.1109/ICCIE. 2009.5223823
53. Mazhar, M.I., Kara, S., Kaebernick, H.: Reusability assessment of components in consumer products—a statistical and condition monitoring data analysis strategy. In: Fourth Australian Life Cycle Assessment Conference—Sustainability Measures For Decision Support, Sydney, Australia (2005)
54. http://www.london2012.com/documents/locog-publications/locog-guidelines-on-carbon-emissions-of-products-and-services.pdf. Accessed on 12 Oct 2013

A Comparative Study of Membership Functions for an Interval Type-2 Fuzzy System Used for Dynamic Parameter Adaptation in Particle Swarm Optimization

Frumen Olivas, Fevrier Valdez and Oscar Castillo

Abstract This paper presents an analysis of the effects in quality results that bring the use of different types of membership functions in an interval type-2 fuzzy system, used to adapt some parameters of particle swarm optimization (PSO). Benchmark mathematical functions are used to test the methods, and a comparative study is performed.

Keywords Type-2 fuzzy logic · Particle swarm optimization · Dynamic parameter adaptation · PSO · Membership functions

1 Introduction

In this paper, we analyze the effect that brings the different types of membership functions, in an interval type 2 fuzzy system used to dynamic parameter adaptation in PSO, which we use in [1]; this system is an extension or generalization from a system we use in [2]. So the main contribution of this paper is the analysis of the effect that brings different types of membership functions to an interval type-2 fuzzy system.

PSO is a bio-inspired method introduced by Kennedy and Eberhart in 1995 [3] and [4], this method maintains a swarm of particles, like a flock of birds or a shoal of fish, and each particle represent a solution to the problem. These particles "fly" in a multidimensional space, using Eq. 1 to update the position of the particle and Eq. 2 to update the velocity of the particle [5]. The equation of the velocity of the

F. Olivas (✉) · F. Valdez · O. Castillo
Tijuana Institute of Technology, Tijuana, Mexico
e-mail: frumen@msn.com

F. Valdez
e-mail: fevrier@tectijuana.mx

O. Castillo
e-mail: ocastillo@tectijuana.mx

particle is affected by a coefficient component and a social component, where coefficient component is affected by a coefficient factor c_1, and social component is affected by a social factor c_2. These parameters are very important because this changes the velocity of the particle; when c_1 is larger than c_2, this improves the swarm exploration of the space of search, and on the other hand, when c_1 is lower than c_2, this improves the swarm exploitation of the best area from the space of search found so far.

$$x_i(t+1) = x_i(t) + v_i(t+1) \tag{1}$$

$$v_{ij}(t+1) = v_{ij}(t) + c_1 r_1(t)\left[y_{ij}(t) - x_{ij}(t)\right] + c_2 r_{2j}(t)\left[\hat{y}_j(t) - x_{ij}(t)\right] \tag{2}$$

where in Eq. 1, x_i is the particle i, t is the current iteration or time; v_i is the velocity of the particle i. And where in Eq. 2, v_{ij} is the velocity of the particle i in the dimension j, c_1 is the cognitive factor or the weight that tell us the importance of the best previous position of the particle, r_1 is a random value in the range of [0, 1], c_2 is the social factor or the weight that tell us the importance of the best global position of the swarm, r_2 is a random value in the range of [0, 1], y_{ij} is the dimension j of the best position found so far by the particle i, x_{ij} is the dimension j of the current position of the particle i, and $\hat{y}_j$ is the dimension j of the best global position of the swarm.

Fuzzy set theory introduced by Zadeh in 1965 [6, 7] helps us to model problems with a degree of uncertainty, by using linguistic information about the problem, and improves the numerical computation by using linguistic labels represented by membership functions. Type-2 fuzzy logic also introduced by Zadeh [8] helps us to model problems with an implicit degree of uncertainty, and this is possible by adding a footprint of uncertainty (FOU) to the membership functions, and this means that even the membership value is fuzzy [9].

Since PSO was introduced in 1995, it has seen some improvements to the convergence and diversity of the swarm using a fuzzy system; for example, Hongbo [10] added a new parameter to the velocity equation, and this new parameter is tuned using a fuzzy logic controller. Shi [11] used a fuzzy system to adjust the inertia weight, where inertia is a parameter added to the velocity formula (Eq. 2), as a weight that affects the previous velocity. Wang [12] used a fuzzy system to determine when to change the velocity of the particles to avoid local minima.

2   Methodology

The change in PSO is by including a type-2 fuzzy system that dynamically adapts the parameters c_1 and c_2; the proposal is shown in Fig. 1, taken from [1].

A fuzzy system can be a very powerful tool [13], but has many parameters to be adjusted to the problem in which it is applied [14]; the realization of this paper helps

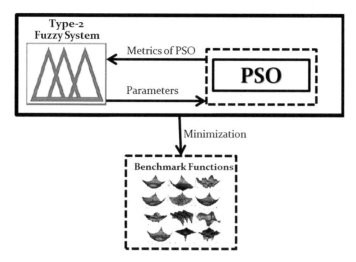

Fig. 1 Proposed method

us to know whether the type of membership function provides a difference in the results in terms of performance. In this case, the membership function types to be compared are triangular, Gaussian, trapezoidal, and generalized bell.

We develop an interval type-2 fuzzy system that dynamically adapts some parameters in PSO, that is, c_1 and c_2, and we choose these parameters because it changes the velocity of the particle, and we use previous knowledge to make the fuzzy rule set, using the idea that in first iterations, PSO will use exploration to search the best area of the space of search, and in final iterations, PSO will use exploitation of the best area found so far.

The fuzzy rule set used for parameter adaptation in PSO is shown in Fig. 2. Taking the idea that when c_1 is higher than c_2, this improves the exploration of the space of search, and when c_1 is lower than c_2, this improves the exploitation of the best area from the space of search found so far.

The variables for the interval type-2 fuzzy system for parameter adaptation are:

1. If (Iteration is Low) and (Diversity is Low) then	(C1 is High) and (C2 is Low)
2. If (Iteration is Low) and (Diversity is Medium) then	(C1 is MediumHigh) and (C2 is Medium)
3. If (Iteration is Low) and (Diversity is High) then	(C1 is MediumHigh) and (C2 is MediumLow)
4. If (Iteration is Medium) and (Diversity is Low) then	(C1 is MediumHigh) and (C2 is MediumLow)
5. If (Iteration is Medium) and (Diversity is Medium) then	(C1 is Medium) and (C2 is Medium)
6. If (Iteration is Medium) and (Diversity is High) then	(C1 is MediumLow) and (C2 is MediumHigh)
7. If (Iteration is High) and (Diversity is Low) then	(C1 is Medium) and (C2 is High)
8. If (Iteration is High) and (Diversity is Medium) then	(C1 is MediumLow) and (C2 is MediumHigh)
9. If (Iteration is High) and (Diversity is High) then	(C1 is Low) and (C2 is High)

Fig. 2 Fuzzy rule set for parameter adaptation in PSO

Inputs:

1. Iteration: this variable is a metric about the iterations of PSO, is a percentage of the iterations elapsed, with a range from [0, 1], and is represented by Eq. 3; this variable is granulated into three membership functions.
2. Diversity: this variable is a metric about the particles of PSO, is the degree of dispersion of the particles in the space of search; also can be seen as the average of the Euclidian distances between each particle and the best particle, with a range of [0, 1] (normalized), and is represented by Eq. 4; this variable is granulated into three membership functions.

Outputs:

1. $C1$: this variable represents the parameter c_1 on the velocity equation (Eq. 2) of PSO, is the cognitive factor or a weight that affects the best position of the particle, with a range of [0, 3], but is granulated so the results in the range of [0.5, 2.5]; this range contains the best possible values for c_1 according to [15], and this variable is granulated into five membership functions.
2. $C2$: this variable represents the parameter c_2 on the velocity equation (Eq. 2) of PSO, is the social factor or a weight that affects the best position of the best particle in the swarm, with a range of [0, 3], but is granulated so the results in the range of [0.5, 2.5]; this range contains the best possible values for c_2 according to [15], and this variable is granulated into three membership functions.

$$\text{Iteration} = \frac{\text{Current Iteration}}{\text{Maximum of Iteration}} \tag{3}$$

$$\text{Diversity}(S(t)) = \frac{1}{n_s} \sum_{i=1}^{n_s} \sqrt{\sum_{j=1}^{n_s} \left(X_{ij}(t) - \bar{X}_j(t) \right)^2} \tag{4}$$

The interval type-2 fuzzy system designed in [1] for the parameter adaptation is shown in Fig. 3, and we develop this system manually, that is, we change the levels of FOU of each point of each membership function, but each point has the same level of FOU; also the input and output variables have only interval type-2 triangular membership functions. However, this system is the best system found by changing the levels of FOU manually. The changes in the levels of FOU are made manually, because we want to use it on a wide variety of problems, so we try to do it in the most general form possible.

Thinking that each point of each membership function should have its own level of FOU, which is why we decide to optimize the levels of FOU using an external PSO [1], the result of this optimization is shown in Fig. 4.

The purpose of this paper was to study performance analysis of using different types of membership functions; we already use interval type-2 triangular membership

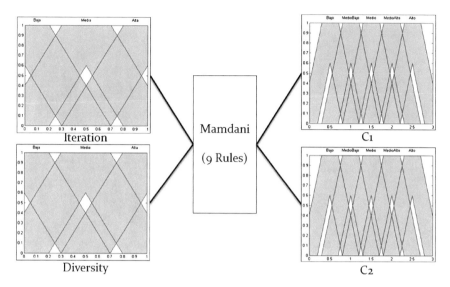

Fig. 3 Interval type-2 fuzzy system with triangular membership functions used for parameter adaptation in PSO (FPSO1)

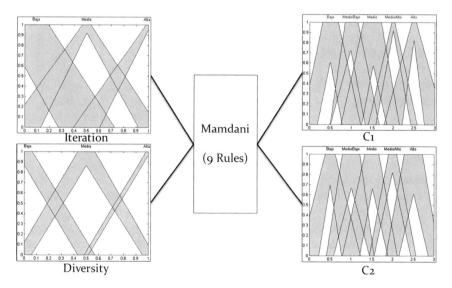

Fig. 4 Optimized interval type-2 fuzzy system used for parameter adaptation in PSO (FPSO5)

functions, so the types used in this analysis are interval type-2 Gaussian membership functions, interval type-2 generalized bell membership functions and interval type-2 trapezoidal membership functions.

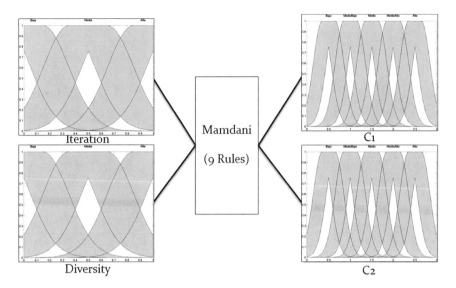

Fig. 5 Interval type-2 fuzzy system with Gaussian membership functions for parameter adaptation in PSO (FPSO2)

To switch between types of membership functions, we try to represent the same level of FOU on each type of membership function, taking as reference the interval type-2 triangular membership functions that we already use.

Figure 5 shows the fuzzy system for parameter adaptation, using interval type-2 Gaussian membership functions. Like the system done manually, now this system has only interval type-2 Gaussian membership functions in input and output variables. We try to maintain the levels of FOU on each membership function as similar to the system that uses triangular membership functions; this is to see only the effect of a different membership function.

Figure 6 shows the fuzzy system for parameter adaptation, using interval type-2 trapezoidal membership functions. And again we try to maintain the levels of FOU on each membership function as similar to the system that uses triangular membership functions.

Figure 7 shows the fuzzy system for parameter adaptation, using interval type-2 generalized bell membership functions. We try to maintain the levels of FOU as in the system that uses triangular membership functions, but more likely the system that uses trapezoidal membership functions.

We also change the type of membership functions from the optimized interval type-2 fuzzy system, trying to maintain the levels of FOU on each point from each membership function. Fig. 8 shows the iteration variable from the optimized interval type-2 fuzzy system and its variations with the different types of membership

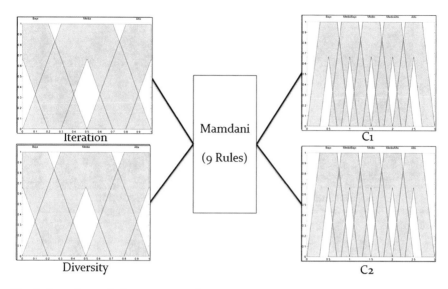

Fig. 6 Interval type-2 fuzzy system with trapezoidal membership functions for parameter adaptation in PSO (FPSO3)

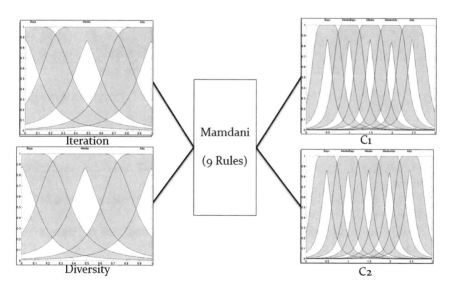

Fig. 7 Interval type-2 fuzzy system with generalized bell membership functions for parameter adaptation in PSO (FPSO4)

functions used. We only show the input iteration from the optimized system because for the other input and outputs, it is the same idea.

At this point, we can consider 8 systems and these are:

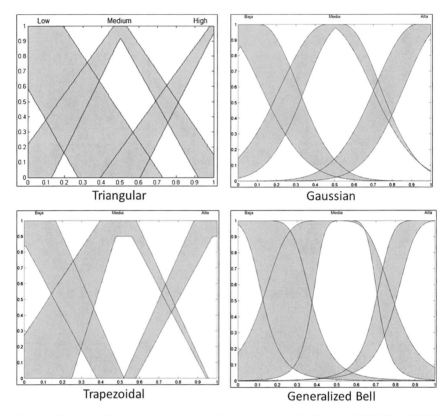

Fig. 8 Different versions of the iteration variable from the optimized system (FPSO5 to FPSO8)

1. FPSO1 the fuzzy system that uses interval type-2 triangular membership functions.
2. FPSO2 the fuzzy system that uses interval type-2 Gaussian membership functions.
3. FPSO3 the fuzzy system that uses interval type-2 trapezoidal membership functions.
4. FPSO4 the fuzzy system that uses interval type-2 generalized bell membership functions.
5. FPSO5 the optimized fuzzy system that uses interval type-2 triangular membership functions.
6. FPSO6 the optimized fuzzy system that uses interval type-2 Gaussian membership functions.
7. FPSO7 the optimized fuzzy system that uses interval type-2 trapezoidal membership functions.
8. FPSO8 the optimized fuzzy system that uses interval type-2 generalized bell membership functions.

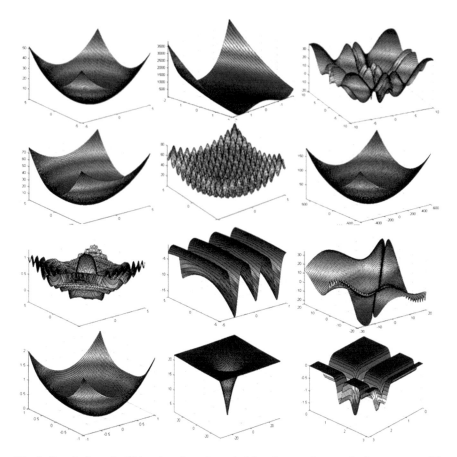

Fig. 9 Sample from the 27 benchmark mathematical functions used to test the fuzzy systems (*f3*, *f4*, *f5*, *f8*, *f27*, *f9*, *f13*, *f14*, *f15*, *f17*, *f18*, *f19*) seen from left to right and line by line

To test these fuzzy systems, we use 27 benchmark mathematical functions obtained from [16] and [17]; Fig. 9 shows a sample of the total of the benchmark functions used to test the systems; we also tested the fuzzy systems using 1000 dimensions on the benchmark functions.

3 Results

Each developed interval type-2 fuzzy system is integrated in a PSO method to dynamically adapt c_1 and c_2 parameters, and each system is tested with the set of benchmark mathematical functions, using the parameters from Table 1. The parameters from Table 1 are used for each fuzzy system integrated in PSO to dynamically adapt the parameters.

Table 1 Parameters for PSO

Parameter	Value
Population	30
Iterations	100
c_1 and c_2	Dynamic
Inertia and constriction factor	1 (no effect)

Table 2 Results obtained by applying the fuzzy systems for the parameter adaptation on PSO in the minimization of benchmark mathematical functions (1)

Fun	FPSO1	FPSO2	FPSO3	FPSO4
$f1$	1	1	1	1
$f2$	2.0e−21	1.0e−20	6.2e−21	2.1e−21
$f3$	1.9e+2	2.0e+2	1.9e+2	2.0e+2
$f4$	1.4e+4	1.5e+4	1.4e+4	1.5e+4
$f5$	1.1e+3	1.2e+3	1.0e+3	1.3e+3
$f6$	−96.64	−98.33	−97.20	−98.90
$f7$	−18.06	−17.83	−17.47	−18.15
$f8$	1.4e−13	1.2e−13	5.9e−14	6.2e−14
$f9$	0.50	0.52	0.48	0.52
$f10$	−0.52	−0.52	−0.52	−0.52
$f11$	0.003	0.003	0.002	0.001
$f12$	−0.335	−0.335	−0.335	−0.335
$f13$	−16.69	−16.47	−16.59	−16.55
$f14$	−23.14	−23.08	−23.18	−23.05
$f15$	4.1e+5	4.1e+5	4.1e+5	4.1e+5
$f16$	1.4e−12	2.4e−12	1.1e−12	2.0e−12
$f17$	2.4e−16	3.8e−16	1.3e−16	3.6e−16
$f18$	3.38	3.41	3.35	3.45
$f19$	−1.80	−1.80	−1.80	−1.80
$f20$	11.34	11.29	11.34	11.34
$f21$	−1	−1	−1	−1
$f22$	3	3	3	3
$f23$	−1.0316	−1.0316	−1.0316	−1.0316
$f24$	2.3694	2.7322	2.5427	2.2306
$f25$	−0.993	−0.996	−0.996	−0.994
$f26$	−186.73	−186.73	−186.73	−186.72
$f27$	8918	8970	8940	9056
Avg	16219.7	16276.9	16252.3	16266.4

The results from Tables 2 and 3 are obtained by applying the 8 fuzzy systems for the parameter adaptation on PSO in the minimization of the benchmark mathematical functions, with 1000 dimensions, and these results are the average of 100 experiments.

Table 3 Results obtained by applying the fuzzy systems for the parameter adaptation on pso in the minimization of benchmark mathematical functions (2)

Fun	FPSO5	FPSO6	FPSO7	FPSO8
f1	1	1	1	1
f2	1.5e−21	7.6e−21	4.1e−21	3.3e−21
f3	2.0e+2	2.0e+2	2.0e+2	2.0e+2
f4	1.4e+4	1.5e+4	1.4e+4	1.5e+4
f5	1.3e+3	1.3e+3	1.2e+3	1.3e+3
f6	−99.09	−98.14	−98.71	−99.09
f7	−18.05	−18.09	−17.58	−17.71
f8	2.8e−13	1.9e−13	1.8e−13	2.3e−13
f9	0.52	0.55	0.52	0.54
f10	−0.52	−0.52	−0.52	−0.52
f11	0.004	0.001	0.002	0.003
f12	−0.335	−0.335	−0.335	−0.335
f13	−16.69	−16.77	−16.63	−16.58
f14	−23.14	−23.23	−23.051	−23.29
f15	4.e+5	4.1e+5	4.1e+5	4.1e+5
f16	1.8e−12	4.2e−12	5.6e−12	6.3e−12
f17	6.3e−16	1.0e−15	1.7e−16	8.8e−16
f18	3.42	3.46	3.43	3.47
f19	−1.80	−1.80	−1.80	−1.80
f20	11.34	11.34	11.34	11.34
f21	−1	−1	−1	−1
f22	3	3	3	3
f23	−1.0316	−1.0316	−1.0316	−1.0316
f24	2.1552	2.3587	2.5278	2.2319
f25	−0.994	−0.995	−0.995	−0.994
f26	−186.72	−186.73	−186.73	−186.73
f27	8986	8963	8991	8980
Avg	16258.5	16268.6	16239.5	16267.4

The results from Tables 2 and 3 show the performance from each interval type-2 fuzzy system for parameter adaptation; in this case, we can see that the results are similar and there is not a big difference between each other. As we know PSO is a meta-heuristic method and has an explicit randomness, and for that reason each result from Tables 2 and 3 is the average of 100 experiments, which is to avoid randomness in the quality results.

The standard deviations from the experiments ranging from 1.5526e−20 to 9.8388 indicate that the results of all experiments are very close to each other. This means that the adaptation of the parameters of PSO, using an interval type-2 fuzzy system, helps PSO to obtain more accurate and consistent results.

4 Conclusions

Tables 2 and 3 show that PSO with the interval type-2 fuzzy system with triangular membership functions has, on average, better results in terms of quality when compared with the other fuzzy systems using other types of membership functions, even when compared with the optimized fuzzy system.

Also from the results shown in Tables 2 and 3, we can conclude that the effect of different types of membership functions, in this kind of problem, makes no difference, so we can continue using the interval type-2 triangular membership function, which is a membership function easy to use, because the parameters are more intuitive and easy to set.

This paper is important because this is the study of the effects that bring different types of membership functions, thus make this comparison help us to avoid an optimization of the types of membership functions, so we can focus on others optimizations for the interval type-2 fuzzy system used for parameter adaptation in PSO.

References

1. Olivas, F., Valdez, F., Castillo, O.: Particle swarm optimization with dynamic parameter adaptation using interval type-2 fuzzy logic for benchmark mathematical functions. In: 2013 World Congress on Nature and Biologically Inspired Computing (NaBIC), pp. 36–40 (2013)
2. Melin, P., Olivas, F., Castillo, O., Valdez, F., Soria, J., Valdez, M.: Optimal design of fuzzy classification systems using PSO with dynamic parameter adaptation through fuzzy logic. Expert Syst. Appl, 3196–3206. Elsevier (2016)
3. Kennedy, J., Eberhart, R.: Particle swarm optimization. In: Proceedings of IEEE International Conference on Neural Networks, IV, pp. 1942–1948. IEEE Service Center, Piscataway
4. Kennedy, J., Eberhart, R.: Swarm Intelligence. Morgan Kaufmann, San Francisco (2001)
5. Engelbrecht, A.: Fundamentals of Computational Swarm Intelligence. Wiley, New York (2006)
6. Zadeh, L.: Fuzzy sets. Inf. Control **8**, 338 (1965)
7. Zadeh, L.: Fuzzy logic. IEEE Comput. 83–92
8. Zadeh, L.: The concept of a linguistic variable and its application to approximate reasoning—I. Inform. Sci. **8**, 199–249 (1975)
9. Liang, Q., Mendel, J.: Interval type-2 fuzzy logic systems: theory and design. IEEE Trans. Fuzzy Syst. **8**(5), 535–550 (2000)
10. Hongbo, L., Ajith, A.: A fuzzy adaptive turbulent particle swarm optimization. Int. J. Innovative Comput. Appl. **1**(1), 39–47 (2007)
11. Shi, Y., Eberhart, R.: Fuzzy adaptive particle swarm optimization. In: Evolutionary Computation, pp. 101–106 (2001)
12. Wang, B., Liang, G., ChanLin, W., Yunlong, D.: A new kind of fuzzy particle swarm optimization FUZZY_PSO algorithm. In: 1st International Symposium on Systems and Control in Aerospace and Astronautics. ISSCAA 2006, pp. 309–311
13. Wang, L.-X.: Fuzzy systems are universal approximators. In: IEEE International Conference on Fuzzy Systems, pp. 1163, 1170. 8–12 Mar (1992)
14. Takagi, T., Sugeno, M.: Fuzzy identification of systems and its applications to modeling and control. IEEE Trans. Syst. Man Cybern. SMC-15(1), 116,132 (1985)

15. Jang, J., Sun, C., Mizutani, E.: Neuro-fuzzy and Soft Computing: A Computational Approach to Learning and Machine Intelligence. Prentice-Hall, Upper Saddle River (1997)
16. Haupt, R., Haupt, S.: Practical Genetic Algorithms, second edn. A Wiley-Interscience publication, New Jersey (2004)
17. Marcin, M., Smutnicki, C.: Test functions for optimization needs (2005)

Part IX
Control and Modeling

Models for Indicating the Period of Failure of Industrial Objects

T.A. Aliev, N.F. Musaeva, O.Q. Nusratov, A.G. Rzayev
and U.E. Sattarova

Abstract We propose a technology for calculating robust correlation matrices and robust normalized correlation matrices to indicate the beginning of the latent period of the emergency state of technological objects. For the same purpose, we also propose a technology for calculating estimates of characteristics of noise and useful signal, which assumes zero values in the normal state. Sets of informative attributes are formed from them in monitoring systems for industrial objects; while the object is in service, the number and value of the nonzero element are used to determine the location and nature of failure.

1 Introduction

Classical probabilistic–statistical methods, methods of stochastic process theory, and statistical methods of signal processing are used these days in solving the problems of diagnostics, identification, and indication of changes in the technical condition of industrial and biological objects. However, the use of these methods for identifying the technical condition of real industrial objects is associated with

T.A. Aliev (✉) · O.Q. Nusratov · A.G. Rzayev
Institute of Control Systems of ANAS, Baku, Azerbaijan
e-mail: telmancyber@gmail.com

O.Q. Nusratov
e-mail: nusratov@cyber.ab.az

A.G. Rzayev
e-mail: asifrzayev48@gmail.com

N.F. Musaeva · U.E. Sattarova
Azerbaijan University of Architecture and Construction, Baku, Azerbaijan
e-mail: musanaila@gmail.com

U.E. Sattarova
e-mail: ulker.rzaeva@gmail.com

© Springer International Publishing Switzerland 2016
L.A. Zadeh et al. (eds.), *Recent Developments and New Direction
in Soft-Computing Foundations and Applications*, Studies in Fuzziness
and Soft Computing 342, DOI 10.1007/978-3-319-32229-2_27

certain difficulties, to overcome which numerous unconventional methods have been developed [1, 2]. In the process, it is not uncommon that signals from the sensors are contaminated by noise [3] and have insufficient volume of experimental information [4], or there is missing observation [5]. In such cases, the researcher first of all endeavors to identify the specific peculiarities of the noise that distinguish it from the useful random signal, for instance, heavy tails [6]. After that, in solving the said problems, correlation and spectral characteristics of noisy signals are calculated and correlation matrices are set up [4–6]. Since noise causes distortion of random signals, the researcher tries to eliminate these effects, using various methods. In most cases, different filtration methods are applied, which take into account the specifics of the process [4], methods of modeling and reconstructing the missing value with application of moving average [5], centering methods [7], etc.

Alongside with these works, new unconventional statistical methods for analysis of random processes are developed, using the estimates of some new characteristics of measurement information, which allow solving such problems as monitoring, control, identification, forecasting, and management of different processes [8–10]. By comparing values of those characteristics obtained from signal analysis, the moment of transition of a system, object, or technological process into failure state or emergency state is determined.

Use of noise as a carrier of diagnostic information is an unconventional method of statistical analysis of random processes [8–10]. For example, Aliev et al. [8] proposed a technology for determining the robust statistical indicators that in the normal state assume zero values to indicate the beginning of the latent imperceptible transition of oil-pumping stations and compressor stations (OPS and CS) from the normal state to the emergency state. The authors of [9] developed noise technologies for indicating and identifying the latent period of transition of an object from the normal state to the emergency state. In [9], methods are proposed for noise indication of change in the dynamic state of industrial objects by means of robust correlation characteristics of noisy signals. In [10], a technology is proposed for determining the estimates of auto- and cross-correlation indicators, noise variances, cross-correlation functions, and coefficients of correlation between the useful signal and the noise that in the normal state assume zero values to indicate the beginning of the latent imperceptible transition of objects from the normal state to the emergency state.

Our analysis of probabilistic–statistical methods used to ensure fault-free operation of technological objects demonstrated that methods for processing of noisy technological parameters are required, which would allow indicating the beginning of the latent period of transition of an object to the emergency state by using noise as a carrier of diagnostic information. The present paper deals with one possible solution to this problem.

2   Problem Statement

It is known that for identifying the beginning of transition of an object from the normal state to the emergency state, one can use the known models that come to solving matrix equations of the type [10]:

$$\vec{R}_{\overset{\circ}{X}\overset{\circ}{Y}}(\mu) = \vec{R}_{\overset{\circ}{X}\overset{\circ}{X}}(\mu)\vec{W}(\mu), \quad \mu = 0,\ \Delta t,\ \ldots,(N-1)\Delta t, \tag{1}$$

where $\vec{R}_{\overset{\circ}{X}\overset{\circ}{X}}(\mu)$ is a square symmetric matrix of the autocorrelation function $R_{\overset{\circ}{X}\overset{\circ}{X}}(\mu)$ of the input signal $\overset{\circ}{X}(t) = X(t) - m_X$ with dimension $N \times N$; $\vec{R}_{XY}(\mu)$ is a column vector of the cross-correlation function $R_{\overset{\circ}{X}\overset{\circ}{Y}}(\mu)$ between the input $X(t)$ and output $Y(t)$; m_X and m_Y are the mathematical expectations of $X(t)$ and $Y(t)$, respectively; $\vec{W}(\mu)$ is a column vector of the impulsive admittance functions [10].

For instance, it is known that matrix Eq. (1) allows solving the problem of identifying the initial stage of transition of a compressor station to the emergency state, the latent period of violation of seismic stability of offshore platforms, as well as the beginning of the latent period of the emergency state of drilling units.

However, this matrix Eq. (1) allows solving the problem of identifying the transition of industrial objects from the normal state to the emergency state if the technological parameters $X(t)$ and $Y(t)$ meet the classical conditions, i.e., if they comply with the normal distribution law and if they are stationary and ergodic.

On the other hand, the analysis of measurement information received from sensors of the said technical objects demonstrates that the real signals $g(t), \eta(t)$ are a mixture of the useful signals $X(t), Y(t)$ and random noises $\varepsilon(t), \varphi(t)$, that is

$$\begin{cases} g(t) = X(t) + \varepsilon(t) \\ \eta(t) = Y(t) + \varphi(t) \end{cases}. \tag{2}$$

The conventional approach to the identification by means of the matrix Eq. (1) suggests that the classical limitations take place; that is, the noisy technological parameters $g(t), \eta(t)$ comply with the normal distribution law and are stationary and ergodic; the random noises $\varepsilon(t), \varphi(t)$ have zero mathematical expectations $m_\varepsilon \approx 0, m_\varphi \approx 0$ and uncorrelated values of the readings as follows:

$$\begin{aligned} \frac{1}{N}\sum_{k=1}^{N} \overset{\circ}{\varepsilon}(k\Delta t)\,\overset{\circ}{\varepsilon}((k+\mu)\Delta t) &\approx 0 \\ \frac{1}{N}\sum_{k=1}^{N} \overset{\circ}{\varepsilon}(k\Delta t)\,\overset{\circ}{\varphi}((k+\mu)\Delta t) &\approx 0 \end{aligned}, \quad \text{when } \mu \neq 0, \tag{3}$$

and the useful signals $X(t)$ and $Y(t)$ are not correlated:

$$\frac{1}{N}\sum_{k=1}^{N} \overset{\circ}{X}(k\Delta t)\, \overset{\circ}{\varepsilon}((k+\mu)\Delta t) \approx 0$$

$$\frac{1}{N}\sum_{k=1}^{N} \overset{\circ}{\varepsilon}(k\Delta t)\, \overset{\circ}{X}((k+\mu)\Delta t) \approx 0$$

$$(4)$$

$$\frac{1}{N}\sum_{k=1}^{N} \overset{\circ}{X}(k\Delta t)\, \overset{\circ}{\varphi}((k+\mu)\Delta t) \approx 0$$

$$\frac{1}{N}\sum_{k=1}^{N} \overset{\circ}{\varepsilon}(k\Delta t)\, \overset{\circ}{Y}((k+\mu)\Delta t) \approx 0$$

$$(5)$$

If we assume the above, the following equality is supposed to hold true

$$\vec{R}_{\overset{\circ}{g}\overset{\circ}{\eta}}(\mu) = \vec{R}_{\overset{\circ}{g}\overset{\circ}{g}}(\mu)\vec{W}(\mu), \tag{6}$$

where $\vec{R}_{\overset{\circ}{g}\overset{\circ}{g}}(\mu)$ is a square symmetric matrix of the autocorrelation function $R_{\overset{\circ}{g}\overset{\circ}{g}}(\mu)$ of the input noisy signal $\overset{\circ}{g}(t) = g(t) - m_g$; $R_{\overset{\circ}{g}\overset{\circ}{\eta}}(\mu)$ is a column vector of the cross-correlation function $R_{\overset{\circ}{g}\overset{\circ}{\eta}}(\mu)$ between the noisy input $\overset{\circ}{g}(t)$ and the output $\overset{\circ}{\eta}(t) = \eta(t) - m_\eta$; m_g and m_η are the mathematical expectations of $g(t)$ and $\eta(t)$, respectively.

However, those ideal conditions do not hold true in the transition of an industrial object from the normal state to the emergency state. In that case, the elements $R_{\overset{\circ}{X}\overset{\circ}{X}}(\mu)$ and $R_{\overset{\circ}{X}\overset{\circ}{Y}}(\mu)$ of the correlation matrices $\vec{R}_{\overset{\circ}{X}\overset{\circ}{X}}(\mu)$ and $\vec{R}_{\overset{\circ}{X}\overset{\circ}{Y}}(\mu)$ contain errors due to influence of noises, and the following inequalities take place:

$$R_{\overset{\circ}{X}\overset{\circ}{X}}(\mu) \neq R_{\overset{\circ}{g}\overset{\circ}{g}}(\mu), \tag{7}$$

$$R_{\overset{\circ}{X}\overset{\circ}{Y}}(\mu) \neq R_{\overset{\circ}{g}\overset{\circ}{\eta}}(\mu). \tag{8}$$

Therefore, the following inequality takes place:

$$\vec{R}_{\overset{\circ}{g}\overset{\circ}{\eta}}(\mu) \neq \vec{R}_{\overset{\circ}{g}\overset{\circ}{g}}(\mu)\vec{W}(\mu). \tag{9}$$

In this case, it is naturally impossible to solve the identification problem for real technological parameters. We must therefore reorganize the inequalities in (9) in the form that would allow solving the problem.

Consider that determining the initial stage of the transition of an object from the normal state to the emergency state requires only an approximate solution to this problem. In many cases, the problem of indicating the transition of an object to the emergency state rather than that of identifying the technical condition can be solved for real technological processes. In this chapter, we therefore consider one possible solution to this problem as well.

In the period T_0 before the beginning of defect origin process, assume that the classical conditions hold true, i.e., the corresponding equalities are true:

$$W_{T_0}[g(t)] = \frac{1}{\sqrt{2\pi D(g)}} e^{-\frac{(g-m_g)^2}{2D(g)}}, \quad D(\varepsilon) \approx 0,$$

$$D(g) \approx D(X), \quad R_{\overset{\circ}{g}\overset{\circ}{g}}(\mu) \approx R_{\overset{\circ}{X}\overset{\circ}{X}}(\mu), \quad m_g \approx m_X \approx 0,$$

$$R_{X\varepsilon}(\mu = 0) \approx 0, \quad r_{X\varepsilon} \approx 0, \tag{10}$$

where $W_{T_0}[g(t)]$ is the signal distribution law for $g(t)$ in the normal technical state of the object; $D(\varepsilon), D(X)$, and $D(g)$ are the variance estimates of the noise, the useful signal, and the summed signal, respectively; $R_{\overset{\circ}{X}\overset{\circ}{X}}(\mu)$ and $R_{\overset{\circ}{g}\overset{\circ}{g}}(\mu)$ are the estimates of the correlation functions at the time shifts $\mu = 0, \mu = \Delta t, \mu = 2\Delta t, \mu = 3\Delta t, \ldots$ of the useful signal $X(t)$ and the summed signal $g(t); m_g$ and m_X are the mathematical expectations of the useful signal and the summed signal; and $R_{X\varepsilon}(\mu = 0)$ and $r_{X\varepsilon}$ are the cross-correlation function and the coefficient of correlation between the useful signal $X(t)$ and the noise $\varepsilon(t)$, respectively.

At the same time, our experimental research demonstrated that in the moment T_1, when the processes of transition of an object to the emergency state begins and the latent period of change in the technical condition of an object comes, the condition (10) is violated and the following inequalities take place:

$$W_{\overset{\circ}{T}}[g(t)] \neq W_{T_1}[g(t)],$$

$$D(\varepsilon) \neq 0, \quad D(g) \neq D(X), \quad R_{\overset{\circ}{g}\overset{\circ}{g}}(\mu) \neq R_{\overset{\circ}{X}\overset{\circ}{X}}(\mu),$$

$$m_g \neq m_X, \quad R_{X\varepsilon}(\mu = 0) \neq 0, \quad r_{X\varepsilon} \neq 0. \tag{11}$$

Thus stated solution to the problem of the beginning of violation of seismic stability of offshore platforms, the latent period of transition of compressor stations and drilling units to the emergency state is reduced to indicating the moment of transition of equality (10) into inequality (11) [8–10].

We consider possible ways to solve the above-mentioned problems in the following paragraphs.

3 Technology for Indicating the Transition of Industrial Objects from the Normal State to the Emergency State

Our numerous experimental research studies demonstrated that in the latent period of transition of an object to the emergency state, noise emerges in the signals due to such defects as tear and wear, cracks, bending, and fatigue; the coefficient of correlation $r_{X\varepsilon}$ between the useful signal and the noise as well as variance and other characteristics change; stationarity, normality, ergodicity, etc., are violated [8–10]. Given the above, let us consider in more detail the possibility of solving the problem of indicating the initial stage of transition of industrial objects from the normal state to the emergency state in the presence of noise. As was indicated in the problem statement, solving numerous important identification problems reduces to the numerical solution of matrix correlation Eq. (1). In that case, correlation matrices $\vec{R}_{\overset{\circ}{X}\overset{\circ}{X}}(\mu)$ and $\vec{R}_{\overset{\circ}{X}\overset{\circ}{Y}}(\mu)$, whose elements are the estimates of the correlation functions $R_{\overset{\circ}{X}\overset{\circ}{X}}(\mu)$ and $R_{\overset{\circ}{X}\overset{\circ}{Y}}(\mu)$ of the useful signals $X(t)$ and $Y(t)$, and the column vector of the impulsive admittance functions $\vec{W}(\mu)$, look as follows:

$$\vec{R}_{\overset{\circ}{X}\overset{\circ}{X}}(\mu) = \left\| \begin{matrix} R_{\overset{\circ}{X}\overset{\circ}{X}}(0) & R_{\overset{\circ}{X}\overset{\circ}{X}}(\Delta t) & \ldots & R_{\overset{\circ}{X}\overset{\circ}{X}}[(N-1)\Delta t] \\ R_{\overset{\circ}{X}\overset{\circ}{X}}(\Delta t) & R_{\overset{\circ}{X}\overset{\circ}{X}}(0) & \ldots & R_{\overset{\circ}{X}\overset{\circ}{X}}[(N-2)\Delta t] \\ \ldots & \ldots & \ldots & \ldots \\ R_{\overset{\circ}{X}\overset{\circ}{X}}[(N-1)\Delta t] & R_{\overset{\circ}{X}\overset{\circ}{X}}[(N-2)\Delta t] & \ldots & R_{\overset{\circ}{X}\overset{\circ}{X}}(0) \end{matrix} \right\|, \quad (12)$$

$$\vec{R}_{\overset{\circ}{X}\overset{\circ}{Y}}(\mu) = \left[R_{\overset{\circ}{X}\overset{\circ}{Y}}(0) \quad R_{\overset{\circ}{X}\overset{\circ}{Y}}(\Delta t) \quad \ldots \quad R_{\overset{\circ}{X}\overset{\circ}{Y}}[(N-1)\Delta t] \right]^{\mathrm{T}}, \quad (13)$$

$$\vec{W}(\mu) = [\, W(0) \quad W(\Delta t) \quad \ldots \quad W((N-1)\Delta t) \,]^{\mathrm{T}}, \quad (14)$$

where

$$R_{\overset{\circ}{X}\overset{\circ}{X}}(\mu) = \frac{1}{N} \sum_{k=1}^{N} \overset{\circ}{X}(k\Delta t) \, \overset{\circ}{X}((k+\mu)\Delta t), \quad (15)$$

$$R_{\overset{\circ}{X}\overset{\circ}{Y}}(\mu) = \frac{1}{N} \sum_{k=1}^{N} \overset{\circ}{X}(k\Delta t) \, \overset{\circ}{Y}((k+\mu)\Delta t). \quad (16)$$

Taking into account that the conditions (3)–(5) hold true for the noisy signals $g(t)$ and $\eta(t)$ and the noises $\varepsilon(t)$ and $\varphi(t)$ and the mean-square value of the noise $\varepsilon(t)$ equals the noise variance

$$\frac{1}{N} \sum_{k=1}^{N} \overset{\circ}{\varepsilon}(k\Delta t) \, \overset{\circ}{\varepsilon}(k\Delta t) = D(\varepsilon), \quad (17)$$

the values of autocorrelation functions and cross-correlation functions take the following form:

$$R_{\overset{\circ}{g}\overset{\circ}{g}}(\mu) = \begin{cases} R_{\overset{\circ}{X}\overset{\circ}{X}}(0) + D(\varepsilon) & \text{when} \quad \mu = 0 \\ R_{\overset{\circ}{X}\overset{\circ}{X}}(\mu) & \text{when} \quad \mu \neq 0 \end{cases}, \quad (18)$$

$$R_{\overset{\circ}{g}\overset{\circ}{\eta}}(\mu) = \frac{1}{N}\sum_{k=1}^{N} \overset{\circ}{g}(k\Delta t)\,\overset{\circ}{\eta}((k+\mu)\Delta t) \approx R_{\overset{\circ}{X}\overset{\circ}{Y}}(\mu).\tag{19}$$

Then, the correlation matrices (12) and (13) $\vec{R}_{\overset{\circ}{g}\overset{\circ}{g}}(\mu)$ and $\vec{R}_{\overset{\circ}{g}\overset{\circ}{\eta}}(\mu)$, whose elements are the estimates of the correlation functions $R_{\overset{\circ}{g}\overset{\circ}{g}}(\mu)$ and $R_{\overset{\circ}{g}\overset{\circ}{\eta}}(\mu)$ of the noisy signals $g(t)$ and $\eta(t)$, are transformed as follows:

$$\vec{R}_{\overset{\circ}{g}\overset{\circ}{g}}(\mu) = \left\| \begin{array}{cccc} R_{\overset{\circ}{g}\overset{\circ}{g}}(0) & R_{\overset{\circ}{g}\overset{\circ}{g}}(\Delta t) & \cdots & R_{\overset{\circ}{g}\overset{\circ}{g}}[(N-1)\Delta t] \\ R_{\overset{\circ}{g}\overset{\circ}{g}}(\Delta t) & R_{\overset{\circ}{g}\overset{\circ}{g}}(0) & \cdots & R_{\overset{\circ}{g}\overset{\circ}{g}}[(N-2)\Delta t] \\ \cdots & \cdots & \cdots & \cdots \\ R_{\overset{\circ}{g}\overset{\circ}{g}}[(N-1)\Delta t] & R_{\overset{\circ}{g}\overset{\circ}{g}}[(N-2)\Delta t] & \cdots & R_{\overset{\circ}{g}\overset{\circ}{g}}(0) \end{array} \right\|$$

$$\approx \left\| \begin{array}{cccc} R_{\overset{\circ}{X}\overset{\circ}{X}}(0)+D(\varepsilon) & R_{\overset{\circ}{X}\overset{\circ}{X}}(\Delta t) & \cdots & R_{\overset{\circ}{X}\overset{\circ}{X}}[(N-1)\Delta t] \\ R_{\overset{\circ}{X}\overset{\circ}{X}}(\Delta t) & R_{\overset{\circ}{X}\overset{\circ}{X}}(0)+D(\varepsilon) & \cdots & R_{\overset{\circ}{X}\overset{\circ}{X}}[(N-2)\Delta t] \\ \cdots & \cdots & \cdots & \cdots \\ R_{\overset{\circ}{X}\overset{\circ}{X}}[(N-1)\Delta t] & R_{\overset{\circ}{X}\overset{\circ}{X}}[(N-2)\Delta t] & \cdots & R_{\overset{\circ}{X}\overset{\circ}{X}}(0)+D(\varepsilon) \end{array} \right\|,\tag{20}$$

$$\vec{R}_{\overset{\circ}{g}\overset{\circ}{\eta}}(\mu) \approx \vec{R}_{\overset{\circ}{X}\overset{\circ}{Y}}(\mu).\tag{21}$$

Thus, when the classical conditions (3)–(5) hold true, the matrix $\vec{R}_{\overset{\circ}{g}\overset{\circ}{\eta}}(\mu)$ of cross-correlation functions between the noisy input signal $g(t)$ and the noisy output signal $\eta(t)$ coincides with the correlation matrix $\vec{R}_{\overset{\circ}{X}\overset{\circ}{Y}}(\mu)$ of the useful signals $X(t)$ and $Y(t)$. On the other hand, the matrix $\vec{R}_{\overset{\circ}{g}\overset{\circ}{g}}(\mu)$ of the noisy input signal $g(t)$ differs from the correlation matrix $\vec{R}_{\overset{\circ}{X}\overset{\circ}{X}}(\mu)$ of the useful input signal $X(t)$ by diagonal elements, which, alongside with the estimates of the correlation function $R_{\overset{\circ}{X}\overset{\circ}{X}}(0)$, also contain the value of the noise variance $D(\varepsilon)$.

Obviously, that necessitates eliminating the influence of the variance $D(\varepsilon)$ of the noise $\varepsilon(t)$ and obtaining the robust correlation matrix $\vec{R}^{R}_{\overset{\circ}{g}\overset{\circ}{g}}(\mu)$, coinciding with the correlation matrix $\vec{R}_{\overset{\circ}{X}\overset{\circ}{X}}(\mu)$ of the useful signals, i.e., making the following condition hold true:

$$\vec{R}^{R}_{\overset{\circ}{g}\overset{\circ}{g}}(\mu) \approx \vec{R}_{\overset{\circ}{X}\overset{\circ}{X}}(\mu)\tag{22}$$

For this purpose, a robust technology for improving conditionality of correlation matrices is proposed in the following paragraphs.

1. Values of the autocorrelation function are determined for the noisy input signal $g(t)$ and the output signal $\eta(t)$:

$$R_{\overset{\circ}{g}\overset{\circ}{g}}(\mu) = \frac{1}{N}\sum_{k=1}^{N}\overset{\circ}{g}(k\Delta t)\,\overset{\circ}{g}((k+\mu)\Delta t). \tag{23}$$

2. The noise variance $D^*(\varepsilon)$ is determined for the noisy input signal $\overset{\circ}{g}(t)$ [10]:

$$D^*(\varepsilon_i) = \frac{1}{N}\sum_{k=1}^{N}[Z(0) - 2Z(1) + Z(2)], \tag{24}$$

where

$$Z(0) = \overset{\circ}{g}_i(k\Delta t)\,\overset{\circ}{g}_i(k\Delta t),$$
$$Z(1) = \overset{\circ}{g}_i(k\Delta t)\,\overset{\circ}{g}_i((k+1)\Delta t),$$
$$Z(2) = \overset{\circ}{g}_i(k\Delta t)\,\overset{\circ}{g}_i((k+2)\Delta t).$$

3. The robust estimates $R^R_{\overset{\circ}{g}\overset{\circ}{g}}(0)$ of the elements of the correlation matrix are calculated:

$$R^R_{\overset{\circ}{g}\overset{\circ}{g}}(0) = R_{\overset{\circ}{g}\overset{\circ}{g}}(0) - D^*(\varepsilon).$$

4. The robust correlation matrix $\vec{R}^R_{\overset{\circ}{g}\overset{\circ}{g}}(\mu)$ is formed:

$$\vec{R}^R_{\overset{\circ}{g}\overset{\circ}{g}}(\mu) \approx \left\|\begin{array}{cccc} R_{\overset{\circ}{g}\overset{\circ}{g}}(0) - D^*(\varepsilon) & R_{\overset{\circ}{X}\overset{\circ}{X}}(\Delta t) & \ldots & R_{\overset{\circ}{X}\overset{\circ}{X}}[(N-1)\Delta t] \\ R_{\overset{\circ}{X}\overset{\circ}{X}}(\Delta t) & R_{\overset{\circ}{g}\overset{\circ}{g}}(0) - D^*(\varepsilon) & \ldots & R_{\overset{\circ}{X}\overset{\circ}{X}}[(N-2)\Delta t] \\ \ldots & \ldots & \ldots & \ldots \\ R_{\overset{\circ}{X}\overset{\circ}{X}}[(N-1)\Delta t] & R_{\overset{\circ}{X}\overset{\circ}{X}}[(N-2)\Delta t] & \ldots & R_{\overset{\circ}{g}\overset{\circ}{g}}(0) - D^*(\varepsilon) \end{array}\right\|. \tag{25}$$

Thus, the robust technology for improving conditionality of correlation matrices allows one, by eliminating the influence of the noise variance, to transform the matrix of noisy technological parameters into the form similar to the matrix, whose elements contain no error from noises, i.e., to obtain a matrix, for which equality (22) holds true.

However, input and output parameters of offshore platforms, compressor stations, and drilling units are various physical quantities (e.g., consumption, pressure, temperature, and velocity) with different range of value change. Therefore, solving the identification problem first of all requires their reduction to a single dimensionless value. As a result of that nondimensionalization, we obtain normalized correlation matrices. The following conversion formulas are used for this purpose:

$$r_{\overset{\circ}{X}\overset{\circ}{X}}(\mu) = R_{\overset{\circ}{X}\overset{\circ}{X}}(\mu)/D(X), \tag{26}$$

$$r_{\overset{\circ}{X}\overset{\circ}{Y}}(\mu) = R_{\overset{\circ}{X}\overset{\circ}{Y}}(\mu)/\sqrt{D(X)\cdot D(Y)}, \tag{27}$$

$$r_{\overset{\circ}{g}\overset{\circ}{g}}(\mu) = R_{\overset{\circ}{g}\overset{\circ}{g}}(\mu)/D(g), \tag{28}$$

$$r_{\overset{\circ}{g}\overset{\circ}{\eta}}(\mu) = R_{\overset{\circ}{g}\overset{\circ}{\eta}}(\mu)/\sqrt{D(g)\cdot D(\eta)}, \tag{29}$$

where $D(X), D(Y), D(g)$ and $D(\eta)$ are the variances of the signals $X(k\Delta t), Y(k\Delta t), g(t),$ and $\eta(t)$:

$$D(X) = \frac{1}{N}\sum_{k=1}^{N} \overset{\circ}{X}(k\Delta t)\, \overset{\circ}{X}(k\Delta t), \tag{30}$$

$$D(Y) = \frac{1}{N}\sum_{k=1}^{N} \overset{\circ}{Y}(k\Delta t)\, \overset{\circ}{Y}(k\Delta t), \tag{31}$$

$$D(g) = \frac{1}{N}\sum_{k=1}^{N} \overset{\circ}{g}(k\Delta t)\, \overset{\circ}{g}(k\Delta t), \tag{32}$$

$$D(\eta) = \frac{1}{N}\sum_{k=1}^{N} \overset{\circ}{\eta}(k\Delta t)\, \overset{\circ}{\eta}(k\Delta t), \tag{33}$$

or in view of the conditions of noncorrelatedness (3):

$$D(g) = \frac{1}{N}\sum_{k=1}^{N} \overset{\circ}{g}(k\Delta t)\, \overset{\circ}{g}(k\Delta t) = D(X) + D(\varepsilon), \tag{34}$$

$$D(\eta) = \frac{1}{N}\sum_{k=1}^{N} \overset{\circ}{\eta}(k\Delta t)\, \overset{\circ}{\eta}(k\Delta t) = D(Y) + D(\varphi), \tag{35}$$

where $D(\varphi)$ is the variance of the noise $\varphi(k\Delta t)$:

$$D(\varphi) = \frac{1}{N}\sum_{k=1}^{N} \overset{\circ}{\varphi}(k\Delta t)\, \overset{\circ}{\varphi}(k\Delta t), \tag{36}$$

Then, the normalized correlation matrices are of the following form:

$$\vec{r}_{\overset{\circ}{X}\overset{\circ}{X}}(\mu) = \left\| \begin{array}{cccc} 1 & \dfrac{R_{\overset{\circ}{X}\overset{\circ}{X}}(\Delta t)}{D(X)} & \cdots & \dfrac{R_{\overset{\circ}{X}\overset{\circ}{X}}[(N-1)\Delta t]}{D(X)} \\[2ex] \dfrac{R_{\overset{\circ}{X}\overset{\circ}{X}}(\Delta t)}{D(X)} & 1 & \cdots & \dfrac{R_{\overset{\circ}{X}\overset{\circ}{X}}[(N-2)\Delta t]}{D(X)} \\[2ex] \cdots & \cdots & \cdots & \cdots \\[1ex] \dfrac{R_{\overset{\circ}{X}\overset{\circ}{X}}[(N-1)\Delta t]}{D(X)} & \dfrac{R_{\overset{\circ}{X}\overset{\circ}{X}}[(N-2)\Delta t]}{D(X)} & \cdots & 1 \end{array} \right\|, \tag{37}$$

$$\vec{r}_{\overset{\circ}{X}\overset{\circ}{Y}}(\mu) = \left[\dfrac{R_{\overset{\circ}{X}\overset{\circ}{Y}}(0)}{\left(\sqrt{D(X)D(Y)}\right)} \quad \dfrac{R_{\overset{\circ}{X}\overset{\circ}{Y}}(\Delta t)}{\left(\sqrt{D(X)D(Y)}\right)} \quad \cdots \quad \dfrac{R_{\overset{\circ}{X}\overset{\circ}{Y}}[(N-1)\Delta t]}{\left(\sqrt{D(X)D(Y)}\right)} \right]^{\mathrm{T}} \tag{38}$$

The corresponding normalized correlation matrices of the noisy signals are transformed into the following form:

$$\vec{r}_{\overset{\circ}{g}\overset{\circ}{g}}(\mu) = \left\| \begin{array}{cccc} \dfrac{R_{\overset{\circ}{g}\overset{\circ}{g}}(0)}{D(g)} & \dfrac{R_{\overset{\circ}{g}\overset{\circ}{g}}(\Delta t)}{D(g)} & \cdots & \dfrac{R_{\overset{\circ}{g}\overset{\circ}{g}}[(N-1)\Delta t]}{D(g)} \\[2ex] \dfrac{R_{\overset{\circ}{g}\overset{\circ}{g}}(\Delta t)}{D(g)} & \dfrac{R_{\overset{\circ}{g}\overset{\circ}{g}}(0)}{D(g)} & \cdots & \dfrac{R_{\overset{\circ}{g}\overset{\circ}{g}}[(N-2)\Delta t]}{D(g)} \\[2ex] \cdots & \cdots & \cdots & \cdots \\[1ex] \dfrac{R_{\overset{\circ}{g}\overset{\circ}{g}}[(N-1)\Delta t]}{D(g)} & \sqrt{\dfrac{R_{\overset{\circ}{g}\overset{\circ}{g}}[(N-2)\Delta t]}{D(g)}} & \cdots & \sqrt{\dfrac{R_{\overset{\circ}{g}\overset{\circ}{g}}(0)}{D(g)}} \end{array} \right\|$$

$$\approx \left\| \begin{array}{cccc} 1 & \dfrac{R_{\overset{\circ}{g}\overset{\circ}{g}}(\Delta t)}{D(X)+D(\varepsilon)} & \cdots & \dfrac{R_{\overset{\circ}{g}\overset{\circ}{g}}[(N-1)\Delta t]}{D(X)+D(\varepsilon)} \\[2ex] \dfrac{R_{\overset{\circ}{g}\overset{\circ}{g}}(\Delta t)}{D(X)+D(\varepsilon)} & 1 & \cdots & \dfrac{R_{\overset{\circ}{g}\overset{\circ}{g}}[(N-2)\Delta t]}{D(X)+D(\varepsilon)} \\[2ex] \cdots & \cdots & \cdots & \cdots \\[1ex] \dfrac{R_{\overset{\circ}{g}\overset{\circ}{g}}[(N-1)\Delta t]}{D(X)+D(\varepsilon)} & \sqrt{\dfrac{R_{\overset{\circ}{g}\overset{\circ}{g}}[(N-2)\Delta t]}{D(X)+D(\varepsilon)}} & \cdots & 1 \end{array} \right\| \tag{39}$$

$$\vec{r}_{\overset{\circ}{g}\overset{\circ}{\eta}}(\mu) \approx \left[\dfrac{R_{\overset{\circ}{g}\overset{\circ}{\eta}}(0)}{\sqrt{D(g)D(\eta)}} \quad \dfrac{R_{\overset{\circ}{g}\overset{\circ}{\eta}}(\Delta t)}{\sqrt{D(g)D(\eta)}} \quad \cdots \quad \dfrac{R_{\overset{\circ}{g}\overset{\circ}{\eta}}[(N-1)\Delta t]}{\sqrt{D(g)D(\eta)}} \right]^{\mathrm{T}}$$

$$= \left[\dfrac{R_{\overset{\circ}{X}\overset{\circ}{Y}}(0)}{\sqrt{A(\varepsilon)A(\varphi)}} \quad \dfrac{R_{\overset{\circ}{X}\overset{\circ}{Y}}(\Delta t)}{\sqrt{A(\varepsilon)A(\varphi)}} \quad \cdots \quad \dfrac{R_{\overset{\circ}{X}\overset{\circ}{Y}}[(N-1)\Delta t]}{\sqrt{A(\varepsilon)A(\varphi)}} \right], \tag{40}$$

where

$$A(\varepsilon) = D(X) + D(\varepsilon),$$
$$A(\varphi) = D(Y) + D(\varphi).$$

Thus, in identification problems [8–10] of identification of the latent period of offshore platforms, compressor stations, and drilling units to the emergency state, diagonal elements of the normalized correlation matrix $\vec{r}_{\overset{\circ}{g}\overset{\circ}{g}}(\mu)$ of the noisy signals $g(t)$ coincide with those of the normalized correlation matrix $\vec{r}_{\overset{\circ}{X}\overset{\circ}{X}}(\mu)$ of the useful signals $X(t)$ and are equal to unit. However, the remaining elements of the normalized correlation matrix $\vec{r}_{\overset{\circ}{g}\overset{\circ}{g}}(\mu)$ of the input signal, as well as all the elements of normalized cross-correlation matrix $\vec{r}_{\overset{\circ}{g}\overset{\circ}{\eta}}(\mu)$ of the noisy input and output signals, contain in the radical expression of the denominator the values of the variances $D(X)$ and $D(Y)$ of the useful signals $X(t)$, $Y(t)$ and the values of the variances $D(\varepsilon)$ and $D(\varphi)$ of the noises $\varepsilon(t)$ and $\varphi(t)$. Consequently, normalization causes additional errors in the elements of normalized correlation matrices. To eliminate the said shortcoming, a robust technology for improving conditionality of normalized correlation matrices is proposed in the following paragraphs.

1. Estimates of the auto- and cross-correlation functions are determined for the noisy input signal $g(t)$ and the output signal $\eta(t)$:

$$R_{\overset{\circ}{g}\overset{\circ}{g}}(\mu) = \frac{1}{N}\sum_{k=1}^{N}\overset{\circ}{g}(k\Delta t)\,\overset{\circ}{g}((k+\mu)\Delta t), \tag{41}$$

$$R_{\overset{\circ}{g}\overset{\circ}{\eta}}(\mu) = \frac{1}{N}\sum_{k=1}^{N}\overset{\circ}{g}(k\Delta t)\,\overset{\circ}{\eta}((k+\mu)\Delta t). \tag{42}$$

2. The noise variances $D^*(\varepsilon)$ and $D^*(\varphi)$ are determined for the noisy input signal $\overset{\circ}{g}(t)$ and the output signal $\overset{\circ}{\eta}(t)$ from (24) and [6, 10]:

$$D^*(\varphi) = \frac{1}{N}\sum_{k=1}^{N}[H(0) - 2H(1) + H(2)], \tag{43}$$

where

$$H(0) = \overset{\circ}{\eta}(k\Delta t)\,\overset{\circ}{\eta}(k\Delta t),$$
$$H(1) = \overset{\circ}{\eta}(k\Delta t)\,\overset{\circ}{\eta}((k+1)\Delta t),$$
$$H(2) = \overset{\circ}{\eta}(k\Delta t)\,\overset{\circ}{\eta}((k+2)\Delta t).$$

3. The robust estimates $r^R_{\overset{\circ}{g}\overset{\circ}{g}}(\mu)$ and $r^R_{\overset{\circ}{g}\overset{\circ}{\eta}}(\mu)$ of the elements of the normalized correlation matrices are calculated:

$$\left.\begin{array}{l} R^R_{\overset{\circ}{g}\overset{\circ}{g}}(\mu) = \dfrac{R_{\overset{\circ}{g}\overset{\circ}{g}}(\mu)}{(D(g)-D^*(\varepsilon))}, \quad \mu \neq 0 \\[2ex] r^R_{\overset{\circ}{g}\overset{\circ}{\eta}}(\mu) = \dfrac{R_{\overset{\circ}{g}\overset{\circ}{\eta}}(\mu)}{\sqrt{(D(g)-D^*(\varepsilon))(D(\eta)-D^*(\varphi))}} \end{array}\right\}, \tag{44}$$

4. The robust normalized correlation matrices $\vec{r}^{R}_{\underset{gg}{\circ\,\circ}}(\mu)$ and $\vec{r}^{R}_{\underset{g\eta}{\circ\,\circ}}(\mu)$ are formed:

$$
\vec{r}^{R}_{\underset{gg}{\circ\,\circ}}(\mu) = \left\| \begin{array}{cccc} 1 & \dfrac{R_{\underset{gg}{\circ\circ}}(\Delta t)}{(D(g)-D^*(\varepsilon))} & \cdots & \dfrac{R_{\underset{gg}{\circ\circ}}[(N-1)\Delta t]}{(D(g)-D^*(\varepsilon))} \\ R_{\underset{gg}{\circ\circ}}(\Delta t) & 1 & \cdots & \dfrac{R_{\underset{gg}{\circ\circ}}[(N-2)\Delta t]}{[D(g)-D^*(\varepsilon)]} \\ \cdots & \cdots & \cdots & \cdots \\ \dfrac{R_{\underset{gg}{\circ\circ}}[(N-1)\Delta t]}{(D(g)-D^*(\varepsilon))} & \dfrac{R_{\underset{gg}{\circ\circ}}[(N-2)\Delta t]}{(D(g)-D^*(\varepsilon))} & \cdots & 1 \end{array} \right\|. \tag{45}
$$

$$
\vec{r}^{R}_{\underset{g\eta}{\circ\,\circ}}(\mu) = \left[\dfrac{R_{\underset{g\eta}{\circ\circ}}(0)}{\sqrt{A^*(\varepsilon)A^*(\varphi)}} \quad \dfrac{R_{\underset{g\eta}{\circ\circ}}(\Delta t)}{\sqrt{A^*(\varepsilon)A^*(\varphi)}} \quad \cdots \quad \dfrac{R_{\underset{g\eta}{\circ\circ}}[(N-1)\Delta t]}{\sqrt{A^*(\varepsilon)A^*(\varphi)}} \right], \tag{46}
$$

where

$$
A^*(\varepsilon) = D(g) - D^*(\varepsilon),
$$
$$
A^*(\varphi) = D(\eta) - D^*(\varphi).
$$

Thus, the robust technology for improving conditionality of normalized correlation matrices allows one, by eliminating the influence of the noise characteristics, to transform the source matrices into the form similar to the matrix, whose elements contain no errors from noises, i.e., to obtain a matrix, for which the following equality holds true:

$$
\vec{r}^{R}_{\underset{gg}{\circ\,\circ}}(\mu) \approx \vec{r}_{\underset{XX}{\circ\,\circ}}(\mu), \tag{47}
$$

$$
\vec{r}^{R}_{\underset{g\eta}{\circ\,\circ}}(\mu) \approx \vec{r}_{\underset{XY}{\circ\,\circ}}(\mu). \tag{48}
$$

In this manner, by means of equalities (47) and (48), we make it possible to solve the problem of identification of the transition of offshore platforms, compressor stations, and drilling units from the normal state to the emergency state. This is due to that fact that the matrices $\vec{R}^{R}_{\underset{gg}{\circ\,\circ}}(\mu), \vec{r}^{R}_{\underset{gg}{\circ\,\circ}}(\mu)$ and $\vec{r}^{R}_{\underset{g\eta}{\circ\,\circ}}(\mu)$, compared with the matrices $\vec{R}_{\underset{gg}{\circ\,\circ}}(\mu), \vec{r}_{\underset{gg}{\circ\,\circ}}(\mu)$, and $\vec{r}_{\underset{g\eta}{\circ\,\circ}}(\mu)$, that appear in Eq. (9) contain significantly less amount of errors.

4 Technology for Indicating the Transition of Industrial Objects to the Emergency State in the Beginning of Its Latent Period

In the following paragraphs, we propose a technology that with a sufficient degree of reliability gives us the full picture of the processes at the starting point of the time interval T_1, when the object enters its emergency state. For this purpose, provision is made for application of the estimates of statistical characteristics of technological parameters.

Assume that the signals $g_1(t), g_2(t), \ldots, g_n(t)$ characterizing the technical condition of the object come from the sensors installed in the vulnerable locations of that industrial object. It follows from expressions (10) and (11) that as the technical condition of an object changes, equalities (10) are violated, with estimates of the noise variances $D(\varepsilon_i)$ and $i = \overline{1, n}$ and the useful signal variance $D(X_i)$ changing in the first place. These estimates can be determined from the following expressions:

$$D^*(\varepsilon_i) = \frac{1}{N} \sum_{k=1}^{N} [Z(0) - 2Z(1) + Z(2)], \tag{49}$$

$$D^*(X_i) = D(g_i) - D(\varepsilon_i). \tag{50}$$

Let us now consider the specifics of calculating the estimates of the auto- and cross-correlation functions $R_{\overset{\circ}{g_i}\overset{\circ}{g_i}}(\mu)$ and $R_{\overset{\circ}{g_i}\overset{\circ}{g_j}}(\mu)$ of the noisy signals $g_i(t), g_j(t)$ and $i, j = \overline{1, n}$ in the periods T_0 and T_1. It is known that as the time shift between $\overset{\circ}{g_i}(k\Delta t)$ and $\overset{\circ}{g_i}((k + \mu)\Delta t)$ grows in the process of calculating those estimates, a moment comes between $\overset{\circ}{g_i}(k\Delta t)$ and $\overset{\circ}{g_j}((k + \mu)\Delta t)$ when the estimates are equal to zero. If we denote that time shift by μ', then the errors of the estimates are obtained from the expressions:

$$R_{\overset{\circ}{g_i}\overset{\circ}{g_i}}(\mu') = \frac{1}{N} \sum_{k=1}^{N} \overset{\circ}{g_i}(k\Delta t) \, \overset{\circ}{g_i}((k + \mu)\Delta t) \approx 0, \tag{51}$$

$$R_{\overset{\circ}{g_i}\overset{\circ}{g_j}}(\mu') = \frac{1}{N} \sum_{k=1}^{N} \overset{\circ}{g_i}(k\Delta t) \, \overset{\circ}{g_j}((k + \mu)\Delta t) \approx 0, \tag{52}$$

which takes the minimum values.

This means that in the stable normal operation mode of an object in the period T_0, various errors unrelated to the beginning of change in the technical condition of an object affect all the estimates of $R_{\overset{\circ}{g_i}\overset{\circ}{g_i}}(\mu)$ and $R_{\overset{\circ}{g}\overset{\circ}{g}}_{i\ j}(\mu)$ at various values of μ, with the exception of the estimates of $R_{\overset{\circ}{g_i}\overset{\circ}{g_i}}(\mu')$ and $R_{\overset{\circ}{g}\overset{\circ}{g}}_{i\ j}(\mu')$.

Thus, both a combination of estimates of the autocorrelation indicators $R_{\overset{\circ}{g_1}\overset{\circ}{g_1}}(\mu'), R_{\overset{\circ}{g_2}\overset{\circ}{g_2}}(\mu'), \ldots, R_{\overset{\circ}{g_i}\overset{\circ}{g_i}}(\mu'), \ldots, R_{\overset{\circ}{g_n}\overset{\circ}{g_n}}(\mu')$ and a combination of estimates of the cross-correlation indicators $R_{\overset{\circ}{g_1}\overset{\circ}{g_2}}(\mu'), R_{\overset{\circ}{g_1}\overset{\circ}{g_3}}(\mu') \ldots, R_{\overset{\circ}{g_1}\overset{\circ}{g_n}}(\mu'), R_{\overset{\circ}{g_2}\overset{\circ}{g_1}}(\mu'),$ $R_{\overset{\circ}{g_2}\overset{\circ}{g_3}}(\mu') \ldots, R_{\overset{\circ}{g_2}\overset{\circ}{g_n}}(\mu'), \ldots R_{\overset{\circ}{g_n}\overset{\circ}{g_1}}(\mu'), R_{\overset{\circ}{g_n}\overset{\circ}{g_2}}(\mu'), \ldots, R_{\overset{\circ}{g_n}\overset{\circ}{g_{n-1}}}(\mu')$ are formed from the characteristics of the signals $g_1(t), g_2(t), \ldots, g_n(t)$ during the operation of an industrial object. Equalities (10) will be violated in the beginning of the latent period of the transition of an object to the emergency state, and many of those estimates will be instantaneously different from zero, which allows using them as reliable indicators.

Let us now consider the possibility of applying estimates of the cross-correlation function $R^*_{X_i\varepsilon_i}(\mu = 0)$ and the coefficient of correlation $r^*_{X_i\varepsilon_i}$ between the useful signal $X_i(t)$ and the noise $\varepsilon_i(t)$ for solving the above-mentioned problems by means of the following formula:

$$R^*_{X_i\varepsilon_i}(\mu = 0) \approx \frac{1}{2}[R(0) - [R(1) + R(2) - R(3)] - R\varepsilon(0)], \qquad (53)$$

where

$$R(0) = R_{\overset{\circ}{g_i}\overset{\circ}{g_i}}(\mu = 0),$$
$$R(1) = R_{\overset{\circ}{g_i}\overset{\circ}{g_i}}(\mu = 1),$$
$$R(2) = R_{\overset{\circ}{g_i}\overset{\circ}{g_i}}(\mu = 2),$$
$$R(3) = R_{\overset{\circ}{g_i}\overset{\circ}{g_i}}(\mu = 3),$$
$$R\varepsilon(0) = R_{\overset{\circ}{\varepsilon_i}\overset{\circ}{\varepsilon_i}}(\mu = 0).$$

In this expression, the estimates $R_{\overset{\circ}{g_i}\overset{\circ}{g_i}}(\mu = 0), R_{\overset{\circ}{g_i}\overset{\circ}{g_i}}(\mu = 1), R_{\overset{\circ}{g_i}\overset{\circ}{g_i}}(\mu = 3)$ are calculated using the conventional algorithm, and the estimate $R_{\overset{\circ}{\varepsilon_i}\overset{\circ}{\varepsilon_i}}(\mu = 0) = D(\varepsilon_i)$ is quite easily calculated from expression (49). Formula (53) can therefore be also considered a reliable and practically convenient indicator. After the estimate of the noise variance $D^*(\varepsilon_i)$ of the noisy signal $g_i(t)$ and the estimate of the cross-correlation function $R^*_{X_i\varepsilon_i}(\mu = 0)$ between the useful signal $X_i(t)$ and the noise $\varepsilon_i(t)$ are determined from expression (49) and expression (53), respectively, we can naturally calculate the estimate of the autocorrelation function $R^*_{\overset{\circ}{X_i}\overset{\circ}{X_i}}(\mu = 0)$ of the useful signal:

$$R^*_{\overset{\circ}{X_i}\overset{\circ}{X_i}}(\mu = 0) \approx R_{\overset{\circ}{g_i}\overset{\circ}{g_i}}(\mu = 0) - R^*_{\overset{\circ}{X_i}\varepsilon_i}(\mu = 0) - D^*(\varepsilon_i). \tag{54}$$

Then, we can calculate the coefficient of correlation $r^*_{X_i\varepsilon_i}$ between the useful signal and the noise quite easily, using the following formula:

$$r^*_{X_i\varepsilon_i} \approx \frac{R^*_{\overset{\circ}{X_i}\varepsilon_i}(\mu = 0)}{\sqrt{R^*_{\overset{\circ}{X_i}\overset{\circ}{X_i}}(\mu = 0) \cdot D^*(\varepsilon_i)}} \tag{55}$$

It is clear that during the operation of an object, using the characteristics of the technological parameters determined from expressions (54) and (55), we can form combinations of estimates of both the cross-correlation indicators $R^*_{X_1\varepsilon_1}(\mu = 0), R^*_{X_2\varepsilon_2}(\mu = 0), \ldots R^*_{X_n\varepsilon_n}(\mu = 0)$ and the coefficients of correlation $r^*_{X_1\varepsilon_1}, r^*_{X_2\varepsilon_2}, \ldots,$ $r^*_{X_n\varepsilon_n}$ that will be different from zero during the transition of the object to the emergency state.

It thus follows that to indicate the beginning of the transition of offshore platforms, compressor stations, and drilling units from the normal state to the emergency state, it is sufficient to calculate the estimates of the noise variances $D^*(\varepsilon_i)$, auto- and cross-correlation functions $R_{\overset{\circ}{g_i}\overset{\circ}{g_i}}(\mu')$ and $R_{\overset{\circ}{g_i}\overset{\circ}{g_j}}(\mu')$ of the noisy signals $g_i(t)$ at the time shift μ', as well as the estimates of the cross-correlation function $R^*_{X_i\varepsilon_i}(\mu = 0)$ and the coefficient of correlation $r^*_{X_i\varepsilon_i}$ between the useful signal $X_i(t)$ and the noise $\varepsilon_i(t)$ from expressions (49)–(55). After this point, we should form the corresponding sets of informative attributes from the obtained estimates:

$$V_\varepsilon = [D^*(\varepsilon_1) \quad D^*(\varepsilon_2) \quad \ldots \quad D^*(\varepsilon_n)], \tag{56}$$

$$V_g = \begin{bmatrix} R_{\overset{\circ}{g_1}\overset{\circ}{g_1}}(\mu') & R_{\overset{\circ}{g_1}\overset{\circ}{g_2}}(\mu') & \ldots & R_{\overset{\circ}{g_1}\overset{\circ}{g_n}}(\mu') \\ R_{\overset{\circ}{g_2}\overset{\circ}{g_1}}(\mu') & R_{\overset{\circ}{g_2}\overset{\circ}{g_2}}(\mu') & \ldots & R_{\overset{\circ}{g_2}\overset{\circ}{g_n}}(\mu') \\ \ldots & \ldots & \ldots & \ldots \\ R_{\overset{\circ}{g_n}\overset{\circ}{g_1}}(\mu') & R_{\overset{\circ}{g_n}\overset{\circ}{g_2}}(\mu') & \ldots & R_{\overset{\circ}{g_n}\overset{\circ}{g_n}}(\mu') \end{bmatrix}, \tag{57}$$

$$V_{X\varepsilon} = \begin{bmatrix} R^*_{X_1\varepsilon_1}(\mu = 0) & R^*_{X_2\varepsilon_2}(\mu = 0) & \ldots & R^*_{X_n\varepsilon_n}(\mu = 0) \\ r^*_{X_1\varepsilon_1} & r^*_{X_2\varepsilon_2} & \ldots & r^*_{X_n\varepsilon_n} \end{bmatrix}. \tag{58}$$

It is obvious that while an object is in the normal technical condition, i.e., when all technological parameters are in the time interval T_0, all elements of the sets of informative attributes $V_\varepsilon, V_g, V_{X\varepsilon}$ will be in the zero state. Equalities (10) will be violated in the beginning of the latent period of transition to the emergency state, and we can determine the location and nature of the failure of the industrial object from the number of the set and the number of the nonzero informative attribute.

5 Conclusion

We demonstrate in this chapter that in the latent period of transition of offshore platforms, compressor stations, and drilling units to the emergency state, noise emerges in the signals due to such defects as tear and wear, cracks, bending, and fatigue; the coefficient of correlation between the useful signal and the noise, their spectrum, variance, and other characteristics change; the stationarity condition and the normal distribution law are violated. Diagnostic information is partially lost, and the useful signal is distorted due to the noise filtration in control system. As a result, the emergency state is detected only in its expressed form. It often proves to be belated, which causes numerous accidents. It is possible to use a matrix equation of type (1) to solve the problem of indicating the beginning of transition of objects to the emergency state. However, the technological parameters being analyzed are noisy signals. Therefore, the adequacy of solving the identification problem is not ensured. To eliminate this obstacle, this article proposes a technology for forming the robust normalized correlation matrices that allow improving the adequacy of the mentioned matrix equations.

Due to this fact, the adequacy of the obtained results improves. However, it is impossible to ensure the absolute adequacy of the obtained model with sufficient degree of reliability. We therefore also propose for wide practical applications a technology for indicating the beginning of the latent period of objects to the emergency state. The convenience of this technology is that its implementation in monitoring systems does not require involvement of highly qualified specialists, using estimates of the noise variance, cross-correlation function, and coefficient of correlation between the useful signal and the noise and the correlation indicators. During the operation of an object, the nonzero state of elements of the sets formed from those estimates is registered in the system as the beginning of the emergency state.

Acknowledgments The work has been fulfilled with the support of the Science Fund of the State Oil Company of the Azerbaijan Republic within the framework of the project "Developing a new generation control, diagnostics and management system based on the robust noise analysis of wattmeter cards of oil wells operated by sucker rod pumps."

References

1. Jin, J., Shi, J.: Feature-preserving data compression of stamping tonnage information using wavelets. Technometrics **41**, 327–339 (1999)
2. Yu, G., Zou, C., Wang, Z.: Outlier detection in functional observations with applications to profile monitoring. Technometrics **54**, 308–318 (2012)
3. Yu, J.: Machine tool condition monitoring based on an adaptive Gaussian mixture model. Trans. ASME J. Manuf. Sci. Eng. **134**(3), 13 (2012)

4. Caballero-Águila, R., García-Garrido, I., Linares-Pérez, J.: Optimal fusion filtering in multisensor stochastic systems with missing measurements and correlated noises. J. Math. Prob. Eng. **2013**, 14 p. (Article ID 418678) (2013)
5. Mouriño, H., Barão, M.I.: Maximum likelihood estimation of the VAR(1) model parameters with missing observations. J. Math. Prob. Eng. **2013**, 13 p. (Article ID 848120) (2013)
6. Musaeva, N.F.: Robust method of estimation with "contaminated" coarse errors. J. Autom. Control Comput. Sci. **37**(6), 50–63 (2003)
7. Wang, X., Sun, S.-L., Ding, K.-H., Xue, J.-Y.: Weighted measurement fusion white noise deconvolution filter with correlated noise for multisensor stochastic systems. J. Math. Prob. Eng. **2012**, 16 p. (Article ID 257619) (2012)
8. Aliev, T.A., Musaeva, N.F., Guluyev, G.A., Sattarova, U.E., Rzaeva, N.E.: System of monitoring of period of hidden transition of compressor station to emergency state. J. Autom. Inf. Sci. **43**, 66–81 (2011)
9. Aliev, T.A., Musaeva, N.F., Guluyev, G.A., Sattarova, U.E.: Noise technology of indication and identification of the latent period of transition of an object from a normal condition to an emergency one. J. Mech. Autom. Control **9**, 13–18 (2010)
10. Aliev, T.A.: Digital Noise Monitoring of Defect Origin. Springer, London (2007). 235 pp

Optimization of an Integrator to Control the Flight of an Airplane

Leticia Cervantes and Oscar Castillo

Abstract In this paper, we show a fuzzy control optimization using genetic algorithms, and this optimization helps us to improve the flight control of an airplane. To control the flight control of the airplane, fuzzy systems were used to control the stability of the airplane. In this paper, the fuzzy systems and the behavior of the airplane are explained to understand the complete work.

Keywords Type-1 fuzzy system · Fuzzy control · Genetic algorithm

1 Introduction

Many problems of control apply different techniques to improve the results. In this paper, the benchmark problems that is the flight control of a DeHavilland Beaver and the flight maneuvers of the plane are explained to understand the problem. The fuzzy systems used in this work have a reason to control the total flight, this will be explained in section problem description. In this work, we used first individual controllers or fuzzy systems to achieve the stability of the plane; fuzzy systems are used to control its stability, and an integrator is used to obtain a better control. Then, a genetic algorithm is applied to improve the fuzzy system and to achieve the total control.

The rest of the paper is organized as follows: In Sect. 2, we present some basic concepts to understand this work; in Sect. 3, we define the proposed method and problem description, and finally conclusions are presented in Sect. 4.

L. Cervantes (✉) · O. Castillo
Tijuana Institute of Technology, Tijuana, Mexico
e-mail: lettyy2685@hotmail.com

O. Castillo
e-mail: Ocastillo@tectijuana.mx

© Springer International Publishing Switzerland 2016   407
L.A. Zadeh et al. (eds.), *Recent Developments and New Direction
in Soft-Computing Foundations and Applications*, Studies in Fuzziness
and Soft Computing 342, DOI 10.1007/978-3-319-32229-2_28

2   Background and Basic Concepts

In this section, some basic concepts are provided to understand this work.

2.1   Genetic Algorithm

Genetic algorithms (GAs) are numerical optimization algorithms inspired by both natural selection and genetics. We can also say that the genetic algorithm is an optimization and search technique based on the principles of genetics and natural selection. A GA allows a population composed of many individuals to evolve under specified selection rules to a state that maximizes the "fitness" [1]. The method is a general one, capable of being applied to an extremely wide range of problems. The algorithms are simple to understand and the required computer code easy to write.

GAs were in essence proposed by John Holland in the 1960s. His reasons for developing such algorithms went far beyond the type of problem solving with which this work is concerned. His book (1975), Adaptation in Natural and Artificial Systems, is particularly worth reading for its visionary approach. More recently others, for example De Jong, in a paper entitled GAs are NOT Function Optimizers, have been keen to remind us that GAs are potentially far more than just a robust method for estimating a series of unknown parameters within a model of a physical system [2].

A typical algorithm might consist of the following:

1. Start with a randomly generated population of n 1 − bit chromosomes (candidate solutions to a problem).
2. Calculate the fitness $f(x)$ of each chromosome x in the population.
3. Repeat the following steps until n offspring have been created:

   - Select a pair of parent chromosomes from the current population, the probability of selection being an increasing function of fitness. Selection is done "with replacement," meaning that the same chromosome can be selected more than once to become a parent.
   - With probability Pc (the "crossover probability" or "crossover rate"), cross over the pair at a randomly chosen point (chosen with uniform probability) to form two offspring. If no crossover takes place, form two offspring that are exact copies of their respective parents. (Note that here the crossover rate is defined to be the probability that two parents will cross over in a single point. There are also "multipoint crossover" versions of the GA in which the crossover rate for a pair of parents is the number of points at which a crossover takes place.)
   - Mutate the two offspring at each locus with probability Pm (the mutation probability or mutation rate) and place the resulting chromosomes in the new population. If n is odd, one new population member can be discarded at random.

- Replace the current population with the new population.
- Go to step 2 [3].

Some of the advantages of a GA include the following:

- Optimizes with continuous or discrete variables,
- Does not require derivative information,
- Simultaneously searches from a wide sampling of the cost surface,
- Deals with a large number of variables,
- Is well suited for parallel computers,
- Optimizes variables with extremely complex cost surfaces (they can jump out of a local minimum),
- Provides a list of optimal values for the variables, not just a single solution,
- Codification of the variables so that the optimization is done with the encoded variables, and
- Works with numerically generated data, experimental data, or analytical functions [2, 4].

2.2 Flight Maneuvers

To control the flight of an airplane, it is necessary to control the 3 axes (pitch, row and yaw).

Attitude is the angular difference measured between an airplane's axis and the line of the Earth's horizon. Pitch attitude is the angle formed by the longitudinal axis, and bank attitude is the angle formed by the lateral axis [5–10]. Rotation about the airplane's vertical axis (yaw) is termed an attitude relative to the airplane's flight path, but not relative to the natural horizon [1, 11–15].

This increased aileron yaws the airplane toward the rising wing, or opposite to the direction of turn. To counteract this adverse yawing moment, rudder pressure must be applied simultaneously with aileron in the desired direction of turn. This action is required to produce a coordinated turn. As airspeed is reduced, the flight controls become less effective and the normal nose-down tendency is reduced [16–20].

The elevators become less responsive and coarse control movements become necessary to retain control of the airplane. The slipstream effect produces a strong yaw so the application of rudder is required to maintain coordinated flight. The secondary effect of applied rudder is to induce a roll, so aileron is required to keep the wings level. This can result in flying with crossed controls. To understand better the maneuvers of an airplane in Fig. 1 shows the maneuvers and its axes (pitch, roll and yaw) [21–24].

Fig. 1 Maneuvers of an
airplane

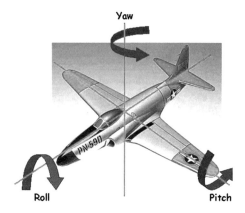

3   Problem Description

The main objective in this case of study was to maintain the stability of the airplane
in flight. The goal was to create the fuzzy system to perform the flight control and
use the simulation tool to test the fuzzy controller with noise. In Fig. 2, we show the
structure of the case os study.

The architecture for the method is shown in Fig. 3.

Fig. 2 Methodology

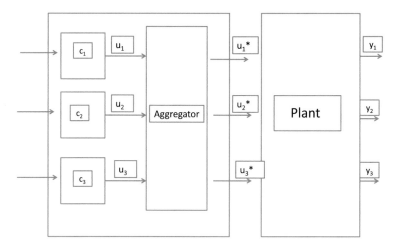

Fig. 3 Architecture of control method

In the last figure, we show the architecture used in this work, where we obtain outputs from individual controllers and we design an integrator with these outputs to obtain new outputs and achieve the control. In the above mentioned figure, we illustrate the methodology with the simulation plant; first reference is established, and then, the Joystick is used to introduce noise (turbulences) after that individual fuzzy systems are used to control the behavior of the airplane, and having results with those controllers, we decide to use an integrator to improve the control of the airplane. As we mentioned, the individual fuzzy systems were designed considering the flight maneuvers to design each fuzzy system [25–30]. To control the flight of an airplane is necessary to control 3 axes (pitch, row and yaw). To obtain the total control of the airplane is necessary to use the wheel and the pedals to maneuver the airplane. The 3 fuzzy systems and rules used in this work are shown in Figs. 4, 5, 6, 7, 8 and 9.

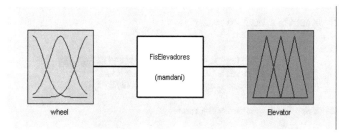

Fig. 4 Fuzzy system to control the elevator

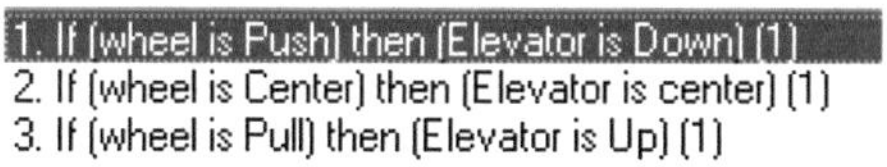

1. If (wheel is Push) then (Elevator is Down) (1)
2. If (wheel is Center) then (Elevator is center) (1)
3. If (wheel is Pull) then (Elevator is Up) (1)

Fig. 5 Rules of the fuzzy system for the elevator

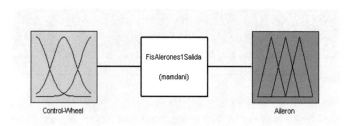

Fig. 6 Fuzzy system to control the aileron

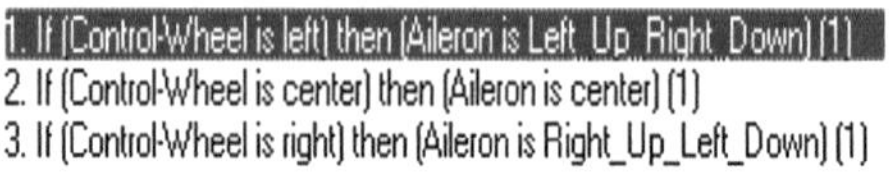

1. If (Control-Wheel is left) then (Aileron is Left_Up_Right_Down) (1)
2. If (Control-Wheel is center) then (Aileron is center) (1)
3. If (Control-Wheel is right) then (Aileron is Right_Up_Left_Down) (1)

Fig. 7 Rules of the fuzzy system for the aileron

FisRudder
(mamdani)
Pedals
Rudder

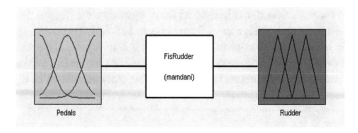

Fig. 8 Fuzzy system to control the rudder

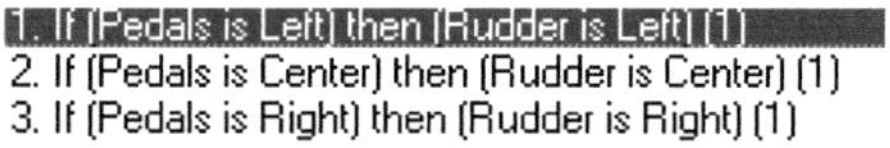

1. If (Pedals is Left) then (Rudder is Left) (1)
2. If (Pedals is Center) then (Rudder is Center) (1)
3. If (Pedals is Right) then (Rudder is Right) (1)

Fig. 9 Rules of the fuzzy system for the rudder

Elevators
Rudder
Aleron
FisIntegrador
(mamdani)
SElevators
SRudder
SAleron

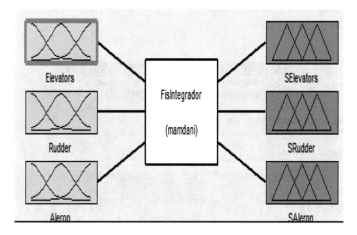

Fig. 10 Integrator

When simulation was performed using the 3 fuzzy systems, we decide to work with an integrator where the output of each fuzzy system will be an input, this means that the integrator has 3 inputs (elevator, rudder and the aileron) to obtain new outputs and achieve the control. The integrator is shown in Fig. 10.

1. If (Elevators is alto) and (Rudder is alto) and (Aleron is bajo) then (SElevators is alto)(SRudder is medio)(SAleron is medio) (1)
2. If (Elevators is alto) and (Rudder is medio) and (Aleron is alto) then (SElevators is medio)(SRudder is alto)(SAleron is medio) (1)
3. If (Elevators is bajo) and (Rudder is medio) and (Aleron is bajo) then (SElevators is bajo)(SRudder is bajo)(SAleron is medio) (1)
4. If (Elevators is medio) and (Rudder is alto) and (Aleron is medio) then (SElevators is medio)(SRudder is medio)(SAleron is medio) (1)
5. If (Elevators is medio) and (Rudder is medio) and (Aleron is alto) then (SElevators is medio)(SRudder is medio)(SAleron is medio) (1)
6. If (Elevators is medio) and (Rudder is bajo) and (Aleron is alto) then (SElevators is bajo)(SRudder is medio)(SAleron is bajo) (1)
7. If (Elevators is bajo) and (Rudder is medio) and (Aleron is alto) then (SElevators is medio)(SRudder is alto)(SAleron is bajo) (1)
8. If (Elevators is bajo) and (Rudder is bajo) and (Aleron is bajo) then (SElevators is medio)(SRudder is medio)(SAleron is alto) (1)
9. If (Elevators is bajo) and (Rudder is alto) and (Aleron is bajo) then (SElevators is medio)(SRudder is medio)(SAleron is medio) (1)
10. If (Elevators is bajo) and (Rudder is medio) and (Aleron is medio) then (SElevators is medio)(SRudder is medio)(SAleron is medio) (1

Fig. 11 Rules of the integrator

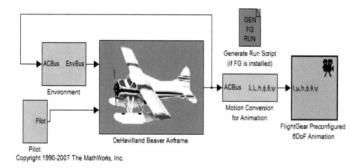

Fig. 12 Simulation plant

The integrator has 3 inputs (elevator, aileron and rudder) and 3 outputs (Selevators, Srudder and Saileron). The main objective to do the integrator is to obtain new outputs to improve the stability and obtain a better control. The rules of the integrator are shown in Fig. 11.

The simulation plant used in this case of study is shown in Fig. 12.

The simulation was performed using triangular membership function, and some results are presented in Table 1: in the first column, the behavior of the aileron is shown when the individual fuzzy controller is applied; in the second column, the behavior of the elevator is shown; and in the third column, the behavior of the rudder is presented with its individual fuzzy controller.

As Table 1 illustrates that the results are not good, we decide to optimize the integrator using a genetic algorithm; in Fig. 13, parameters of the genetic algorithm are presented.

Table 1 Results for simulation plant with a individual fuzzy controllers

Aileron	Elevator	Rudder
0.9142	1.0481	0.8881
0.9457	1.1064	0.932
0.8901	1.1133	0.9515
0.9121	1.0851	0.9446
0.8918	1.117	0.9098
0.8598	1.0906	0.8914
0.9441	1.0878	0.9258
0.9592	1.1512	0.8962
0.9045	1.1158	0.9305
0.9033	1.0465	0.9521

Fig. 13 Behavior of elevator, rudder and aileron

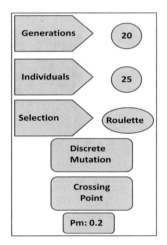

The genetic algorithm was applied to optimize the membership function of each input and each output. When the genetic algorithm was used in the integrator, the results were better, these results are shown in Table 2; in this table, the behavior of the aileron, elevator and rudder is shown, but in this case, an integrator is used to improve the control in the simulation.

When genetic algorithm was used in the integrator, the error was improved and the control was better. This algorithm helped to the integrator to obtain the better parameters in each membership function to obtain a good control, and behavior of the elevators, aileron and rudder is shown in Fig. 14. Figure 15 shows the convergence of the genetic algorithm.

Table 2 Results for simulation plant using an integrator

Aileron	Elevator	Rudder
0.4185	0.3881	0.1877
0.5047	0.6688	0.3047
0.5031	0.5033	0.502
0.4073	0.3611	0.1626
0.5052	0.5053	0.5049
0.3600	0.374	0.2684
0.2468	0.2417	0.2964
0.2413	0.2654	0.4503
0.3595	0.3784	0.5396
0.5043	0.5041	0.5079
0.5039	0.5045	0.508
0.5042	0.5063	0.5046
0.3529	0.3785	0.3601
0.5071	0.5076	0.5121
0.2288	0.243	0.5022

Fig. 14 Behavior of elevator, rudder and aileron

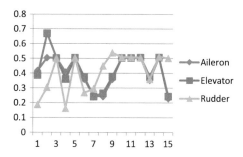

Fig. 15 Convergence of the genetic algorithm

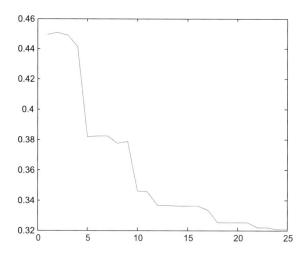

4 Conclusions

After applying the integrator, the results were better in the behavior of the airplane, but when the genetic algorithm optimized the membership function of the integrator, the error was decreased and the behavior of the airplane was improved. The simulation plant has a block that generates turbulences, but at the same time, the joystick introduced noise-generating abrupt movements on airplane, and with this situation, we can say that when an integrator is used using the individual controllers as inputs with genetic algorithm is a good alternative to improve the results and achieve the total control of the plane.

References

1. Holmes, T.: US Navy F-14 Tomcat Units of Operation Iraqi Freedom. Osprey Publishing Limited, Oxford (2005)
2. Mitchell, M.: An Introduction to Genetic Algorithms. Massachusetts Institute of Technology, MA (1999)
3. Coley, A.: An Introduction to Genetic Algorithms for Scientists and Engineers. World Scientific Book, Singapore (1999)
4. Haupt, R., Haupt, S.: Practical Genetic Algorithm. Wiley Interscience Book, NY (2004)
5. Abusleme, H., Angel, C.: Control difuso de un vehiculo volador no tripulado, Ph.d. Thesis, Pontificia University of Chile (2000)
6. Blakelock, J.: Automatic Control of Aircraft and Missiles. Prentice-Hall, NJ (1965)
7. Dwinnell, J.: Principles of Aerodynamics. McGraw-Hill Book Company, NY (1929)
8. Engelen, H., Babuska, R.: Fuzzy logic based full-envelope autonomous flight control for an atmospheric re-entry spacecraft. Control Eng. Pract. J. 11(1), 11–25 (2003)
9. Federal Aviation Administration: Airplane Flying Handbook. Department of Transportation Federal Aviation Administration, US (2007)
10. Federal Aviation Administration. Pilot's Handbook of Aeronautical Knowledge. Department of Transportation Federal Administration, US (2008)
11. Gardner, A.: US Warplanes The F-14 Tomcat. The Rosen Publishing Group, US (2003)
12. Gibbens, P., Boyle, D.: Introductory Flight Mechanics and Performance. University of Sydney, Australia (1999)
13. Kadmiry, B., Driankov, D.: A fuzzy flight controller combining linguistic and model-based fuzzy control. Fuzzy Sets Syst. J. 146(3), 313–347 (2004)
14. Keviczky, T., Balas, G.: Receding horizon control of an F-16 aircraft: A comparative study. Control Eng. Pract. J. 14(9), 1023–1033 (2006)
15. Liu, M., Naadimuthu, G., Lee, E.S.: Trajectory tracking in aircraft landing operations management using the adaptive neural fuzzy inference system. Comput. Math Appl J, 56(5), 1322–1327 (2008)
16. McLean, D.: Automatic Flight Control System. Prentice Hall, NJ (1990)
17. McRuer, D., Ashkenas, I., Graham, D.: Aircraft Dynamics and Automatic Control. Princeton University Press, NJ (1973)
18. Nelson, R.: Flight Stability and automatic control, 2nd edn. Department of Aerospace and Mechanical Engineering, University of Notre Dame, McGraw Hill (1998)
19. Reiner, J., Balas, G., Garrard, W.: Flight control design using robust dynamic inversion and time-scale separation. Automatic J. 32(11), 1493–1504 (1996)

20. Sanchez, E., Becerra, H., Velez, C.: Combining fuzzy, PID and regulation control for an autonomous mini-helicopter. J. Inf. Sci. **177**(10), 1999–2022 (2007)
21. Song, Y., Wang, H.: Design of flight control system for a small unmanned tilt rotor aircraft. Chin. J. Aeronaut. **22**(3), 250–256 (2009)
22. Walker, D.J.: Multivariable control of the longitudinal and lateral dynamics of a fly-by-wire helicopter. Control. Eng. Pract. **11**(7), 781–795 (2003)
23. Rachman, E., Jaam, J., Hasnah, A.: Non-linear simulation of controller for longitudinal control augmentation system of F-16 using numerical approach. Inf. Sci. J. **164**(1–4), 47–60 (2004)
24. Sefer, K., Omer, C., Okyay, K.: Adaptive neuro-fuzzy inference system based autonomous flight control of unmanned air vehicles. Expert Syst. Appl. J. **37**(2), 1229–1234 (2010)
25. Cervantes, L., Castillo, O.: Design of a fuzzy system for the longitudinal control of an F-14 Airplane. Soft Computing for Intelligent Control and Mobile Robotics, vol. 318, pp. 213–224. Springer, Berlin (2011)
26. Cervantes, L. Castillo, O.: Intelligent control of nonlinear dynamic plants using a hierarchical modular approach and type-2 fuzzy logic. Lecture Notes in Computer Science, vol. 7095, pp. 1–12. Springer, Berlin (2011)
27. Cervantes, L., Castillo, O.: Hierarchical genetic algorithms for optimal type-2 fuzzy system design. In: Annual Meeting of the North American Fuzzy Information Processing Society, pp. 324–329 (2011)
28. Cervantes, L., Castillo, O.: Automatic design of fuzzy systems for control of aircraft dynamic systems with genetic optimization. In: World Congress and AFSS International Conference, pp. OS-413-1–OS-413-7 (2011)
29. Cervantes, L., Castillo, O.: Comparative study of type-1 and type-2 fuzzy systems for the three tank water control problem. LNAI, vol. 7630, pp. 362–373. Springer, Berlin (2013)
30. Jamshidi, M., Vadiee, N., Ross, T.: Fuzzy Logic and Control: Software and Hardware Applications, vol. 2. Prentice-Hall, University of New Mexico (1993)

Comparative Study of Bio-inspired Algorithms Applied in the Design of Fuzzy Controller for the Water Tank

Leticia Amador-Angulo and Oscar Castillo

Abstract Recently, bio-inspired methods have become powerful optimization algorithms to solve complex problems. We also mention alternative approaches without optimization techniques for obtaining the controller. Swarm intelligence is the part of artificial intelligence based on the study of actions of individuals in various decentralized systems. The main objective of the work is based on the main reasons for the optimization of the classical control of type Mamdani in the fuzzy controller, specifically in tuning membership functions of the fuzzy controller for the benchmark problem known as the water tank using two methods of optimization a simple ant colony optimization (S-ACO) and the bee colony optimization (BCO) for membership functions' parameters of a fuzzy logic controller (FLC) in order to find the optimal intelligent controller for a benchmark problem known as the water tank. Finally, we provide a comparison of both methods for the case of designing of the classical control of type Mamdani in the fuzzy controllers.

Keywords Ant colony optimization · Bee colony optimization · Uncertainty · Fuzzy controller · Type-1 fuzzy logic · Convergence

1 Introduction

Fuzzy control systems combine information and knowledge of human experts (natural language) with measurements and mathematical models. Fuzzy systems transform the knowledge base in a mathematical formulation that has proven to be very efficient in many applications [1, 2].

L. Amador-Angulo (✉) · O. Castillo
Division of Graduate Studies, Tijuana Institute of Technology, Tijuana, Mexico
e-mail: leticia.amadorangulo@yahoo.com.mx

O. Castillo
e-mail: ocastillo@tectijuana.mx

© Springer International Publishing Switzerland 2016
L.A. Zadeh et al. (eds.), *Recent Developments and New Direction in Soft-Computing Foundations and Applications*, Studies in Fuzziness and Soft Computing 342, DOI 10.1007/978-3-319-32229-2_29

Fuzzy models have emerged as an interesting generalization of mathematical models based on fuzzy sets [10, 20]. Fuzzy systems transform human knowledge into a mathematical formulation. That is why it has been shown that bio-inspired algorithms in nature, for this research S-ACO and BCO, allow finding the values of the membership functions of the fuzzy controller of an intelligent and collaborative simulation by the way ants behave in their natural environment. With this technique, one can design and find a better fuzzy controller for the problem.

The rest of the paper is organized as follows. Section 2 describes the theoretical basis of the paper, Sect. 3 describes the problem statement, Sect. 4 describes the concept of swarm intelligence, Sect. 5 describes the simulations and results, and finally Sect. 6 describes the conclusion and the future work.

2 Theoretical Basis

A fuzzy logic system (FLS) that is defined entirely in terms of type-1 fuzzy sets is known as type-1 fuzzy logic system (Type-1 FLS) [5], and its elements are defined in the following Fig. 1 [19–22].

A fuzzy set in the universe U is characterized by a membership function $u_A(x)$ taking values on the interval [0,1] and can be represented as a set of ordered pairs of an element and the membership value to the set:

$$A = \{(x, u_A(x))|x \in U\} \tag{1}$$

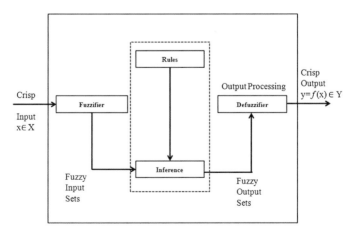

Fig. 1 Architecture of a type-1 fuzzy logic system

Fig. 2 Examples of the types of membership functions

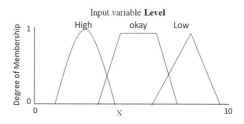

A variety of types of membership functions exist, but one of them is typically known as the bell of Gauss (Gaussian). The mathematical function is defined with the following equation [2].

$$f(x) = \exp\left(\frac{-0.5(x-c)^2}{\sigma^2}\right) \qquad (2)$$

where c is the parameter representing the mean and σ is the parameter representing the variance. The Gaussian distribution is defined by its mean c and standard deviation $\sigma > 0$, and it is satisfied that the lower the σ, the narrower the "bell." An example of a Gaussian-type membership function is shown in Fig. 2 for the linguistic value high.

The distribution and type of membership functions help to assess the values of the linguistic variables differently in each type of membership function. For this research, experiments were performed by changing the membership functions of the fuzzy controller inputs, and we experimented with Gaussian and triangular because in practice they give better results [20]. Figure 2 shows graphically the three types of membership functions with which we realized tests in the model of fuzzy controller.

In this paper, the optimization of the parameter values of membership functions was made with the implementation of fuzzy systems obtained with the classical control of Mamdani style with fuzzy logic for the benchmark problem of the water tank, which are presented in the following section.

3 Problem Statement

3.1 Description of the Problem

The problem to be considered is known as the water tank controller, which aims at control-ling the water level in a tank, therefore, based on the actual water level in the tank the controller has to be able to provide the proper activation of the valve.

3.2   Model Equations of the Water Tank

The process of filling the water tank is presented as a differential equation for the height of the water in the tank, H, given by:

$$\frac{d}{dt}\text{Vol} = A\frac{dH}{dt} = bV - a\sqrt{H} \tag{3}$$

where Vol is the volume of water in the tank, A is the cross-sectional area of the tank, b is a constant related to the flow rate into the tank, and a is a constant related to the flow rate out of the tank. The equation describes the height of water, H, as a function of time, due to the difference between flow rates into and out of the tank. Figure 3 shows graphically the mathematical model.

To evaluate the valve opening in a precise way, we rely on fuzzy logic, which is implemented as a fuzzy controller that performs automated tasks considered as the water level and how fast it be entered to, thereby maintaining the level of water in the tank in a better way.

3.3   Characteristics of the Fuzzy Controller

We present the characteristics of the classical control of type Mamdani in the fuzzy controller, besides the results of model evaluation.

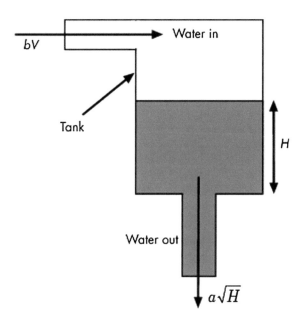

Fig. 3 Graphical representation of the mathematical equation of the water tank filler

3.4   Membership Functions

Membership functions are for the two inputs of the fuzzy system: The first is called **level**, which has three membership functions with linguistic value of *high, okay,* and *low*. The second input variable is called **rate** with three membership functions with linguistic value of *negative, none,* and *positive*, as shown in Fig. 4, representations of fuzzy variables. The names of the linguistic labels are assigned based on the empirical process of filling behavior of a water tank.

The classical control of type Mamdani using the fuzzy inference system has an output called **valve**, which is composed of five triangular membership functions with the following linguistic values: *close_ fast, close_slow, no_change, open_slow,* and *open_ fast*, representation shown in Fig. 5.

Fig. 4 Type-1 fuzzy inference system input variable

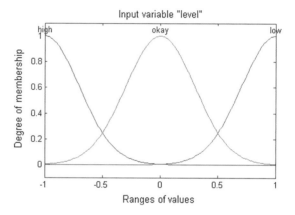

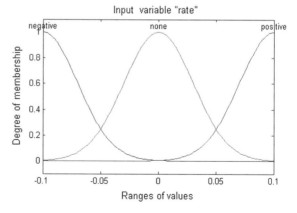

Fig. 5 Type-1 fuzzy
inference system output
variable

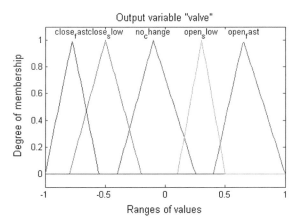

3.5 Rules

The simulation shows that five rules are sufficient, which are detailed below:

- If (level is okay) then (valve is not_change).
- If (level is low) then (valve is open_fast).
- If (level is high) then (valve is close_fast).
- If (level is okay) and (rate is positive) then (valve is close_slow).
- If (level is okay) and (rate is negative) then (valve is open_slow).

The rules are based on the behavior that the water tank is to be filled. All the combinations of rules were taken from experimental knowledge according to how the process is performed in a tank filled with water. We start with 5 rules to visualize the behavior of the classical control of Mamdani style in the fuzzy controller [13].

3.6 Mean Square Error (MSE)

The metric on which the evaluation is being made for the fuzzy controller is by using the mean square error for measuring the behavior of the controller in reference to the error tends to be 0 (zero), and the errors for the proportional integral derivative (PID) controller and fuzzy controller are used. The MSE is the sum of the variance and the squared deviation of the estimator (reference). The mathematical definition is presented in the following equation.

$$\text{MSE} = \frac{1}{n} \sum_{i=1}^{n} (\bar{Y}_i - Y_i)^2 \tag{4}$$

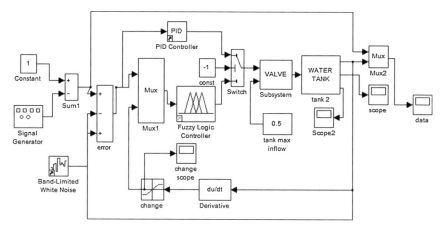

Fig. 6 Block diagram for the simulation of the FLC

3.7 Evaluation of Control Diagram

Figure 6 shows the block diagram used for the fuzzy logic controller (FLC) that obtained the best results of the water tank benchmark problem. Generally, the fuzzy controller is a closed-loop control, the aim is to make the plant output to follow the input r, the adder is applied to the system, it is used for a controller in the first, the output is connected directly to one of two inputs of the adder, and in the second situation, the output and the model are perturbed with noise in order to introduce uncertainty in the data feedback. The noise is a disturbance that tells the model with the objective that the ACO and BCO algorithms further explore its search space and show better results.

Finally, at the output of the adder, we have the error signal, which is applied to the fuzzy controller together with a signal derived from this, which is a change in the error signal over time [16].

Figure 7 shows the simulation model where the black line denotes the reference model and the pink line the output of the model using a classical control of Mamdani style in fuzzy controller. These simulations are made without noise.

3.8 Implementation of Noise in the Model

Different noise levels were applied as a disturbance in the signal processing to evaluate the target type-1 fuzzy controller and to visualize the results in the model. Figure 8 shows the representation of the simulation with a noise level of 0.3, and this is displayed in blue for the PID controller behavior and in pink for the classical control of Mamdani style in fuzzy controller with the above disturbance.

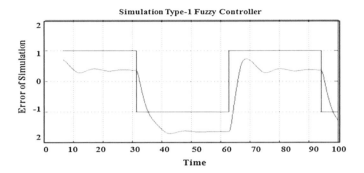

Fig. 7 Simulation model using the type-1 fuzzy controller

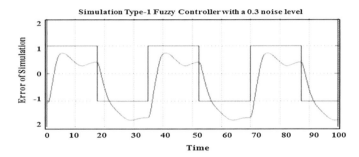

Fig. 8 Simulation model using the type-1 fuzzy controller with a noise level of 0.3

The bio-inspired algorithms in nature have proved to be a technique of optimization good in the design of fuzzy controllers.

4 Swarm Intelligence

Many species in the nature are characterized by swarm behavior. Fish schools, flocks of birds, and herds of land animals are formed as a result of *biological need* to stay together. Individual in herd, fish school, or flock of birds has a higher probability to stay alive, since predator usually assaults only individual. A collective movement characterizes flocks of birds, herds of animal, and fish schools. Herds of animals respond quickly to changes in the directions and speed of their neighbors. Swarm behavior is also one of the main characteristics of social insects (bees, wasps, ants, and termites). Communication between individual insects in a colony of social insects has been well known. The communication systems between individual insects contribute to the configuration of the "collective intelligence" of the social insect colonies. The term "swarm intelligence" that denotes this "collective intelligence" has come into use [4, 9, 18].

4.1 The Theory of Graphs

Let us define the graph $G = (V, E)$, where V is the set of nodes and E is the matrix of the links between nodes. G has $n_G = |V|$ nodes. Let us define L^K as the number of hops in the path built by the ant k from the origin node to the destiny node. Therefore, it is necessary to find:

$$Q = \{q_a, \ldots, q_f | q_1 \in C\} \tag{5}$$

where Q is the set of nodes representing a continuous path with no obstacles, q_a, …, q_f are former nodes of the path, and C is the set of possible configurations of the free space. If $x^k(t)$ denotes a Q solution in time t, $f(x^k(t))$ expresses the quality of the solution.

BCO and S-ACO are algorithm implementation that adapts the behavior of real bees to solutions of minimum cost path problems on graphs [12].

- *Ant Colony Optimization Algorithm*

S-ACO is inspired by ants and their behavior for finding shortest paths from their nest to sources of food. Without any leader that could guide the ants to optimal trajectories, the ants manage to find these optimal trajectories over time in a distributed fashion. In the S-ACO algorithm, the metaphorical ants are agents programmed to find an optimal combination of elements of a given set that maximizes some utility function [3]. The key ingredient in S-ACO and its biological counterpart are the pheromones. With real ants, these are chemicals deposited by the ants and their concentration encodes a map of trajectories, where stronger concentrations represent better trajectories [6, 7, 14].

The chemical that they leave a trace on the track allows a highly collaborative play to seek for food [7]. The ants collaborate together to explore all possible ways to find the solution to the problem (food). Graphically shown in Fig. 9 is smart mechanism that ants use to forage for food.

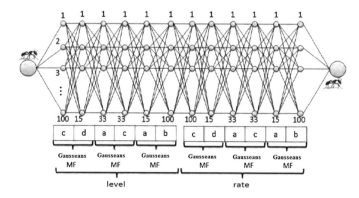

Fig. 9 S-ACO architecture

The S-ACO is an algorithm implementation that adapts the behavior of real ants to solutions of minimum cost path problems on graphs [6, 11]. A number of artificial ants build solutions for a certain optimization problem and exchange information about the quality of these solutions making allusion to the communication system of real ants.

The S-ACO algorithm is based on Eqs. (6)–(8):

$$
p_{ij}^k(t) = \begin{cases} \dfrac{\tau_{ij}^k}{\sum_{j \in N_{ij}^k} \tau_{ij}^{\alpha}(t)} & \text{if} \quad j \in N_i^k \\ 0 & \text{if} \quad j \notin N_i^k \end{cases}
\tag{6}
$$

$$
\tau_{ij}(t) \leftarrow (1 - \rho)\tau_{ij}(t)
\tag{7}
$$

$$
\tau_{ij}(t+1) = \tau_{ij}(t) + \sum_{k=1}^{n_k} \tau_{ij}(t)
\tag{8}
$$

Equation (6) represents the probability of an ant k located on a node i selects the next node denoted by j, where N_i^k is the set of feasible nodes (in a neighborhood) connected to node i with respect to ant k, τ_{ij} is the total pheromone concentration of link ij, and, α is a positive constant used as a gain for the pheromone influence.

Equation (7) represents the evaporation pheromone update, where $\rho \in [0, 1]$ is the evaporation rate value of the pheromone trail. The evaporation is added to the algorithm in order to force the exploration of the ants and avoid premature convergence to suboptimal solutions [7]. For $\rho = 1$, the search becomes completely random [8]. Equation (7) represents the concentration pheromone update, where $\Delta\tau_{ij}^k$ is the amount of pheromone that an ant k deposits in a link ij in a time t.

The general steps of S-ACO are indicated in Table 1.

Table 1 Basic steps of the S-ACO algorithm

Pseudocode of ACO
1. Set a pheromone concentration τ_{ij} to each link (i, j)
2. Place a number $k = 1, 2, \ldots, n_k$ in the nest
3. Iteratively build a path to the food source (destiny node), using Eq. (6) for every ant Remove cycles and compute each route weight $f(x^k(t))$. A cycle could be generated when there are no feasible candidates nodes, that is, for any i and any k, $N_i^k = \emptyset$ then the predecessor of that node is included as a former node of the path
4. Apply evaporation using Eq. (7)
5. Update of the pheromone concentration using Eq. (8)
6. Finally, finish the algorithm in any of the three different ways When a maximum number of epochs has been reached When it has found an acceptable solution, with $f(x_k(t)) < \varepsilon$ When all ants follow the same path

- *Bee Colony Optimization Algorithm*

The BCO is inspired by the bees' behavior in the nature. The basic idea behind the BCO is to create the multiagent system (colony of artificial bees) capable to successfully solve difficult combinatorial optimization problems. The artificial bee colony behaves partially alike and partially differently from bee colonies in nature [18].

The population of agents (artificial bees) consisting of bees collaboratively searches for the optimal solution. Every artificial bee generates one solution to the problem. The algorithm is divided into forward pass and backward pass. The existence of a large number of different social insect species, and variation in their behavioral patterns, The S-ACO and BCO have the characteristic of being able to performing a variety of complex tasks [18]. The best example is the collection and processing of nectar, the practice of which is highly organized. Each bee decides to reach for the nectar source by following a nestmate who has already discovered a path of nectar source dance, in that way trying to convince their nestmate to follow them. If a bee decides to leave the hive to get nectar, she follows the bee dancers to one of the nectar areas. The mechanisms by which the bee decides to follow a specific dancer are not well understood, but it is considered that "the recruitment among bees is always a function of the quality of the food source" [16].

Graphically shown in Fig. 10 is the representation of the smart mechanism that bees use to forage for food in the example of third forward pass.

The bee colony optimization (BCO) metaheuristic [15, 17, 18] has been introduced fairly recently by *Lučić* and *Teodorović* as new direction in the field of swarm intelligence and has not previously applied in type-1 fuzzy controller design.

The BCO algorithm is based on Eqs. (9)–(13):

$$p_{ij,n} = \frac{\left[\rho_{ij,n}\right]^{\alpha} \cdot \left[\frac{1}{d_{ij}}\right]^{\beta}}{\sum_{j \in A_{i,n}} \left[\rho_{ij,n}\right]^{\alpha} \cdot \left[\frac{1}{d_{ij}}\right]^{\beta}} \tag{9}$$

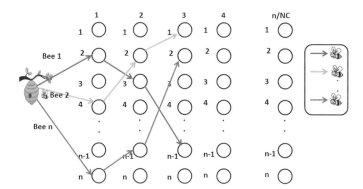

Fig. 10 BCO representation

$$\rho_{ij,n} = \begin{cases} & ,j \in F_{i,n}, \ |A_{i,n}| > 1 \\ \frac{1-\lambda |A_{i,n} \cap F_{i,n}|}{|A_{i,n} - F_{i,n}|} & ,j \notin F_{i,n}, \ |A_{i,n}| > 1 \\ & , \ |A_{i,n}| = 1 \end{cases} \begin{matrix} \forall j \in A_{i,n}, \\ 0 \le \lambda \le 1 \end{matrix} \qquad (10)$$

$$D_i = K \cdot \frac{Pf_i}{Pf_{colony}} \qquad (11)$$

$$Pf_i = \frac{1}{L_I}, L_i = \text{Tour length} \qquad (12)$$

$$Pf_{colony} = \frac{1}{N_{Bee}} \sum_{i=1}^{N_{Bee}} Pf_i \qquad (13)$$

Equation (9) represents the probability of a bee k located on a node i that selects the next node denoted by j, where N_i^k is the set of feasible nodes (in a neighborhood) connected to node i with respect to bee k and ρ_{ij} is the probability of visiting the following node. Note that the β is inversely proportional to the city distance; d_{ij} represents the distance of node i until node j, for this algorithm indicating the total dance that a bee travelled in this moment. Finally, α is a binary variable that is used to find better solutions in the algorithm.

For the representation of S-ACO and BCO algorithm in fuzzy controller, the Eq. (9) and Eq. (10) are replaced for the Eq. (4), previously described in the section 3.

Equation (11) represents that a waggle dance will last for certain duration, determined by a linear function, where K denotes the waggle dance scaling factor, Pf_i denotes the profitability scores of bee i as defined in Eq. (12), and Pf_{colony} denotes the bee colony's average profitability as in Eq. (13) and is updated after each bee completes its tour. For this research, the waggle dance is represented for the mean square error in the model, the intensity of waggle dance indicates that a bee is close to a good solution, control is represented as the minimization of error is found with the MSE.

The general steps of BCO are indicated in Table 2.

5   Simulations and Results

Various type-1 fuzzy logic system were designed, we changing the types of membership functions in inputs, to observe the behavior of the controller; the results will be discussed in the section of results and comparisons. Were performed several experiments would changing the type of membership function of fuzzy controller. The types of membership functions used are Gaussian, triangular, and trapezoidal in the two input variables of the controller, namely level and rate. The fuzzy controller performance is shown in Table 3.

Table 2 Basic steps of the BCO algorithm

Pseudocode of BCO
1. Initialization: an empty solution is assigned to every bee
2. For every bee: //the forward pass
(a) Set $k = 1$; //counter for constructive moves in the forward pass;
(b) Evaluate all possible constructive moves;
(c) According to evaluation, choose on move using the roulette wheel;
(d) $k = k + 1$; if $k \leq$ NC goto step (b)
3. All bees are back to the hive; //backward pass stars
4. Evaluate (partial) objective function value for each bee
5. Every bee decide randomly whether to continue its own exploration and become a recruiter, or to become a follower
6. For every follower, choose a new solution from recruiters by the roulette wheel
7. If solutions are not completed goto step 2
8. Evaluate all solutions and find the best one
9. If stopping condition is not met goto step 2
10. Output the best solution found

B Represents the number of bees in the hive
NC Represents the number of constructive moves during one forward pass

Table 3 Results for type-1 fuzzy logic controller

#	Experiments of type-1 fuzzy logic controller						
	Characteristics of the fuzzy controller					Error	Noise 0.3
	Type of FMs		Number of inputs	Number of outputs	Number of rules	# of Iterations = 100	
	Input	Output					
1	**Gaussian**	**Triangular**	**2**	**1**	**5**	**0.0986**	**No**
2	Triangular	Triangular	2	1	5	1.2770	No
3	Gaussian	Triangular	2	1	9	0.1178	No
4	Triangular	Triangular	2	1	9	1.2629	No
5	Trapezoidal	Triangular	2	1	5	0.1003	No
6	Trapezoidal	Triangular	2	1	9	0.1289	No
7	Gaussian	Triangular	2	1	5	0.1772	Yes
8	Triangular	Triangular	2	1	5	0.9287	Yes
9	Gaussian	Triangular	2	1	9	0.2084	Yes
10	Triangular	Triangular	2	1	9	0.9417	Yes
11	Trapezoidal	Triangular	2	1	5	0.9076	Yes
12	Trapezoidal	Triangular	2	1	9	0.9005	Yes

To obtain the errors shown in Table 1, we considered **100** iterations of use of the fuzzy controller, and we obtained the corresponding average. Table 1 shows that the best experiment is number 1 with an error of 0.0986, where in the controller is provided with type-1 fuzzy logic system with 5 rules and without disturbance noise applyed in the model.

In empirical related questions, we could make several experiments to find the fuzzy controller design that minimizes the error and show a better performance of the solution, but for them there are techniques that simulate the behavior of insects (ants, birds, fireflies, bees, among others), which allow to solve combinatorial problems with collaborative mechanisms that each present in their environment, which is why for this research; the goal is to analyze the behavior of the S-ACO and BCO algorithms have to find the design of a Fuzzy Controller that minimize the error in the model.

- *Results of the S-ACO Algorithm*

Simulations for 30 experiments have been performed with different character-istics of the algorithm, by changing the kind of membership functions at the input and the number of rules in the fuzzy controller, and the best experiments are presented in Table 4.

Table 4 shows the results in which the computational time increases when there are more iterations and the number of ants is increased in the search space, and experiments were performed by changing the value of the pheromone, number of ants, and number of iterations, thereby obtaining the experiment 3 which is the one with the minimum error for all testing, a behavior that showed the S-ACO algorithm is that in the first iterations the minimum error is found.

Figure 11 shows on the left the first input of fuzzy controller called **level** and on the right the second input of fuzzy controller called **rate** that the S-ACO algorithm found as the best of all experiments performed, and the membership functions are Gaussian because they were the ones that showed the error of simulation lowest of all.

The behavior of the model using S-ACO algorithm is shown in Fig. 12 (a), where the yellow line represents the reference value and pink line represents the output to find by S-ACO algorithm. The convergence S-ACO is presented below in Fig. 12 (b). The simulation was performed with 100 iterations.

Table 4 Results for type-1 FLC optimizing by S-ACO algorithm

No.	Ants	Iteration	Alpha α	Beta β	Pheromone	Average error	S-ACO time (min)
1	10	50	2	5	0.5	0.085	18:15
2	30	100	2	5	0.7	0.096	25:20
3	**30**	**50**	**2**	**5**	**0.7**	**0.049**	**20:45**
4	10	100	2	5	0.5	0.084	23:03
5	20	50	2	5	0.5	0.053	25:30
6	20	100	2	5	0.7	0.051	23:40
7	50	50	2	5	0.5	0.091	49:58
8	50	100	2	5	0.7	0.085	59:39
9	60	50	2	5	0.5	0.065	1:02:45
10	60	100	2	5	0.7	0.080	59:39

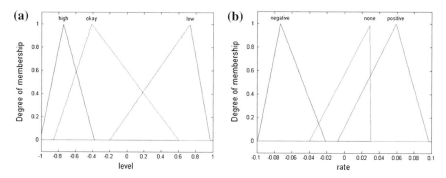

Fig. 11 Distribution of membership functions of the type-1 fuzzy controller found by the S-ACO

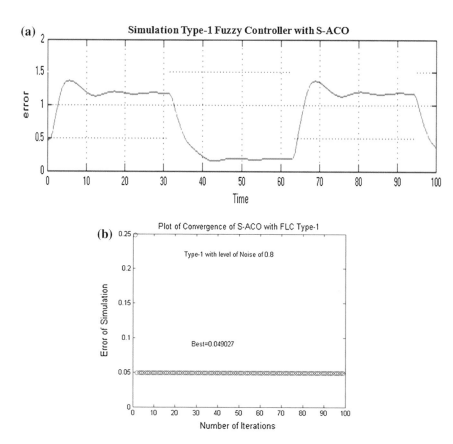

Fig. 12 Simulation and convergence of S-ACO. **a** Simulation model using type-1 fuzzy controller with S-ACO, **b** Optimization behaviour for the S-ACO on type-1 FLC

Table 5 Comparison the results with fuzzy controllers optimized and without optimizing

No	Type of FMs (input)	No. rules	Without optimizing		Optimized	
			Average error	Time (s)	Average error	S-ACO Time (min)
1	Gaussian	5	**0.0986**	05:45	0.085	43:15
2	Gaussian	9	0.1178	08:15	0.096	25:45
3	Triangular	5	1.2770	09:13	**0.049**	22:45
4	Triangular	9	1.2629	12:36	0.084	23:03
5	Trapezoidal	5	0.1003	15:30	0.067	32:40
6	Trapezoidal	9	0.1289	18:42	0.052	28:39

Table 4 shows a comparison with the optimized experiments performed with the fuzzy controller that solves the problem water tank and without optimization.

Table 5 shows that having triangular membership functions gives a better performance in the fuzzy controller, and the results that were found when the fuzzy controller is optimized are less error due to the intelligent ants together to find a better solution. All experiments were done for 100 iterations of the fuzzy controller.

It is shown that by increasing the number of fuzzy rules in the controller, it tends to explore much more expert by itself, which is why experiments show that without optimizing the controller we get the minimum error in the triangular membership functions with 5 rules, whereas the optimized controller has better results with only 5 rules, and this is because the controller tank water with 5 rules is sufficient for evaluation. That is why, based on experimentation, it can be shown that the simulation error is less when using bio-inspired S-ACO algorithm in the fuzzy controller design.

With the results shown in Table 4, it is envisioned that the S-ACO algorithm computing time increases if the controller is not being optimized, yet it is much less error than is obtained with S-ACO optimization in the fuzzy controller. These results show that the best error when NOT optimized is **0.0986**, whereas the best optimization error with S-ACO is **0.049**. Another variant is changing the number of rules in the fuzzy controller, but does not present a major impact on the outcome. However, to further explore the different ways to design a fuzzy controller leads to better results in their evaluation.

- *Results of the BCO Algorithm*

Simulation for 15 experiments has also been performed with different characteristics of the algorithm, by changing the size of population, number of iterations, number of follower bees, and values of alpha and beta, and the experiments are presented in Table 6.

Table 6 shows that the computational time increases when the iterations are high and the number of follower bees is increased. We changed the number of follower

Table 6 Results of type-1 fuzzy controller using BCO without noise

Experiment	Population	Iterations	Follower bees	Alpha α	Beta β	Average error	BCO time (min)
1	90	20	10	0.4	0.5	0.080920	26:24
2	80	20	10	0.4	0.5	0.076116	06:57
3	70	25	20	0.4	0.5	0.072430	18:23
4	60	30	25	0.4	0.5	0.07576	21:36
5	50	30	30	0.4	0.5	0.080448	23:14
6	70	25	15	0.4	0.5	0.098945	12:43
7	70	25	20	0.4	0.5	0.075186	15:04
8	70	25	25	0.4	0.5	0.084104	18:57
9	70	25	30	0.5	0.6	0.072270	31:52
10	90	30	60	0.5	0.6	0.070211	58:22
11	90	25	70	0.5	0.6	**0.068650**	39:25
12	90	20	80	0.5	0.6	0.072144	19:05
13	80	20	80	0.5	0.6	0.068682	01:45:18
14	80	25	85	0.5	0.6	0.072575	26:54
15	70	20	70	0.5	0.6	0.071754	26:41

bee, size of population and iterations for to note that both the algorithm affects these parameters, It is observable the experiment 11 with MSEof 0.068650, population of 90, iterations of 25 and follower bee of 70.

Figure 13 shows on the left the first input of fuzzy controller called **level** and on the right the second input of fuzzy controller called **rate** that the BCO algorithm found as the best of all experiments performed, and the membership functions are Gaussian because they were the ones that showed the error of simulation lowest of all.

The simulation of the fuzzy controller is shown in Fig. 14 (a), where the yellow line represents the reference value and pink line represents the behavior with BCO

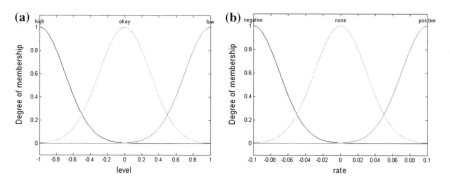

Fig. 13 Distribution of membership functions of the type-1 fuzzy controller found by the BCO

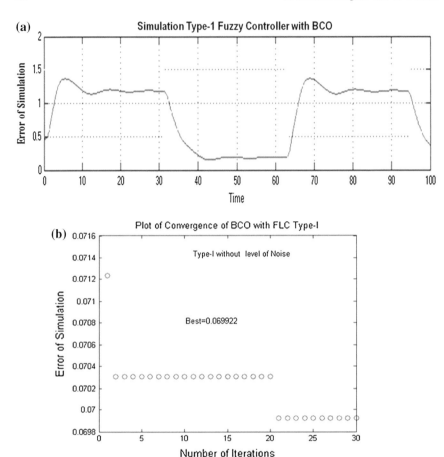

Fig. 14 Simulation and convergence of BCO. **a** Simulation model using type-1 fuzzy controller with BCO, **b** Optimization behaviour for the BCO on type-1 FLC

algorithm. The convergence BCO algorithm is presented below in Fig. 14 (b). The simulation was performed with 100 iterations.

Finally, Table 7 shows the comparison the results to find the S-ACO and BCO. We presented the average error in the simulation and the time. We to cant observed

Table 7 Comparison of the results of S-ACO and BCO

Comparison of the results of S-ACO and BCO				
	Without optimizing		S-ACO	BCO
	With noise	Without noise		
Average error	0.1772	0.0986	0.0491	0.0686
Time of simulation (min)	09:20	05:20	22:45	39:25

that S-ACO algorithm is better compared to BCO algorithm in the minimization of the error for the studied case presented.

6 Conclusions

This paper proposes a comparison the bio-inspired algorithm that solves combinatorial problems, called S-ACO and BCO for the design of optimal Fuzzy Controllers. These algorithms allow finding the appropriate distributions of parameters in the membership functions for obtained a better performance in the studied case presented.The results show that S-ACO to find the better error in the simulations. The water tank problem was analyzed and has a low complexity level this is why the best results obtained in this study are shown when the Type-1 Fuzzy Controller was used without level of noise applied in the model, also, for to find better results we apply the robustness of two techniques inspired in the nature for solve problems; such as S-ACO and BCO algorithms in the design of Fuzzy Controller.

Future work includes changing the parameters of the BCO and S-ACO algorithm such as values of alpha, beta, and population size and number of follower bees to BCO and number of ants, values of alpha, beta, and pheromone to ACO, thereby finding better design fuzzy controllers. An interesting aspect is to perform the dynamic adjustment of these parameters with the help of fuzzy logic.

Acknowledgment We would like to express our gratitude to the CONACYT and Tijuana Institute of Technology for the facilities and resources granted for the development of this research.

References

1. Abraham, A., Grsan, C., Ramos, V.: Swarm Intelligent Data Mining, pp. 75–99 (2006)
2. Amador-Angulo, L., Castillo, O.: Comparison of Fuzzy Controllers for the Water Tank with Type-1 and Type-2 Fuzzy Logic, NAFIPS 2013, Edmonton, Canada, pp. 1–6 (2013)
3. Amador-Angulo, L., Castillo, O.: Comparison of the optimal design of fuzzy controllers for the water tank using ant colony optimization. In: Transactions on Engineering Technologies: International Multi Conference of Engineers and Computer Scientists 2013, Tijuana, B.C., Mexico, pp. 1–2 (2013)
4. Bonabeau, E., Dorigo, M., Theraulaz, G.: Swarm Intelligence. Oxford University Press, Oxford (1997)
5. Castillo, O., Martinez-Marroquin, R., Melin, P., Valdez, F., Soria, J.: Comparative study of bio-inspired algorithms applied to the optimization of type-1 and type-2 fuzzy controllers for an autonomous mobile robot. Inf Sci (2010)
6. Dorigo, M., Birattari, M., Stützle, T.: Ant colony optimization. In: IEEE Computational Intelligence Magazine, Nov 2006, pp. 28–39 (2006)
7. Dorigo, M., Stützle, T.: Ant Colony Optimization. Cambridge Massachusettes Institute of Technology, Bradford (2004)

8. Dorigo, M., Blum, C.: Ant colony optimization theory: a survey. Theor. Comput. Sci. **344**(2–3), 243–278 (2005)
9. Engelbrecht, A.P.: Fundamentals of Computational Swarm Intelligence. Wiley, England (2005)
10. Klir, G.J., Yuan, B.: Fuzzy Sets and Fuzzy Logic: Theory and Applications. Practice Hall, New Jerey (1995)
11. Galea, M., Shen, Q.: Simultaneous ant colony optimization algorithms for learning linguistic fuzzy rules (2006)
12. Martinez, R., Castillo, O., Soria, J.: Optimization of Membership Functions of a Fuzzy Logic Controller for an Autonomous Wheeled Mobile Robot Using Ant Colony Optimization, pp. 4–6 (2006)
13. Mendel, J.: Uncertain Rule-Based Fuzzy Logic Systems, Introduction and new directions. PH PTR (2001)
14. Meyer, B.: Convergence control in ACO. In: Genetic and Evolutionary Computation Conference (GECCO), Seattle, WA (2004)
15. Naredo, E., Castillo, O.: ACO-Tuning of a Fuzzy Controller for the Ball and Beam Problem. MICAI, pp. 58–69 (2011)
16. Teodorović, D., Lućić, P., Marcković, G., Dell'Orco, M.: Bee colony optimization: principles and applications. In: Reljin, B., Stanković, S. (eds.) Proccedings of the Eighth Seminar on Neural Network Applications in Electrical Engineering-NEUREL, pp. 151–156. University of Belgrade, Belgrade (2006)
17. Teodorović, D.: Transport modeling by multi-agent systems: a swarm intelligence approach. Transp. Plann. Technol. **26**, 289–312 (2003)
18. Teodorović, D.: Swarm intelligence systems for transportation engineering: pinciples and applications. Transp. Res. Part C Emerging Technol. **16**, 651–782 (2008)
19. Yen, J., Langari, R.: Fuzzy Logic: Intelligence, Control and Information. Prentice Hall, USA (1999)
20. Zadeh, L.A.: The concept of a lingüistic variable and its application to approximate reasoning, part II. Inf. Sci. **8**, 301–357 (1975)
21. Zadeh, L.A.: Fuzzy sets. Inf. Control **8**, 338–353 (1965)
22. Zadeh, L.A.: Toward a theory of fuzzy information granulation and its centrality in human reasoning and fuzzy logic. Fuzzy Sets Syst. **90**, 117–117 (1997) (Elsevier)

Mathematical Model of Ecopyrogenesis Reactor with Fuzzy Parametrical Identification

Y.P. Kondratenko and O.V. Kozlov

Abstract This paper presents the development of the mathematical model with fuzzy parametrical identification of the ecopyrogenesis (EPG) complex reactor as a temperature control object. The synthesis procedure of the fuzzy parametrical identification system of Mamdani type is presented. The analysis of computer simulation results in the form of static and dynamic characteristic graphs of the reactor as a temperature control object confirms the high adequacy of the developed model to the real processes. The developed mathematical model with fuzzy parametrical identification gives the opportunity to investigate the behavior of the temperature control object in steady and transient modes, in particular, to synthesize and adjust the temperature controller of the reactor temperature automatic control system (ACS).

1 Introduction

The EPG technology is widely used for the utilization of municipal solid wastes (MSWs) and low-grade coal in order to obtain alternative liquid fuels and generator gas [1]. It provides a simplified sorting of organic solid waste into two categories: the first category—dried organic waste, which includes all of the polymer waste, including polyvinylchloride (PVC) but not more than 2 %, worn tires, rubber, oil sludge, and paper, and the second category—waste with high humidity, which includes food waste, shredded wood, paper, and cardboard. The first category of waste is disposed by a multiloop circulatory pyrolysis (MCP) to obtain from the

Y.P. Kondratenko (✉)
Department of Intelligent Information Systems, Petro Mohyla Black
Sea State University, Mykolaiv 54003, Ukraine
e-mail: y_kondrat2002@yahoo.com

O.V. Kozlov
Department of Computerized Control Systems, Admiral Makarov National
University of Shipbuilding, Mykolaiv 54025, Ukraine
e-mail: oleksiy.kozlov@nuos.edu.ua

© Springer International Publishing Switzerland 2016
L.A. Zadeh et al. (eds.), *Recent Developments and New Direction
in Soft-Computing Foundations and Applications*, Studies in Fuzziness
and Soft Computing 342, DOI 10.1007/978-3-319-32229-2_30

mass of raw materials up to 60–85 % of the liquid fuel of light fractions with characteristics comparable to diesel fuel. The second category of waste is utilized by the method of multiloop bizonal circulatory gasification (MCG) by obtaining generator gas, which has a calorific value 1100–1250 kcal/m^3 [1]. For realization of the EPG technology, specific technological complexes are used, which are, in turn, complicated multicomponent technical objects. Automation of such technological complexes allows to significantly increase the operation efficiency and economic parameters.

Stabilization of the pyrolysis reactor set temperature value is one of the important tasks of automatic control of the EPG process [1]. The possibility of reactor temperature mode control with high-quality indicators allows controlling of the thermal destruction process inside the reactor with high accuracy. This gives the opportunity to obtain the high-quality liquid fractions of alternative fuel at the outlet and, in turn, requires to use a special kind of temperature ACS.

To study the ACS effectiveness at the stage of its design, it is reasonable to use the mathematical and computer modeling methods that are quite effective and low cost, compared with experimental and other approaches, especially while studying the behavior of thermal power objects and their control systems [2–5]. In particular, development and adjustment of pyrolysis reactor ACS temperature controller require an availability of an adequate mathematical model. Also reactor temperature ACS quality indicators significantly depend on the accuracy of the synthesized model and its adequacy to the real processes.

Therefore, the aim of this work is development and research of the highly adequate mathematical model of the EPG complex pyrolysis reactor as a temperature control object.

2 EPG Complex Reactor Heating Temperature Control System

The functional structure of the EPG complex pyrolysis reactor heating temperature control system is presented in Fig. 1 [6].

The set value of the reactor heating temperature T_{SR} is established by the setting device SD. The setting device signal u_{SD} is compared with the temperature sensor TS signal u_{TS} with the help of the summator, and control error ε is calculated. In turn, the signal u_{TS} corresponds to the real value of the reactor heating temperature T_{RR}. The temperature controller TC produces the control signal u_{TC} that goes to the linear flow regulator LFR. The linear flow regulator is a gas control valve with a DC servo drive, which has linear characteristics of the gas flow depending on the input voltage. The gas flow rate value Q_G that corresponds to LFR input DC voltage signal (from 0 to 10 V) goes to the gas burner GB. The gas burner power P_{GB} is used to heat the reactor, which is also influenced by the disturbances $f(t)$. If the real value of reactor heating temperature T_{RR} deviates from the set value T_{SR}, the

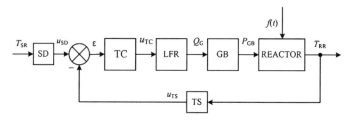

Fig. 1 Functional structure of the EPG complex pyrolysis reactor heating temperature control system

temperature controller produces the control signal u_{TC}, according to the control law that changes the gas flow rate and corresponding gas burner power that provides the corresponding heating of the reactor.

3 Structure of the Reactor Mathematical Model with Fuzzy Parameter Identification

The typical transfer function of the heat power control object, which consists of proportional link, aperiodic link, delay link, and inertial link of nth order [7], can be used in modeling of the EPG complex pyrolysis reactor as a temperature control object that has heating device power as an input and heating temperature as an output.

$$W_R(p) = \frac{U_{out}(p)}{U_{in}(p)} = \frac{ke^{-\tau p}}{(T_1 p + 1)(T_2 p + 1)^n}, \tag{1}$$

where $W_R(p)$—transfer function of the control object; $U_{out}(p)$—image of the controlled coordinate $U_{out}(t)$ (reactor heating temperature); $U_{in}(p)$—image of the control action $U_{in}(t)$ (gas burner heat power); k—gain; τ—time delay; T_1, T_2—time constants of the aperiodic and inertial links, respectively.

Transient process characteristic $h(t)$ and other dynamic characteristics of the generalized heat power control object [6–8], which has transfer function (1), are given in Fig. 2, where $h(t)$—control object transient process characteristic, which is obtained experimentally; $h''(t)$—second derivative nature of the transient process characteristic $h(t)$; $h(t_i)$—value of the transient process characteristic at the extreme point; h_s—steady value of the transient process characteristic; T_0 and τ_0—time constant and time delay that can be found from the experimental transient process characteristic.

The problem of the EPG complex reactor mathematical model is reduced to the identification of the transfer function (1) parameters and inertial link order n. In this case, the approximation criterion is the requirement of coincidence of the

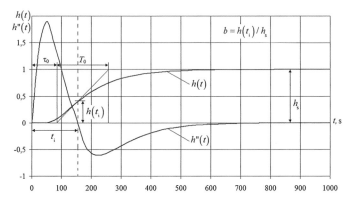

Fig. 2 Transient process characteristics of the generalized heat power control object

experimental transient characteristic of a real object $h(t)$ with the approximate transient characteristic $h_a(t)$ of the mathematical model at the points $t = 0$, $t = \infty$, and inflection point t_i, which is determined from the requirement $h''(t) = 0$. In addition, the characteristics $h(t)$ and $h_a(t)$ must have the same slope at the inflection point. Considering all the above said, the criterion of the approximation has the following form shown in Eq. (2)

$$\left.\begin{array}{l} h_a(0) = h(0) \\ h_{a,\text{steady}} = h_\text{steady} \\ h_a(t_i) = h(t_i) \\ h'_a(t_i) = h'(t_i) \end{array}\right\}. \tag{2}$$

Based on the analytical solution of the differential equation corresponding to the transfer function (1) at $n = 1$, we can create a parameter identification algorithm for approximating the transfer function. The approximation criterion (2) after substituting the results of the analytical solution of the differential equation becomes

$$\left.\begin{array}{l} xe^{-y} = e^{-y/x} \\ (1+x)e^{-y} = 1 - b \\ T_1/T_0 = e^{-y} \end{array}\right\}, \tag{3}$$

where $x = T_1/T_2$ and $y = t_{i,\,a}/T_1$—dimensionless coefficients.

It is possible to calculate the variables x and y as well as parameters T_1, T_2, $t_{i,\,a}$ (at $\tau = 0$, $h_a(t_i) = h(t_i)$, $h'_a(t_i) = h'(t_i)$) by using known experimental transfer function values b and T_0 and numerically solving transcendental Eq. (3). The calculation of the variables x and y, and respectively time constants T_1, T_2 and moment of inflection $t_{i,\,a}$ of the transient characteristic of the mathematical model (1) at $\tau = 0$, for which the condition $h_a(t_i) = h(t_i)$ and $h'_a(t_i) = h'(t_i)$ performed is possible. To bring the point of inflection of the approximating function to the real value of t_a, it is necessary to enter a delay $\tau = t_{i,\,a} - t_a$.

The reverse of further increasing the degree of adequacy of the reactor mathematical model is the optimization of its structure, which is done by varying the exponent n of Eq. (1). Let us consider the algorithm of structural and parametric identification of the above-stated mathematical model using the methods of nonlinear programming, in particular optimization gradient methods [9], and also based on the structure and parameter identification of the complex transient objects, which are considered in [7, 8].

The problem formulation of the nonlinear programming concedes the objective function choice, the definition of the optimized parameter set, the set of constraints, and also the formation of the primary hypothesis of the optimal parameter values.

The quadratic integral functional of experimental transfer function $h(t)$ deviation from the identified approximating transfer function $h_a(t)$ is proposed to be used as the objective function.

$$I[h_a(t), h(t)] = \int_0^{T_{max}} (h_a(t) - h(t))^2 dt. \tag{4}$$

Herewith, reactor model transfer function $h_a(t)$ is uniquely determined by the parameters and structure of the approximating transfer function (1).

As the exponent n of the Eq. (1) can take only integer nonnegative values, so it is not advisable to include n to the set of optimization parameters when we use the nonlinear optimization algorithm. In particular, the definition of the n value is possible by means of complete listing of the set of its admissible values $n \in \{1, 2, 3, \ldots, n_{max}\}$, where n_{max} is the limit of the aperiodic link order, $n_{max} = 8$. Thus, the optimized parameter set for the nonlinear optimization algorithm is reduced to the form: $\mathbf{P} = \{T_1, T_2, \tau\}$. The set of conditions [8, 9] with the additionally imposed requirements of the time constants T_1, T_2 positivity and time delay τ nonnegativity is a constraint for the nonlinear optimization process: $T_1 > 0; T_2 > 0; \tau \geq 0$. The primary hypothesis is proposed to be formulated on the basis of the transfer function (1) at $n = 1$. Let us accomplish the identification of the transfer function (1) parameters on the basis of the given above approach and EPG complex reactor heating experimental transient process characteristics (Fig. 3).

The experimental transient process characteristics of heating are obtained for the reactor capacity of 14 l, and the following are values of gas burner power P_{GB} and reactor load level L_R: (1) $P_{GB} = 16$ kW, $L_R = 0.2$ m; (2) $P_{GB} = 16$ kW, $L_R = 0.3$ m; (3) $P_{GB} = 16$ kW, $L_R = 0.5$ m; (4) $P_{GB} = 20$ kW, $L_R = 0.2$ m; (5) $P_{GB} = 20$ kW, $L_R = 0.3$ m; (6) $P_{GB} = 20$ kW, $L_R = 0.5$ m; (7) $P_{GB} = 25$ kW, $L_R = 0.2$ m; (8) $P_{GB} = 25$ kW, $L_R = 0.3$ m; and (9) $P_{GB} = 25$ kW, $L_R = 0.5$ m.

The transfer function (1) parameters of the EPG complex pyrolysis reactor are found in the identification process and presented in Table 1.

As Table 1 shows, the parameters k, T_1, and T_2 of pyrolysis reactor transfer function (1) change at different values of the gas burner power P_{GB} and reactor load level L_R. Therefore, for the synthesis of a universal mathematical model of the

Fig. 3 Experimental
transient process
characteristics of the EPG
complex reactor heating

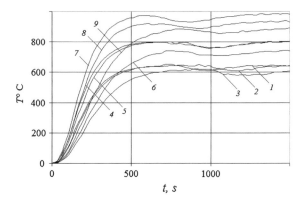

Table 1 Transfer function
parameters of the EPG
complex pyrolysis reactor

Curve number	P_{GB} (kW)	L_R (m)	k	T_1	T_2	τ	n
1	16	0.2	0.043	85.2	34.3	0	2
2	16	0.3	0.042	92.8	56.8	0	2
3	16	0.5	0.04	99.3	103.2	0	2
4	20	0.2	0.041	84.9	33.6	0	2
5	20	0.3	0.04	92.5	56.4	0	2
6	20	0.5	0.038	99.1	102.5	0	2
7	25	0.2	0.041	84.5	32.8	0	2
8	25	0.3	0.037	92.4	55.3	0	2
9	25	0.5	0.035	99.1	102	0	2

reactor on the basis of the data presented in Table 1, it is advisable to use the
specific identification system that determines the coefficients k, T_1, and T_2 of the
reactor transfer function (1), depending on the parameters P_{GB} and reactor load
level L_R.

The analysis of dependences $k = f(P_{GB}, L_R)$, $T_1 = f(P_{GB}, L_R)$, and $T_2 = f(P_{GB}, L_R)$ presented in Table 1 shows the reasonability of the parametrical identification
system development on the basis of fuzzy logic principles and algorithms that are
widely used in the synthesis of mathematical models and control devices of objects
with significant uncertainties and nonlinearities [10–16]. The mathematical models
and control systems based on fuzzy logic are developed and successfully introduced
in the following fields: technological processes control, transport control, medical
and technical diagnostics, financial management, stock forecast, pattern recognition, etc. [17–25].

The mathematical model of the pyrolysis reactor with fuzzy parametrical identification system is presented in Fig. 4.

Let us consider the synthesis procedure particularities of fuzzy identification
system of Mamdani type in detail.

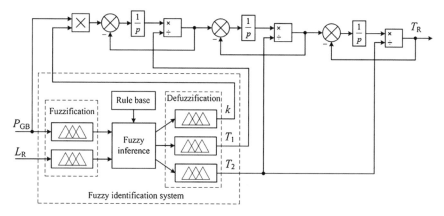

Fig. 4 Mathematical model of the pyrolysis reactor with fuzzy parametrical identification system

4 Synthesis Procedure of the Mamdani-Type Identification System

The main stages of the Mamdani-type fuzzy logic inference are fuzzification, aggregation, activation, accumulation, and defuzzification [11]. The corresponding linguistic meaning and degree of fuzzy set membership are determined for each input variable on the fuzzification stage [12]. In this case, it is advisable to choose the following linguistic terms for the input and output variables, whose parameters are presented in Fig. 5.

For input variables P_{GB} and L_R, range of values is determined in relative units. In Fig. 5, the following notations are accepted: VS—very small; S—small; M—middle; B—big; L—low; and H—high.

The knowledge base is formed to implement the fuzzy inference. The rules of the knowledge base according to the Mamdani algorithm are the linguistic statements in the form:

$$\text{IF } ``P_{GB} = a" \text{ AND } ``L_R = b" \text{ THEN } ``k = c" \text{ AND } ``T_1 = d" \text{ AND } ``T_2 = e",$$

where a, b, c, d, e—corresponding linguistic terms' values.

For this case, the knowledge base consists of 12 rules, which correspond to all possible combinations of two input fuzzy variables. The identification system knowledge base is presented in Table 2.

The truth degree is determined for every rule of the fuzzy inference system at the next stage (aggregation), and truth degree-finding procedure for each fuzzy output rule subconclusion is implemented on the activation stage.

Fig. 5 Fuzzy identification
system linguistic terms'
parameters

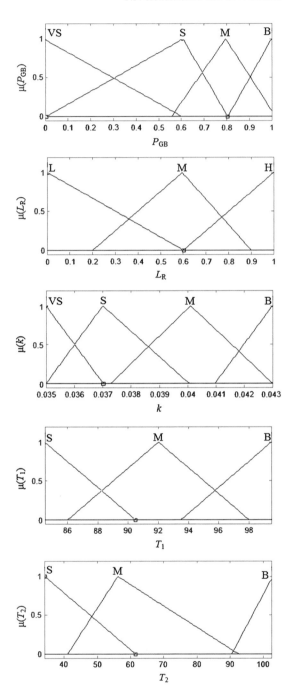

Table 2 Identification system knowledge base

Rule number	Input variables		Output variables		
	P_{GB}	L_R	k	T_1	T_2
1	VS	L	B	S	S
2	VS	M	B	M	M
3	VS	H	B	B	B
4	S	L	B	S	S
5	S	M	B	M	M
6	S	H	M	B	B
7	M	L	M	S	S
8	M	M	M	M	M
9	M	H	S	B	B
10	B	L	B	S	S
11	B	M	M	M	M
12	B	H	S	B	B

The further stage of the fuzzy logic inference is accumulation that is the membership function-finding procedure for every output linguistic variable [13]. The aim of accumulation is to combine all output linguistic terms with corresponding truth degrees of each rule for obtaining the output variable membership function.

Thus, the resultant membership function for the fuzzy decision is formed at the accumulation stage, which is necessary to be converted to precise output signal value.

The procedure of finding output signal u_{out} precise numerical value is the defuzzification procedure.

There are several methods of defuzzification: the gravity center method, the square center method, the left modal value method, the right modal value method, and the other [11, 14]. In this case, the gravity center method is chosen, according to which the output signal value is calculated by the formula (5)

$$u_{out} = \frac{\sum_{i=1}^{n} u_i \cdot \mu(u_i)}{\sum_{i=1}^{n} \mu(u_i)}, \tag{5}$$

where u_{out}—one of the identification system output signals (k, T_1, T_2), n—the number of output linguistic variable values; u_i—ith value of the corresponding output linguistic variable; $\mu(u_i)$—the number of the resultant membership functions for the corresponding value u_i.

The characteristic surface of the developed fuzzy identification system is presented in Fig. 6.

Fig. 6 Fuzzy identification
system characteristic surface

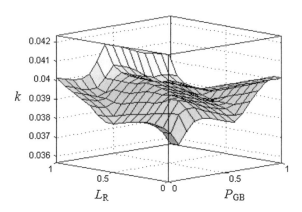

5 Computer Simulation and Adequacy Evaluation of the Reactor Mathematical Model with Fuzzy Parameter Identification System

The computer simulation of the pyrolysis reactor heating transients was carried out for the experimental EPG complex using two models: (a) developed in this paper mathematical model with fuzzy parametrical identification system and (b) well-tested model (based on heat transfer processes equations) which was presented in [26]. The experimental EPG complex has the following parameters: the displacement volume of the pyrolysis reactor is 14 l, the maximum power of gas burner is 25 kW. The simulation was carried out at P_{GB} = 18 kW, L_R = 0.4 m, constant temperature of ambient T_A = 0 °C. Simulation results as the EPG reactor heating transients are graphically shown in Fig. 7, where the following notations are accepted: 1—transient of the real object, 2—transient of the model based on the heat exchange processes equations from [26], 3—transient of the developed model with fuzzy parametrical identification system.

Fig. 7 EPG reactor heating
transient processes

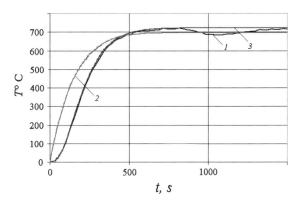

Adequacy and comparative analysis of (a) the developed mathematical model with fuzzy parametrical identification system and (b) model from [26] (based on heat transfer processes equations) is carried out with application of the hypotheses assessing methods of mathematical statistics [27], namely:

1. The sum of squared errors (SSE), which shows the total deviation of values of the mathematical model $T_m(t)$ from the corresponding values of the experimental data $T_e(t)$

$$\text{SSE} = \sum_{i=1}^{k} [T_{ei} - T_{mi}]^2 \rightarrow 0; \tag{6}$$

2. The coefficient of determination (R^2), which is a part of the variance of the variable deviation dependent from its average value. In other words, R^2 is the square of mixed correlation between experimental values and the values of the synthesized mathematical model

$$R^2 = 1 - \frac{\sum_{i=1}^{k} [T_{ei} - T_{mi}]^2}{\sum_{i=1}^{k} [T_{ei} - \overline{T_{ei}}]^2} \rightarrow 1, \tag{7}$$

where $\overline{T_{ei}} = \frac{1}{k} \sum_{i=1}^{k} T_{ei}$ is the arithmetic mean of the experimental sample;

3. The root mean square error (RMSE) is an estimation of the standard deviation of the random component between the data of the synthesized regression model and the experimental values

$$\text{RMSE} = \sqrt{\frac{1}{k} \sum_{i=1}^{k} [T_{ei} - T_{mi}]^2} \rightarrow 0. \tag{8}$$

The calculation results of statistical evaluations of the adequacy of EPG reactor mathematical models are shown in Table 3.

Based on the statistical data (Table 3), we can conclude that the mathematical model developed in this work has better results of experimental samples' compliance than the model based on heat transfer processes equations that are presented in [26].

Table 3 Comparative analysis of the adequacy of EPG reactor mathematical models

Mathematical model type	SSE	R^2	RMSE
Based on the heat exchange processes equations	319726.7	0.854	79.177
With fuzzy parametrical identification	13891.6	0.993	16.504

6 Conclusions

In this work, the mathematical model with fuzzy parametrical identification of the EPG complex reactor as a temperature control object is developed.

The obtained model gives the opportunity to study the behavior of the given temperature control object in the steady and transient modes, particularly to synthesize and adjust the temperature controller of the reactor ACS. The application of the mathematical apparatus of the fuzzy logic under the development of this model parametrical identification system allows us to take into account the specific features of EPG complex pyrolysis reactor as the control object with essentially nonlinear and undefined parameters.

The analysis of the computer simulation results and calculation results of statistical evaluations of the adequacy of EPG reactor mathematical models shows that the mathematical model with fuzzy parametrical identification developed in this work has higher adequacy than the model based on heat transfer processes equations.

References

1. Markina, L.M.: Development of new energy-saving and environmental safety technology at the organic waste disposal by ecopyrogenesis. J. Collect. Works NUS **4**(8) (2011) (in Ukrainian)
2. Štemberk, P., Lanska, N.: Heating system for curing concrete specimens under prescribed temperature. In: 13th Zittau Fuzzy Colloquium, Proceedings of East-West Fuzzy Colloquium, pp.82–88. Zittau, Hochschule Zittau, Goerlitz, Germany (2006)
3. Han, Z.X., Yan, C.H., Zhang, Z.: Study on robust control system of boiler steam temperature and analysis on its stability. J. Zhongguo Dianji Gongcheng Xuebao, Proc. Chin Soc. Electr. Eng. **30**(8), 101–109 (2010)
4. Fiss, D., Wagenknecht, M., Hampel, R.: Modeling a boiling process under uncertainties. In: 19th Zittau Fuzzy Colloquium, Proceedings of East-West Fuzzy Colloquium, pp. 141–22. Zittau, Hochschule Zittau, Goerlitz, Germany (2012)
5. Chaikin, B.S., Mar'yanchik, G.E., Panov, E.M., Shaposhnikov, P.T., Vladimirov, V.A., Volovik, I.S., Makarevich, B.A.: State-of-the-art plants for drying and high-temperature heating of ladles. Int. J. Refract. Ind. Ceram. **47**(5), 283–287 (2006)
6. Kondratenko, Y.P., Kozlov, O.V.: Fuzzy controllers in reactors control systems of multiloop pyrolysis plants. In: 19th Zittau Fuzzy Colloquium, Proceedings of East-West Fuzzy Colloquium, pp. 15–22. Zittau, Hochschule Zittau, Goerlitz, Germany (2012)
7. Rotach, V.Y.: Automatic control theory of heat and power processes. M. Energoatomizdat 296p, (1985) (in Russian)
8. Kondratenko, Y.P., Sydorenko, S., Kravchenko, D.: Fuzzy control systems of non-stationary plants with variable parameters. In: 12th Zittau East-West Fuzzy Colloquium, Conference Proceedings, Heft 84/2005, Nr.2090-2131, Wissenschaftliche Berichte, Institut fur Prozesstechnik, Prozessautomatisierung und Messtechnik, Zittau, pp.140–152 (2005)
9. Himmelblau, D.: Applied nonlinear programming (Trans. from English. a. Bihovckiy M.L. (ed.), M.: Mir), 534p., 1974 (in Russian)
10. Zadeh, L.A.: Information and control. Fuzzy Sets **8**, 338–353 (1965)

11. Zimmermann, H.-J.: Fuzzy Set Theory—and Its Applications. Kluwer Academic Publishers, Boston (1992)
12. Zadeh, L.A.: The role of fuzzy logic in modeling, identification and control, modeling identification and control. Model. Identif. Control 15(3), 191–203 (1994)
13. Yager, R.R., Filev, D.P.: Essentials of Fuzzy Modeling and Control. John Wiley, New York (1994)
14. Hampel, R., Wagenknecht, M., Chaker, N.: Fuzzy Control: Theory and Practice. Physika-Verlag, Heidelberg (2000)
15. Piegat, A.: Fuzzy Modeling and Control. Physica-Verlag, Heidelberg, New York (2001)
16. Jamshidi, M., Vadiee, N., Ross, T.J. (eds.): Fuzzy logic and control: software and hardware application. In: Jamshidi, M. (ed.) Prentice Hall Series on Environmental and Intelligent Manufacturing Systems, vol. 2. Prentice Hall, Englewood Cliffs (1993)
17. Kacprzyk, J., Yager, R.R., Zadrożny, S.: A fuzzy logic based approach to linguistic summaries of databases. Int. J. Appl. Math. Comput. Sci. 10(4), 813–834 (2000)
18. Takagi, T., Sugeno, M.: Fuzzy identification of systems and its applications to modeling and control. IEEE Trans. Syst. Man Cybern. SMC-15(1) (1985)
19. Vachkov, G., Kiyota, Y., Komatsu, K.: Identification of dynamic cause-effect relations for systems performance evaluation. Appl. Sci. Soft Comput. Adv. Soft Comput. 24, 187–194 (2004)
20. Yager, R.R., Filev, D.P.: Unified structure and parameter identification of fuzzy models. Sys. Man Cybern. 23(4) (1993)
21. Suna, Q., Li, R., Zhang, P.: Stable and optimal adaptive fuzzy control of complex systems using fuzzy dynamic model. J. Fuzzy Sets Syst. 133, 1–17 (2003)
22. Skrjanc, I.: Design of fuzzy model-based predictive control for a continuous stirred-tank reactor. In: 12th Zittau Fuzzy Colloquium, Proceedings of East-West Fuzzy Colloquium, pp.126–139. Zittau, Hochschule Zittau, Goerlitz, Germany (2005)
23. Kondratenko, Y.P., Al Zubi E.Y.M.: The optimisation approach for increasing efficiency of digital fuzzy controllers. In: Annals of DAAAM for 2009 & Proceeding of the 20th Int. DAAAM Symp. Intelligent Manufacturing and Automation, Published by DAAAM International, Vienna, Austria, pp. 1589–1591 (2009)
24. Hayajneh, M.T., Radaideh, S.M., Smadi, I.A.: Fuzzy logic controller for overhead cranes. Eng. Comput. 23(1), 84–98 (2006)
25. Kondratenko, Y.P., Kozlov, O.V., Klymenko, L.P., Kondratenko, G.V.: Synthesis and research of neuro-fuzzy model of ecopyrogenesis multi-circuit circulatory system. In: Advance Trends in Soft Computing, Studies in Fuzziness and Soft Computing, vol. 312, pp. 1–14. Springer-Verlag, Berlin (2014)
26. Kondratenko, Y.P., Kozlov, O.V.: Mathematic modeling of reactor's temperature mode of multiloop pyrolysis plant. In: Modeling and Simulation in Engineering, Economics and Management, K.J. Engemann, A.M. Gil-Lafuente, J.L. Merigo (Eds.), International Conference MS 2012, New Rochelle, NY, USA (May 30–June 1, 2012), Proceedings. Lecture Notes in Business Information Processing, vol. 115, pp. 178–187 (2012)
27. Kondratenko, Y.P., Shishkin, A.S.: Nonlinear regression mathematical model of magnetic systems for signal registration slip. J. NTU KPI Ser. Instrum. 33, 127–134 (2007). (in Russian)

Synthesis and Optimization of Fuzzy Controller for Thermoacoustic Plant

Y.P. Kondratenko, O.V. Korobko and O.V. Kozlov

Abstract The paper is devoted to the synthesis of digital system for control of thermoacoustic plant. Based on the analysis of thermoacoustic systems, the main tasks of the synthesized system are shown. Its structure and main components are described. Using the created system, the traditional PD and fuzzy Mamdani and Sugeno controllers are implemented and compared. Best regulator is then additionally tuned using the described input terms optimization procedure. The comparative analysis of initial and optimized controllers is shown using graphs.

1 Introduction

One of the important components of developed industry is the use of heat engines that convert heat energy into mechanical or electrical. Conventionally, heat machines can be divided into two groups: direct effect mechanisms (heat engines) and reverse action mechanisms (heat pumps, refrigerators). The most common are the mechanical heat machines in which the mutual energy conversion is based on the use of special mechanical devices, such as piston mechanisms (internal combustion engines, steam engines, stirling engines) and rotary devices (gas and steam turbines).

Thermoacoustic devices [1] (TAD) are the newest type of unconventional heat machines. Their work is based on the mutual transformation of heat and acoustic

Y.P. Kondratenko (✉)
Intelligent Information Systems Department, Petro Mohyla Black Sea State University, Mykolayiv, Ukraine
e-mail: y_kondratenko@rambler.ru

O.V. Korobko · O.V. Kozlov
Computer-Aided Control Systems Department, National University of Shipbuilding, Mykolayiv, Ukraine
e-mail: oleksii.korobko@nuos.edu.ua

O.V. Kozlov
e-mail: oleksiy.kozlov@nuos.edu.ua

© Springer International Publishing Switzerland 2016
L.A. Zadeh et al. (eds.), *Recent Developments and New Direction in Soft-Computing Foundations and Applications*, Studies in Fuzziness and Soft Computing 342, DOI 10.1007/978-3-319-32229-2_31

453

energies [2]. The main feature of thermoacoustic systems is that unlike the other heat machines, the powerful acoustic pulses [3], generated by TAD, are the carrier of mechanical energy and the "executive mechanism" of thermodynamic cycle.

Considering the reversibility of thermoacoustic effect, TAD can be subdivided into two main types:

- thermoacoustic engines (TAE), which consume supplied heat energy and produce acoustic wave inside the resonator [4, 5];
- thermoacoustic refrigerators (TAR) and heat pumps (TAHP) which consume energy from supplied acoustic wave and produce cooling or heating power, respectively [1, 6].

The schematic of simple thermoacoustic refrigerator is shown in Fig. 1. It consists of sound wave electromechanical generator, hollow resonator filled with gas medium, heat exchange surfaces (T_C, T_H) for heat adding and subtraction, and a special heat exchange surface (the stack), which is the main catalyst of acoustic and thermal energies mutual conversion in TAD.

Absence of moving parts [1] in such heat engines increases their reliability and reduces energy losses in the mechanical connections.

However, for high-power thermoacoustic systems, the acoustic pressure level of 150–180 dB [4, 7] must be maintained in resonator, which leads to the appearance of parasitic nonlinear effects [2]. Therefore, such heat machines are characterized by high sensitivity to the working fluid properties (viscosity, density, thermal conductivity) as well as the environment effects (temperature, pressure, etc.). Furthermore, all TADs that are driven by thermoacoustic engines consume external heat during their working process. This external power characteristic (power, temperature) can wary and depend on outer uncontrolled conditions, such as the working regime of heat energy source and surrounding temperature.

An effective work of TAD in such conditions is possible only if the different nature values of the internal parameters (acoustic, hydrodynamic, thermal) are stabilized. For that reason, it is reasonable to develop the automatic system for thermoacoustic processes control.

The main motivation of this research is to design the digital control system that is capable of output temperature stabilization of thermoacoustic engine-driven plant under the unstable conditions caused by the both internal and external factors.

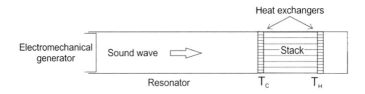

Fig. 1 Structure of the thermoacoustic refrigerator with electromechanical generator of sound waves

2 Synthesis of Discrete Control System for the Thermoacoustic Engine-Driven Plants

2.1 Structure and Main Components of a TAD Control System

Figure 2 shows the schematic of created experimental control system for the thermoacoustic engine-driven plant. In this plant, the supplied heat Q_{IN} is converted into the acoustic wave energy by the TAE branch. This acoustic energy is then used by the heat pump (TAHP) to produce output heat Q_{OUT} with higher temperature. Q_{ATM} is the cooling power with atmospheric temperature.

Due to the thermoacoustic process features, the computer system combines rapid data (acoustic pressure actuations, current, and voltage oscillations) and slow data signals (temperature variations).

The main elements of the system [7] are personal computer (PC); pressure (PS1, …, PS4), temperature (TS1, …, TS4), current (CS), and voltage (VS) sensors; a programmable logic controller (PLC) ICPDAS μPAC 7186EX-SM that collects the thermal behavior data of TAD and microcontroller (MCU) STM32F407VGT6, that is used to collect rapidly changing data from the system pressure, current, and voltage sensors and transfer it to PC.

STM32F4 microcontroller has three analog-to-digital converters, which are characterized by 12-bit resolution and conversion frequency of up to 2.4 Msps and hardware implemented floating-point unit and therefore is suitable for digital signal processing tasks [8]. This provides high performance and applicability of the proposed computer system for any thermoacoustic unit, since the operating frequency range of TAD does not exceed tens of kHz. PLC performs data acquisition

Fig. 2 Structure of the computer system for control of thermoacoustic plants

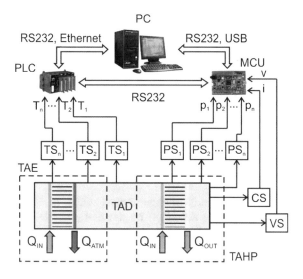

of the status of TAD peripheral equipment and sends it to PC by the Ethernet network.

The designed system allows the communication between all components via RS232 network; also to enable the transfer of large amounts of data, PLC supports the Ethernet connection with PC and MCU can communicate with PC via USB. This grants maximum flexibility to the designed control system for the thermoacoustic engine-driven plants.

Created software for PC [6] allows storing the received from the controller information in the archive database, execution of the necessary algorithms for digital processing of measured signals, and displaying all available information on the user's screen.

As a result, the proposed structure of created control system allows the registration and ongoing monitoring of the thermoacoustic processes and is suitable for the real-time control algorithms implementation for both thermoacoustic engines and refrigerators.

2.2 Experimental Investigation of the TAE-Driven Thermoacoustic Plants

The tests proposed by the authors discrete control system were conducted using the experimental thermoacoustic engine (Fig. 8) [7, 19], with a resonator length of 1010 mm and diameter of 46 mm, which was filled with air at atmospheric pressure. Engine was coupled with thermoacoustic heat pump, which acted as a payload of the system.

Thermoacoustic engine-starting process is presented in Fig. 3 where on the top are shown rooted mean square values of calculated [8] pressure $p_0(t)$ and particle velocity $v_0(t)$ and on the bottom temperature changes in points TS1 and TS2 (Fig. 4).

Further research showed that in given configuration the maximum pressure magnitude inside the TAE resonator depends on the length of TAHP resonator branch. Thus, the control of TAD working process can be made by the adjustments in TAHP resonator length. Therefore, let us consider the synthesis of TAD control system based on the change of the TAHP resonator length.

2.3 Synthesis of the Control System for Thermoacoustic Plant

Based on the results of experimental studies [7, 9] (Fig. 3), it can be concluded that the TAE can be described by the first-order transfer functions that represent the pressure $p(t)$ and particle velocity $v(t)$ which are generated due to the supplied power $P_{supl}(t)$.

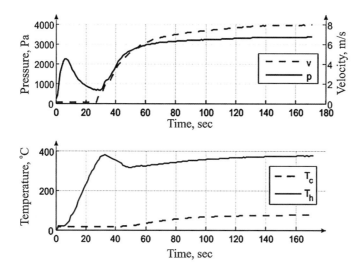

Fig. 3 Results of experimental studies of the TAE parameters

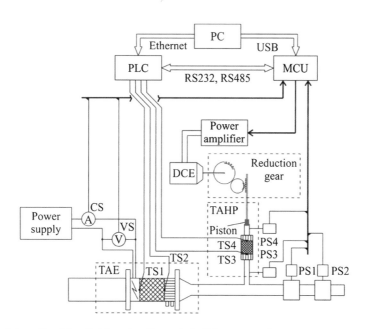

Fig. 4 Schematic of the digital system for control of thermoacoustic processes

The exact values of pressure (1) and velocity (2) transfer functions (TF) were identified using the least squares regression method [10] with the desired functionality in the form (3).

$$W_p(p) = \frac{14.66}{48.72p + 1}, \tag{1}$$

$$W_U(p) = \frac{5 \times 10^{-5}}{41.36p + 1}, \tag{2}$$

$$\min_{k,T} J(k, T) = \min_{k,T} \sum_{j=1}^{N} \left[\frac{k}{T} e^{\frac{-t}{T}} - y_j \right]^2, \tag{3}$$

where k, T—parameters of first-order link, y—experimental sample, and N—experimental sample length.

It should be noted that TF $W_U(p)$ (4) represents volume velocity in TAE resonator and it can be calculated as $U(t) = v(t)S$ with S is the cross-sectional area of TAE resonator.

TAHP branch (Fig. 4) is represented by the transfer matrix $W_{TAHP}(p)$ that is based on the Rott's representation of continuity and momentum equations [1] and identified model (6) of temperature difference $W_T(p)$ between the stack hot TS4 and cold TS3 ends (Fig. 4).

$$W_T(p) = \frac{7.18 \times 10^{-4} p + 6.05 \times 10^{-6}}{p^2 + 0.12p + 5.49 \times 10^{-4}}. \tag{4}$$

The change of the TAHP resonator length is simulated by the system that transforms DC engine $W_{DCE}(p)$ rotational movement into linear using the reduction gear. The effect of resonator length change is simulated by the experimentally obtained nonlinear dependency of pressure magnitude $p(t)$ from the acoustic wave frequency inside the TAHP resonator. The relation between TAHP resonator length L_{res} and acoustic wave frequency f_{res} is described by Eq. (5) [2].

$$f_{res} = \frac{a}{\lambda} = \frac{\sqrt{\gamma kT/m}}{nL}, \tag{5}$$

where f—sound wave frequency; a—speed of sound; λ—sound wave length; k—Boltzmann's constant; γ—the adiabatic coefficient of gasses; T—absolute temperature; m—molecular mass of gas; L—resonator length; and n—acoustic wave length multiplier.

2.4   Resulting Control System Simulation

The resulting model of TAD is shown in Fig. 5. The input values for the system are supplied to TAE electric power P_{supl}, starting length of TAHP resonator L_{TAHP}, value of cold end of TAHP stack T_C, value of sound speed $a = 343$ m/s, and desired temperature T_{des} of TAHP stack hot end.

The acoustic wave length multiplier n for current setup is equal to 2 (half wave length resonator), k_R is the reduction gear coefficient which transform rotational movement of the DC engine into linear movement of the TAHP resonator end piston. The acoustic pressure magnitude is adjusted by the value of the nonlinear dependency $p_m(f_{\text{res}})$ which is normalized by the pressure value at the resonant frequency (5). The output of the system is the value of the TAHP stack hot end $T_H(p)$.

The desired value of TAHP stack hot end is set to $T_{\text{des}} = 45$ °C. It should be noted that synthesized system uses the proportional-derivative (PD) regulator, whose transfer function in continuous form defined as

$$W_{\text{REG}}(p) = 0.05 + 0.3\frac{p}{0.01p + 1} \tag{6}$$

Taking into account that the designed system uses the digital controllers for all control operations, authors implemented digital PD controller [11] using the method of substitutions. This is one of the approximate methods for the discrete systems synthesis. Approximate conversion provides less accurate results that the direct Z-

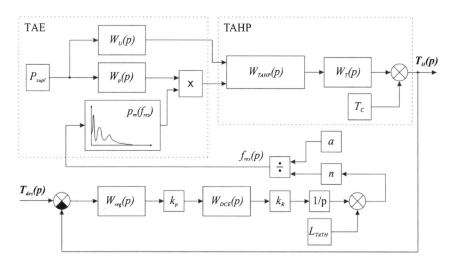

Fig. 5 The resulting model of TAD for indirect control by TAHP resonator length adjustment

transform approach, but can be calculated much easier. Another advantage of using the approximate methods for continuous dynamic objects transform is that, unlike the precise Z-transformation, the total discrete transfer function connected in series is the product of discrete TFs found for each of the continuous transfer functions separately.

When determining the approximate discrete transfer function, better accuracy results of the discretized model can be achieved by using the Tustin substitutions method

$$H(z) = H(p)\big|_{p=\frac{2(z-1)}{T_0(z+1)}}. \tag{7}$$

Applying the Tustin transformation (7) to the synthesized continuous controller (6), authors obtain its equivalent discrete form (8) with the discretization period $T_0 = 0.1$ s due to the sample time of thermocouple data.

$$W_{\text{reg}}(z)\big|_{T_0=0.1} = \frac{6.037z - 5.374}{z - 0.06542}. \tag{8}$$

The simulation results of synthesized TAD model working process are shown in Fig. 6. They demonstrate that the implementation of control system decreases the transient response time (with 5 % setting bounds) from $t_{\text{trans}}^{\text{TAD}} = 365.6$ to $t_{\text{trans}}^{\text{PD}} = 154.73$ s.

Thermoacoustic systems can be characterized by a number of unpredictable factors, which depends on surrounding temperature, pressure, and particular qualities of thermoacoustic plant, such as the roughness of the resonator material and exact qualities of working fluid. Therefore, it is reasonable to consider the implementation of fuzzy controllers [12] to the control system of thermoacoustic plant.

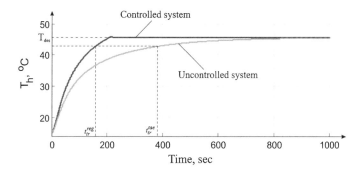

Fig. 6 Simulation results of synthesized TAD model

3 Synthesis of Fuzzy Control System for the Thermoacoustic Engine-Driven Plant

Fuzzy control provides a formal methodology [13] for representing, manipulating, and implementing a human's heuristic knowledge about how to control a system.

Fuzzy controllers are of interest primarily to manage objects that either could not be described or could be described with great difficulties. However, even for control objects for which could be obtained mathematical model, these regulators are often better than others, because they allow obtaining higher quality (fewer errors in transient and steady state) of automatic control.

Since the control algorithms based on fuzzy logic can be implemented using only a computer, the automatic control system with fuzzy controller [14, 15] for thermoacoustic plants is digital.

The microcontroller-based fuzzy controller block diagram for thermoacoustic plant is given in Fig. 7, where we show a fuzzy controller embedded in a closed-loop control system.

The fuzzy controller has four main components [16]: The "rule-base" holds the knowledge, in the form of a set of rules, of how best to control the system; the inference mechanism evaluates which control rules are relevant at the current time and then decides what the input to the plant should be; the fuzzification interface simply modifies the inputs so that they can be interpreted and compared to the rules in the rule-base; and the defuzzification interface converts the conclusions reached by the inference mechanism into the inputs to the plant.

Authors designed [17] fuzzy controllers of Mamdani and Sugeno types for the control system of thermoacoustic plant. Comparison of their transient responses is given in Fig. 8.

It should be noted that both fuzzy controllers have two input variables (temperature error and its first derivative). Each input is characterized by the five triangle input terms (Fig. 9). Output of Mamdani controller has seven terms. The output of Sugeno controller consists of the first-order polynomial. The resulting rule-base for Mamdani controller is given in Table 1.

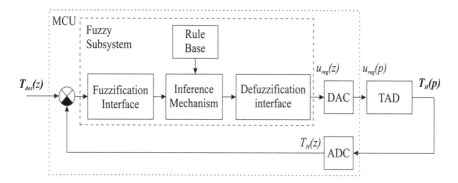

Fig. 7 Block diagram for the TAD control system with fuzzy controller

Fig. 8 Comparison of designed fuzzy and continuous controllers

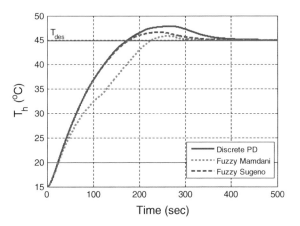

Fig. 9 Input terms of Sugeno-type fuzzy regulator

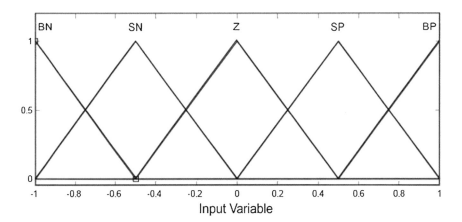

Table 1 Rule base for designed Mamdani controller

		Error derivative, dε/dt				
		BN	SN	Z	SP	BP
Error, ε	BN	BN	BN	BN	MN	SN
	SN	BN	MN	MN	SN	SP
	Z	SN	SN	Z	SP	SP
	SP	SN	SP	MP	MP	BP
	BP	SP	MP	BP	BP	BP

Comparison of transient responses of designed fuzzy and discrete PD controllers shows that best results of designed control system can be achieved with Sugeno type of fuzzy controller.

4 Fuzzy Controllers Optimization

One of the advantages of fuzzy logic systems is their flexibility in terms of tuning procedures for designed controllers: You can choose different types of controllers, implement different types and shapes of their terms, etc.

In this article, authors demonstrate the approach of input terms optimization based on the gradient descend algorithm [18]. The goal (9) of optimization procedure was to minimize the integral of squared error $f(t, P)$ between the simulated $y_{sim}(t, P)$ and desired $y_{des}(t, P)$ transient responses.

$$\min_{P} f(t, P) = \min_{P} \left[\int e(t, P)^2 dt \right]$$
$$= \min_{P} \left[\int (y_{des}(t, P) - y_{sim}(t, P))^2 dt \right]. \tag{9}$$

The vector of parameters P is the array of fuzzy controller terms vertexes for each input (10).

$$P = \left\{ p_{i,j}^k \right\}, k = \{1, 2\}, i = \{1, \ldots, 5\}, j = \{1, 2, 3\}. \tag{10}$$

where k is the number of input, i is the number of term, and the j is the number of vertex of term i.

For optimization of the thermoacoustic plant fuzzy controller, authors formulate the desired transient response by the formula (11).

$$y_{des}(t) = 15 + 30 \left(1 - e^{-t/2} \right). \tag{11}$$

Based on the results of previous section for optimization, authors selected the Sugeno fuzzy controller as the one with the best response characteristics.

The appearance of optimized [19] input terms of Sugeno fuzzy controller is given in Fig. 10.

The transient responses of the designed control systems of the TAE-driven thermoacoustic plant with initial and optimized Sugeno fuzzy controllers [20, 21] along with the desired transient response are given in Fig. 11.

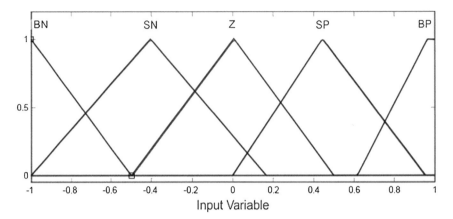

Fig. 10 Optimized input terms of Sugeno-type fuzzy regulator

Fig. 11 Transient responses of initial and optimized fuzzy control systems for thermoacoustic plant

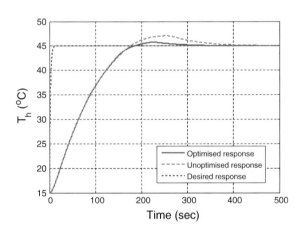

Figures 12 and 13 show the characteristic surfaces of the designed Sugeno fuzzy controller before and after optimization, respectively.

Figure 11 clearly shows that the optimization of the input terms of designed Sugeno fuzzy controller allowed to lower the overshot of system from 4.6 to 1.71 % and also decreases the setting time (in terms of 3 % setting bounds) of control system transient response from $t_{\text{init}}^{\text{sug}} = 290.11$ to $t_{\text{opt}}^{\text{sug}} = 156.76$ s.

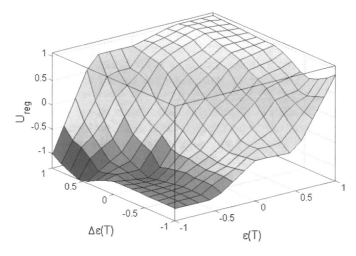

Fig. 12 Characteristic surface of initial Sugeno-type fuzzy controller for digital control systems of thermoacoustic plant

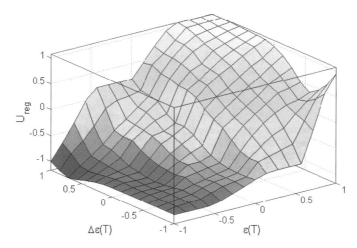

Fig. 13 Characteristic surface of optimized Sugeno-type fuzzy controller for digital control systems of thermoacoustic plant

5 Conclusions

Thermoacoustic plants are the new type of heat machines. Their work is based on the mutual transformations of the acoustic and heat energies. Since their work is based on the energy transfers by the acoustic waves, their working process is characterized by the strong dependency on the external uncontrolled factors and also internal unpredicted characteristics.

To accomplish specified goal, authors designed the mathematical model that describes the working process of thermoacoustic plant. Experimental setup used by the authors combines the thermoacoustic engine for acoustic energy generation and heat pump for its transformation into the useful external energy.

Mathematical modeling of thermoacoustic system transient response with designed conventional PD, fuzzy Mamdani, and Sugeno controllers showed that the created fuzzy controller of Sugeno type demonstrates best results.

For further tuning of designed fuzzy controller, authors implement the procedure of the controllers input terms optimization based on the gradient descend minimization of goal function. The goal function used the integral of squared error between simulated and ideal transient responses of designed control system of thermoacoustic engine-driven experimental plant.

Comparative analysis of the control systems with initial and optimized Sugeno-type fuzzy controllers shows that the optimum tuning of the input terms of fuzzy controller allowed decreasing the transient response time from 290.11 to 156.76 s and system overshot from 4.6 to 1.71 %.

Thus, the main contribution of authors includes the development of simplified, suitable for simulation mathematical model of TAD, use of fuzzy control methods in the stabilization system of TAHP output temperature, and implementation of fuzzy controller's optimization algorithm for overall system transient response improvement.

References

1. Swift, G.W.: Thermoacoustics—A Unifying Perspective for Some Engines and Refrigerators. Acoustical Society of America (2002)
2. Rayleigh, L.: The Theory of Sound, vol. 2, 2nd edn. Dover, NewYork (1945)
3. Bailliet, H., Lotton, P., Bruneau, M., Gusev, V., et al.: Acoustic power flow measurement in a thermoacoustic resonator by means of laser Doppler anemometry (L.D.A.) and microphonic measurement. Appl Acoust. **60**, 1–11 (2000)
4. Wheatley, J.C., Swift, G.W., Migliori, A.: The natural heat engines. Los Alamos Sci. **14**(2), 2–33 (1986)
5. Mehta, S.M., Desai, K.P., Naik, H.B., Atrey, M.D.: Design of standing wave type thermoacoustic prime mover for 300 hz operating frequency. In: Proceedings of the 16th International Cryocooler Conference, USA, pp. 343–352 (2011)
6. Ryan, T.S.: Design and Control of a Standing-Wave Thermoacoustic Refrigerator. University of Pittsburgh (2006)
7. Kondratenko, Y., Korobko, V., Korobko, O.: Distributed Computer System for Monitoring and Control of Thermoacoustic Processes. In: Proceedings of the 7th IEEE International Conference IDAACS'2013, vol. 1, pp. 249–253. Berlin, Germany, 12–14 Sept 2013
8. Kondratenko, Y., Korobko, V., Korobko, O.: Microprocessor system for thermoacoustic plants efficiency analysis based on a two-sensor method. Sens. Transducers **24**(Special Issue), 35–42 (2013)
9. Biwa, T., Tashiro, Y., Nomura, H., Ueda, Y.: Experimental verification of a two-sensor acoustic intensity measurement in lossy ducts. J. Acoust. Soc. Am. **124**(3), 1584–1590 (2008)

10. Wolberg, J.: Data Analysis Using the Method of Least Squares: Extracting the Most Information from Experiments. Springer (2005)
11. Maler, O., Pnueli, A., Sifakis, J.: On the synthesis of discrete controllers for timed systems. Lect. Notes Comput. Sci. **900**, 229–242 (1995)
12. Tanaka, K., Wang, H.O.: Fuzzy Control Systems Design and Analysis: A Linear Matrix Inequality Approach. Wiley, New York, USA (2001)
13. Zadeh, L., Kacprzyk, J.: Fuzzy Logic for the Management of Uncertainty. Wiley, New York (1992)
14. Banks, W., Hayward, G.: Fuzzy Logic in Embedded Microcomputers and Control Systems. Byte Craft Limited, North Waterloo, Ontario Canada (2002)
15. Amador-Angulo, L., Castillo, O., Pulido, M.: Comparison of fuzzy controllers for the water tank with type-1 and type-2 fuzzy logic. In: IFSA World Congress and NAFIPS Annual Meeting (IFSA/NAFIPS), 2013 Joint, pp. 1062–1067 (2013)
16. Pedrich, W.: On generalized fuzzy relational equations and their applications. J. Math. Anal. Appl. **107**, 520–536 (1985)
17. Jamshidi, M., Vadiee, N., Ross, T.J.: Fuzzy Logic and Control. Software and Hardware Applications. PTR Prentice-Hall, Englewood Cliffs, NJ (1993)
18. Nesterov, Y.: Introductory Lectures on Convex Optimization. A Basic Course. Springer (2004)
19. Shapiro, J.F.: Mathematical Programming: Structures and Algorithms. Wiley-Interscience, New York (1979)
20. Yager, R.R., Filev, D.P.: Essentials of Fuzzy Modeling and Control. Wiley (1994)
21. Hampel, R., Wagenknecht, M., Chaker, N.: Fuzzy Control: Theory and Practice. Springer, Berlin and Heidelberg (2000)

Analytical Models of WLAN Standard IEEE 802.11

F.H. Mammadov and M.Y. Orudjova

Abstract Wireless LAN (WLAN) is considered having "point-to-point" mode, consisting of an information maintenance device (IMD) and subscriber stations distributed across multiple identical subnets interacting by a common wireless radio link. Basing on Laplace transform, analytical models of service and information delivery processes at the stations of the network subnets have been developed. Methods for calculating probability-time characteristics of the service and information delivery processes at the stations of the subnets and in the network as a whole have been proposed.

Keywords Wireless LAN · Probability characteristics · Process of service · Process of delivery · Service interval · Delivery interval · The Laplace transform · Distribution density

Wireless LAN (WLAN) is considered having ad hoc "point-to-point" mode, consisting of an information maintenance device (IMD), wireless radio channel (WRC), and stations divided into 3 subnets different for the intensity of incoming message. Physical structure of the network under consideration and timing diagrams of its subnets have been developed. Basing on Laplace transform, analytical models of service and information delivery processes at the subnets have been developed. Methods for calculating probability-time characteristics of the subnets and in the network as a whole have been proposed.

Consider a WLA [1, 2] having ad hoc "point-to-point" mode, consisting of N stations and IMD, which interact with each other by a WRC with the length of Dm (Fig. 1).

Besides, the stations of this network are divided into three (first, second, and third) subnets, three logical groups of stations using time division of general WRC

F.H. Mammadov (✉) · M.Y. Orudjova
Azerbaijan Technical University, Baku, Azerbaijan
e-mail: famil_mammadov@mail.ru

M.Y. Orudjova
e-mail: abdullayevam@yahoo.com

© Springer International Publishing Switzerland 2016   469
L.A. Zadeh et al. (eds.), *Recent Developments and New Direction
in Soft-Computing Foundations and Applications*, Studies in Fuzziness
and Soft Computing 342, DOI 10.1007/978-3-319-32229-2_32

Fig. 1 The physical structure of the wireless LAN

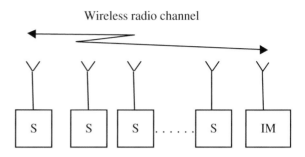

and differing in the intensity of incoming message stream. Message servicing coming from the second and third subnets is carried out by IMD without backlog with the retention of WRC.

The subnet logging into each station which receives the message flow with great intensity λ_f is called the first subnet, with an average intensity λ_s is called the second subnet, and the low intensity λ_t is called the third subnet.

The first subnet consists of N_f stations, divided into M_f similar subnets. The second subnet consists of N_s stations, divided into M_s similar subnets. Stations of the first and second subnets use a protocol for synchronous random-temporary access, respectively, at the intervals T_f and T_s, besides $T_f/T_s = \delta$-integer. The thirds subnet is allocated with a separate window in a time division in cycle intervals T_s. The length of M_i cycle in T_s intervals equals to $M_i = \delta M_f + M_s + 1$.

In the process, the message transfer as frames for the second subnet is invested consistently in discrete time intervals n_{serv} for service maintenance of the third subnet in cycle intervals T_{serv}.

Since in a wireless radio channel, there exists error probability in the transmitted packet, error checking errors in it are carried out by using jam-resistant code and a system solving feedback with the expectation of DFB-EXP, by the way of transmitting acknowledgment character with the first and second subnets in n_{acf} and n_{acs} discharges, respectively.

Message transmitted from the stations of the second and third subnets is considered as maintained if no errors are found in them, no conflicts are observed and service unit returns them the response command. Otherwise, delivery and service are repeated.

Message transfer in the form of a frame between the stations I and j of the first subnet is performed in the following sequence. Station i, which has a data frame in its buffer, starts to transfer it to the station j of the given subnet.

The frame transmitted at V_s while being distributed through the wireless channel, on which t_{dij} time is spent comes into the station j, where over time t_{dacf} its decoding is carried out. If during decoding errors are observed in the station j, accordingly a positive acknowledgment is sent on the station i, on the distribution of which t_{sya} t_{dji} time where during t_{dacf} its decoding is performed. Timing diagram of message transmitting in the station of the first subnet with an available access is shown in Fig. 2.

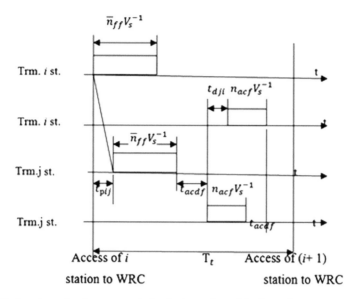

Fig. 2 Timing chart of a frame transmission in the station of the first subnet

From the timing diagram of a frame transmission in the station of the first subnet shown in Fig. 2, the duration of the time interval T_t for transmitting a frame of the said subnet is defined by the following expression:

$$T_t = (\bar{n}_{ff} + n_{fd} + n_{acf})V_s^{-1} + t_{fdf} + t_{acdf}, \quad T_t = (\bar{n}_{ff} + n_{fd} + n_{acf})V_s^{-1} + t_{fdf} + t_{acdf},$$
$$\bar{n}_{ff} = r_{prf} + r_{ff} + r_{af} + r_{cf} + r_{ckf} + k_f,$$
$$n_{ff} = t_{ff}V_s, \quad t_{ff} = t_{dijs} + t_{djif}, \quad t_{dijf} = t_{df}/2,$$
$$n_{acf} = r_{af} + r_{cf} + k_f, \quad r_{af} = \log_2 N_{\Pi} + 1\text{-integer}, \quad t_{dijf} = n_{dff}V_s^{-1}, \quad t_{acdf} = n_{dacf}V_s^{-1},$$
$$(1)$$

where V_s—transmission rate in the wireless radio channel (Mbit/s), t_{fds} and t_{acds}—time of frame decoding and the acknowledgment character of the first subnet, $\bar{n}_{ff}$—the average length of the frame in the first subnet, n_{fd}—discrete frame distribution time of the first subnet at WRC, n_{acf}—length of the acknowledgment character of the first subnet, r_{prf}—the length of the first frame preamble of the first subnet, r_{ff}—length of the flag field of the frame of the first subnet, r_{af}—the length of the address field of the frame of the first subnet, r_{cf}—length of the control field of the frame of the first subnet, r_{ckf}—length of the check digit of the frame of the first subnet, k_f—length of the information part of the package for the first subnet, t_{dijf} and t_{djif}—time for frame distribution from station i to j and from j to i for the first subnet, n_{dff} and n_{dacf}—discrete time for frame decoding and the acknowledgment character of the first subnet.

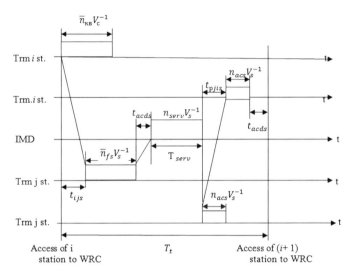

Fig. 3 Timing chart of a frame transmission in the station of the second and the third subnets

Let us consider now the sequence of frame transmission between the station i and j of the second and third subnets (Fig. 3). The station i of the second subnet, which has data frame in its buffer, starts to transfer it to the station j of the given subnet. The frame transmitted at V_s speed on the distribution through wirelessly radio channel of which t_{dijt} time is spent comes for the station j, where after decoding it arrives at the IMD. After frame servicing, IMD at the station j on the same subnet transmits a positive acknowledgment character to the station i, on the distribution of which t_{jit} time is spent, where its decoding is performed.

Following from the timing chart of a frame transmission in the stations of the second and the third subnets, the duration of the time interval T_t for transmitting a frame of the said subnet is defined by the following expression:

$$
\begin{aligned}
T_s &= T_{ds} + n_{\text{serv}} T_{\text{serv}}, \\
T_{ds} &= (\bar{n}_{fs} + n_{ds} + n_{\text{acs}}) V_s^{-1} + t_{fs} + t_{\text{acs}}, \\
\bar{n}_{fs} &= r_{\text{prs}} + r_{fs} + r_{\text{as}} + r_{\text{cs}} + r_{\text{cks}} + k_s, \\
n_{fs} &= t_{fs} V_s, t_{fd} = t_{\text{dijs}} + t_{\text{djis}}, t_{\text{dijs}} = t_{ds}/2, \\
&= r_{\text{as}} + r_{\text{cs}} + k_s, r_{\text{as}} = \log_2 N_{\Pi} + 1\text{-integer}, \\
t_{\text{dijs}} &= n_{\text{dfs}} V_s^{-1}, t_{\text{acds}} = n_{\text{dacs}} V_s^{-1},
\end{aligned}
\tag{2}
$$

where T_{ds}—duration of frame delivery interval on the second subnet, t_{fds} and t_{acds}—time of frame decoding and the acknowledgment character of the second subnet, $\bar{n}_{fs}$—the average length of the frame in the second subnet, n_{acs}—length of the acknowledgment character of the second subnet, n_{ds}—frame distribution time of the second subnet at WRC, r_{prs}—the length of the frame preamble of the second subnet, r_{fs}—length of the flag field of the frame of the second subnet, r_{as}—the length of the address field of the frame of the second subnet, r_{cs}—length of the control field of the frame of the second subnet, r_{cks}—length of the check digit of the frame of the second subnet, k_s—length of the information part of the package for the second subnet, t_{dijs} and t_{djis}—time for frame distribution from station i to j and from j to i for the second subnet, n_{dfs} and n_{dacs}—discrete time for frame decoding and the acknowledgment character of the second subnet.

Let us assume that Poisson flow enters the input of the buffer of each station at the first subnets with intensity λ_p at the interval T_p, and Poisson flow enters the inputs of the buffer at the stations of different subnets with intensity λ_v at the interval T_p, the service of which is conducted in discrete time. Such systems can be described [3] by a stochastic system.

Then, Laplace transform of the distribution density of the service interval of a delivery process of the first subnet at the interval T_f has the form:

$$
\begin{aligned}
&g_p(s) = Q_{fs}/(e^{-STM_a} - P_{fs}), \\
&Q_{fs} = q_a Q_{fc} Q_{fm}, \quad P_{fs} = 1 - Q_{fs}, \\
&Q_{fc} = (1-p)^n_{fc}, \quad Q_{fm} = (1 - q_a \rho_f)^{(N_f/M_f)-1}, \\
&n_{fc} = r_{fa} + r_{fc} + k_f, \quad T = V_s^{-1}, N_f/M_f\text{-integer},
\end{aligned}
\tag{3}
$$

where q_a—parameter random access to WRC, ρ_f—probability of buffer occupancy at the station of the first subnet, p—error probability in WRC, s—Laplace operator.

Laplace transform of the distribution density of the interval of a delivery and service of the second subnet at the interval T_s is determined by the following expression:

$$
\begin{aligned}
&g_s(s) = Q_{ss}/(e^{-STM_{int}} - P_{ss}), \quad Q_{ss} = q_s Q_{sc} Q_{sm} \bar{\Pi}_{serv}, \quad P_{ss} = 1 - Q_{ss}, \\
&Q_{sc} = (1-p)^n_{sc}, n_{sc} = r_{sa} + r_{sc} + k_s, \quad Q_{sm} = (1 - q_v \rho_s)^{(N_s/M_s)-1}, N_s/M_s\text{-integer},
\end{aligned}
\tag{4}
$$

wherein ρ_v—probability of buffer occupancy at the station of the second subnet, $\bar{\Pi}_{serv}$ probability of timely frame service [4], T, s, and p—are defined by the expression (3).

Laplace transform of the distribution density of the interval of a delivery and service of the third subnet at the interval T_t is defined as follows:

$$
\begin{aligned}
&g_t(s) = Q_{ts} g_{ts}(s)/1 - P_{ts} g_{ts}(s)), \\
&Q_{ts} = Q_{tc} \bar{\Pi}_{serv}, g_{st}(s) = e^{-STM_{int}}, \\
&P_{ts} = 1 - Q_{ts}, Q_{tc} = (1-p)^n_{tc}, \quad n_{tc} = r_{ta} + tc + k_t.
\end{aligned}
\tag{5}
$$

The probability of buffer occupancy of the stations at the first, the second, and the third subnets is, respectively, determined based on the equation of interference obtained in [4] work by the following system of equations:

$$
\begin{aligned}
&\rho_f = q_{if}\bar{n}_{sf}; \bar{n}_{sf} = (d/ds)\,g_f(s)|_{s\to 0},\\
&q_{if} = fT_p, \quad \rho_f < 1.\\
&\rho_s = q_{is}\bar{n}_{ss}; \bar{n}_{ss} = (d/ds)\,g_s(s)|_{s\to 0},\\
&q_{is} = \lambda_s T_s, \quad \rho_s < 1.\\
&\rho_t = q_{it}\bar{n}_{st}; \bar{n}_{st} = (d/ds)\,g_t(s)|_{s\to 0},\\
&q_{it} = \lambda_t T_s, \quad \rho_t < 1.
\end{aligned}
\tag{6}
$$

wherein T_f, T_s and $g_f(s)$, $g_s(s)$, $g_t(s)$ are determined by the expressions (1)–(5), respectively.

For the corresponding network, Laplace transform of the distribution density for frame delay at the stations of the first, the second, and the third subnets on the basis of the model obtained in [4] work are determined by the following expressions:

$$
\begin{aligned}
&f_{qf}(s) = ((1 - \rho_f)(1 - s)g_f(s))/(1 - s(1 - q_{if}) - q_{if}sg_f(s))\\
&f_{qs}(s) = ((1 - \rho_s)(1 - s)g_s(s))/(1 - s(1 - q_{is}) - q_{is}sg_s(s)),\\
&f_{qt}(s) = ((1 - \rho_t)(1 - s)g_t(s))/(1 - s(1 - q_{it}) - q_{it}sg_t(s)),
\end{aligned}
\tag{7}
$$

where $g_p(s)$, $g_s(s)$, $g_t(s)$ and ρ_p, ρ_s, ρ_t, q_{ip}, q_{is} and q_{it} are determined by (3)–(6), respectively.

We now turn to the probabilistic-time characteristics of the considered network. Since the average delay at the stations of the first, the second, and the third subnets will, respectively, has the form [4]:

$$
\begin{aligned}
&\bar{t}_{qf} = -(d/ds)f_{qf}(s)|_{s\to 0},\\
&\bar{t}_{qs} = -(d/ds)f_{qs}(s)|_{s\to 0},\\
&\bar{t}_{qt} = -(d/ds)f_{qt}(s)|_{s\to 0},
\end{aligned}
\tag{8}
$$

wherein $f_{qf}(s)$, $f_{qs}(s)$ и $f_{qt}(s)$ are defined by the expression (7).

Solving Eq. (8) with taking into consideration (3)–(5) and (7), we obtain final expressions for calculating the average delay time at the stations of the first, the second, and the third subnets, respectively :

$$
\begin{aligned}
&\bar{t}_{qf} = (M_{int}T_s/2(Q_{sf} - q_{if}M_{int}))(2 - q_{if}(M_{int} + 1)),\\
&\bar{t}_{qs} = (M_{int}T_s/2(Q_{ss} - q_{is}M_{int}))(2 - q_{is}(M_{int} + 1)),\\
&\bar{t}_{qt} = (M_{int}T_s/2(Q_{st} - q_{it}M_{int}))(2 - q_{it}(M_{int} + 1)),
\end{aligned}
\tag{9}
$$

wherein T_p is defined by the expressions (2), Q_{sf}, Q_{ss}, Q_{st}—(3)–(5), and q_{if}, q_{is}, q_{it}—(6).

Probability of timely delivery of packets is calculated in a stochastic timed out message in the first, second, and third subnets, which is given by the aging time of the geometric distribution with parameter γ_a will be, respectively:

$$\overline{\Pi}_{qf} = f_{qf}(\gamma_{sf}), \quad \gamma_{sf} = 1 - T_p/\overline{T}_{df},$$
$$\overline{\Pi}_{qs} = f_{qs}(\gamma_{ds}), \quad \gamma_{ss} = 1 - T_p/\overline{T}_{ds}, \tag{10}$$
$$\overline{\Pi}_{qt} = f_{qt}(\gamma_{dt}), \quad \gamma_{st} = 1 - T_p/\overline{T}_{dt},$$

where T_p is determined by the expression (2), $\overline{T}_{df}, \overline{T}_{ds}$ and $\overline{T}_{dt}$—the average allowed time for message aging in the first, the second, and the third subnets $f_{qf}(\gamma_{sf}), f_{qs}(\gamma_{ds}), f_{qt}(\gamma_{dt})$ are defined by the expression (7) with the replacement of them by $\gamma_{sf}, \gamma_{ss}, \gamma_{st}$, respectively, i.e.:

$$\overline{\Pi}_{qf} = ((1 - \rho_f)(1 - \gamma_{sf})g_f(\gamma_{sf}))/(1 - \gamma_{sf}(1 - q_{if}) - q_{if}\gamma_{sf}g_f(\gamma_{sf})),$$
$$\overline{\Pi}_{qf} = ((1 - \rho_s)(1 - \gamma_{ss})g_s(\gamma_{ss}))/(1 - \gamma_{ss}(1 - q_{is}) - q_{is}\gamma_{ss}g_s(\gamma_{st})), \tag{11}$$
$$\overline{\Pi}_{qf} = ((1 - \rho_t)(1 - \gamma_{st})g_t(\gamma_{st}))/(1 - \gamma_{st}(1 - q_{it}) - q_{it}\gamma_{st}g_t(\gamma_{st})),$$

wherein γ_{sf}, γ_{ss}, and γ_{st}—determined by the expression (10).

Data rate in the first, the second, and third, respectively, subnets of common application will be determined by the following expressions:

$$R_{sf}^{O\Pi} = \lambda_f k_f N_f,$$
$$R_{ss}^{O\Pi} = \lambda_s k_s N_s, \tag{12}$$
$$R_{st}^{O\Pi} = \lambda_t k_t,$$

where k_f, k_s, k_t are defined expressions (3)–(5), respectively. Data rate in the first, the second, and the third subnets of real time correspondingly equals to:

$$R_{sf}^{PB} = R_{sf}^{O\Pi}\overline{\Pi}_{qf},$$
$$R_{ss}^{PB} = R_{ss}^{O\Pi}\overline{\Pi}_{qs}, \tag{13}$$
$$R_{st}^{PB} = R_{st}^{O\Pi}\overline{\Pi}_{qt},$$

wherein $\overline{\Pi}_{qf}, \overline{\Pi}_{qs}, \overline{\Pi}_{qt}$ and $R_{sf}^{O\Pi}, R_{ss}^{O\Pi}, R_{st}^{O\Pi}$ are determined by (11) and (12) respectively.

Data rate of WLAN of general application and the real time is overall determined by the following expressions, respectively:

$$R_{c\Sigma}^{O\Pi} = \sum_{j=1}^{N_n} \lambda_{fj}k_{fj} + \sum_{i=1}^{N_E} \lambda_{si}k_{si} + \lambda_t k_t, \quad j = \overline{2,6}, \ i = \overline{1,5},$$

$$R_{c\Sigma}^{O\Pi} = \sum_{j=1}^{N_n} \lambda_{fj}k_{fj}\overline{\Pi}_{qfj} + \sum_{i=1}^{N_E} \lambda_{si}k_{si}\overline{\Pi}_{qsi} + \lambda_t k_t\overline{\Pi}_{qt} \tag{14}$$

wherein $\overline{\Pi}_{qfj}, \overline{\Pi}_{qsi}$ and $\overline{\Pi}_{qt}$, determined by the expression (10).

Using these expressions and taking into account the internal parameters of the used protocol, we can analyze probabilistic-time characteristics of a wireless local area network as a whole, as well as its subnets.

References

1. Stallings, W.: Wireless Communication Lines and Networks, Moscow, 640p. ISBN: 5-8459 0409-9 (2003)
2. Vishnevsky, V., Semenova, O.: Polling Systems: Theory and Applications in the Broadband Wireless Networks, p. 312. Technosphere, Moscow (2007)
3. Ivchenko, G.I., Kashimanov, V.A., Kovalenko, I.N.: Queueing Theory, 304p. ISBN №: 978-5-02119 (2012) (Iss..2, Corr. and add)
4. Mamedov, F.G., Orudzheva, M.J.: Basic analytical models of wireless LANs. J. Inf. Commun. Technol. **16**, 25–31 (2012)

Neural Network-Based Approach for Design and Modeling Evolution Processes of Economic Clusters

E.A. Babkin, N.A. Klimova and O.R. Kozyrev

Abstract We present here the new approach for design and modeling evaluation of economic clusters based on artificial neural networks platform. We show here the basic principles and discuss the application of the approach for Hopfield networks.

1 Introduction

The task of intentional design of economic and innovation clusters can be considered as one of the major activities of policy makers and regional business developers [1–3]. Large number of contributing agents and complex nonlinear dynamics of cluster networking system makes thorough analysis of alternatives during cluster design quite difficult and require application of computer-based decision support systems at the stage of design or the forecasting. Traditionally to automate the analysis, different analytical methods or simulation modeling are used. We believe that the design of innovative clusters in general and analysis of alternative scenarios of agglomeration and distribution in particular require application of soft methods and models which reflect evolutionary design and robust accommodation to the changing conditions. Traditional methods of operation research cannot provide needed means. It is known that application of branch-and-bound algorithms may lead to dramatic changes in solutions' structure even if the modification of initial conditions is minimal.

E.A. Babkin (✉) · N.A. Klimova · O.R. Kozyrev
Intrafab Pte. Ltd, 120 Lower Delta Road #09-07 Cendex Centre,
Singapore 169208, Singapore
e-mail: babkin@hrvaas.com

N.A. Klimova
e-mail: klimova@hrvaas.com

O.R. Kozyrev
e-mail: kozyrev@hrvaas.com

© Springer International Publishing Switzerland 2016
L.A. Zadeh et al. (eds.), *Recent Developments and New Direction
in Soft-Computing Foundations and Applications*, Studies in Fuzziness
and Soft Computing 342, DOI 10.1007/978-3-319-32229-2_33

477

Taking such a viewpoint, we offer a new approach to the design of innovative clusters and to the modeling of the processes of agglomeration and distribution based on dynamics of artificial neural networks.

2 Main Approach

In our studies of agglomeration and distribution of cluster's components, we use a generic structural model of a distributed innovation cluster of the regional or trans-regional scale. Multiple physical facilities of the innovation regional or trans-regional cluster are linked with each other and form the cluster network infrastructure. Different consumers of the cluster products and services are linked with particular physical facilities by logistic services. The consumers expose multiple demand queries specifying individual or combined services or products. In our model, products and services of the cluster correspond to the economic agents that produce a particular demanded product or service. Individual entrepreneurs, innovative firms, research institutions, and other kinds of organization represent economic agents. From the systemic point of view, agglomeration of some subset of the agents may have holistic effects, bring extra added value, and increase entrepreneurial opportunities. In such situation, design of alternative structures of the agglomeration agents' groups and scenarios of their distribution among the cluster's physical facilities stands for the key research issue. In our model, we specifically emphasize such systemic phenomenon and call joint reinforcement and business improvement effects of agent's agglomeration as business contiguity and actively exploit it in during modeling.

To design agglomeration and distribution scenarios and model their dynamics, we offer the using of two-stage approach. The first stage of the approach is design of agglomeration groups from a set of individual economic agents. The second stage is distribution of the designed agglomeration groups among the physical facilities of the cluster network infrastructure. For each stage, we define various domain-specific constraints, including business contiguity of the economic agents' business security constraints, limitations of facility's capacities as well as specify optimum criteria. Our principal objective was to minimize the total transaction costs of consecutive provisioning of a set of consumers demand queries. From the mathematical point of view, the specified problem represents a kind of NP-complete nonlinear optimization problems of integer programming. So far, different task-specific approaches were proposed such as branch-and-bound method with a set of heuristics [4] and probabilistic algorithms etc. However, not many of them exploit benefits of parallel processing and grid technologies. In such circumstances, our strategic goal was developing expressive mathematical model and computationally efficient methods of optimization which will enforce next-generation decision support systems for design of the structure of innovation clusters.

The following mathematical model is proposed to define precisely different elements and constraints of the problem considered (Table 1).

Table 1 Formalization of the problem

The name	The designation
Characteristics of the cluster network infrastructure	
The set of economic agents	$\mathbf{D^G} = \left\{ d_i^g / i = \overline{1,I} \right\}$
The vector of agents' average establishment size	$\boldsymbol{\rho} = \{\rho_i\}$
The vector of agent's instantiation numbers	$\boldsymbol{\pi} = \{\pi_i\}$
The matrix of business contiguity of agents	$\mathbf{A^G} = \left\| a_{ii'}^g \right\|$, where $a_{ii'}^g = 1$ if combination of ith and i'th agents increase entrepreneurial opportunities, $a_{ii'}^g = 0$—otherwise
Characteristics of consumers' demand queries	
The set of consumers' demand queries	$\mathbf{Q} = \left\{ q_p / p = \overline{1, P_0} \right\}$
The matrix of agents involvement during provisioning demand queries	$\mathbf{W^Q} = \left\| w_{pi}^Q \right\|$, where $w_{pi}^Q = 1$ if demand query p uses product or service provisioned by the ith agent, $w_{pi}^Q = 0$—otherwise
The matrix of frequencies of demand queries issued by the consumers	$\mathbf{\Lambda^Q} = \left\| \xi_{kp}^Q \right\|$, where ξ_{kp}^Q is the frequency of the issuing of the pth demand query by the consumer k
Characteristics of the consumers	
The set of the consumers	$\mathbf{U} = \left\{ u_k / k = \overline{1, K_0} \right\}$
The matrix of an attachment of the consumers to physical facilities in the cluster infrastructure	$\mathbf{v} = \| v_{kr} \|$, where $v_{kr} = 1$ if the kth consumer is attached to the facility r of the cluster infrastructure, $v_{kr} = 0$—otherwise
The matrix of demand queries issuing by the consumers	$\mathbf{\Phi^Q} = \left\| \varphi_{kp}^Q \right\|$, where $\varphi_{kp}^Q = 1$ if consumer k issues the demand query p, $\varphi_{kp}^Q = 0$,—otherwise
The matrix of an attachment of the demand queries to the consumer's local facilities	$\mathbf{\Lambda^Q} = \left\| \delta_{pr}^Q \right\|$, where $\delta_{pr}^Q = 1$ if $\sum_{k=1}^{k_0} v_{kr}\varphi_{kp}^Q \geq 1$; $\delta_{pr}^Q = 0$ if $\sum_{k=1}^{k_0} v_{kr}\varphi_{kp}^Q = 0$

In the given mathematical framework, the stated problem of the innovation cluster design using the criteria of a minimum of average transaction costs during consecutive provisioning a set of consumers' demand queries might be formulated as follows for different transaction costs t:

$$\min_{\{x_{ir}, y_{ir}\}} \sum_{k=1}^{K_0} \sum_{p=1}^{P_0} \xi_{kp}^Q \cdot \varphi_{kp}^Q \cdot \left\{ \sum_{r_1=1}^{R_0} v_{kr_1} \cdot \left[\sum_{r_2=1}^{R_0} z_{pr_2} \cdot \left(t^{\text{dis}} + t_{r_1 r_2}^{\text{ser}} + t_{r_1 r_2}^{\text{trf}} \cdot \left(1 + \sum_{t=1}^{T} z_{pr_2}^t \right) \right) + t^{\text{ass}} \right] + \sum_{r_2=1}^{R_0} \sum_{t=1}^{T} z_{pr_2}^t \cdot \left(t_{r_2}^{\text{srh}} + t_{r_2} \right) \right\}$$

subject to

1. the number of economic agents in the agglomeration group $\sum_{i=1}^{I} x_{it} \leq F_t, \forall t = \overline{1, T}$, where F_t—maximum number of agents in the group t;
2. single agent inclusion in the group $\sum_{i=1}^{I} x_{it} = 1, \forall i = \overline{1, I}$;
3. the required level of information and business security of the design $x_{it} x_{i't} = 0$ for given agents d_i and $d_{i'}$;
4. irredundant allocation of agglomeration groups $\sum_{r=1}^{R_0} y_{tr} = 1, \forall t = \overline{1, T}$;
5. the size of the formed agglomeration group $\sum_{i=1}^{I} x_{it} y_{tr} \rho_i \psi_0 \leq \theta_{tr}$, $\forall t = \overline{1, T}, \forall r = \overline{1, R_0}$, where θ_{tr}—the greatest allowable size of the group t determined by characteristics of the facility r;
6. the total number of the designed agglomeration group located at the facility r: $\sum_{t=1}^{T} y_{tr} \leq h_r, \forall r = \overline{1, R_0}$, where h_r—the maximum number of groups allowed in the facility r;
7. the total establishment size of the facilities $\sum_{t=1}^{T} \sum_{i=1}^{I} \psi_0 \rho_i \pi_i x_{it} y_{tr} \leq \eta_r^{\text{EMD}}, \forall r = \overline{1, R_0}$;
8. the total transaction costs for production or provisioning services and products $\sum_{r=1}^{R_0} \sum_{t=1}^{T} z_{pr}^t \cdot (t_r^{\text{srh}} + t_r) \leq T_p, \forall p = \overline{1, P_0}$ for given $Q_p \in Q$, where T_p—the allowable transaction cost p needed for production or provisioning.

To solve the problem stated, we offer to extend the generic principles of neuro-algorithms [4, 5]. Many combinatorial optimization problems are NP-complete, which require state space explorations leading to $O(a^N)$ computations for a system with N degrees of freedom. Different kinds of heuristic methods are therefore often used to find reasonably good solutions. The generic paradigm of artificial neural networks (ANN) falls within this category. Whereas the use of ANN for pattern recognition and prediction problems is a nonlinear of conventional linear interpolation/extrapolation methods, ANN in the optimization domain really brings something new to the "table." In contrast to existing search and heuristics methods, the ANN approach does not fully or partly explore the different possible configurations. This is done in a way that allows for a statistical interpretation of the result. For feedback NN, the concept of network energy (Lyapunov function) is entered.

The main property of energy function is that during neural network states evolution it decreases and achieves a local minimum, in which it keeps constant energy. It allows solving combinatorial optimization tasks, if they can be formulated as a task of energy minimization. The neural approach is particularly transparent for graph bisection problem [5] due to its binary nature. This problem considers a set of N nodes to be partitioned into two halves such that the connectivity (cutsize) between the two halves is minimized. Traveling salesman problem (TSP) is also the optimization task frequently arising in practice. It was proved that this task belongs to the large set of tasks named NP-complete.

Described in [6], decision (the Hopfield method), based on feedback networks, is typical concerning a finding acceptable, though also of not optimum decisions. Nevertheless, answer turns out so quickly that in the certain cases the method can appear useful. The convergence of the decisions received on a Hopfield method for traveling salesman problem strongly depends on coefficients of energy function, and there is no regular method of definition of their values.

In [6], a general method called the stable state analysis technique was developed to determine constraints that the weights in the Hopfield energy function must satisfy so that valid solutions of high quality can be always obtained. In [6], the effectiveness of this method is demonstrated through a reinvestigation of the capability of the Hopfield neural network to solve the TSP. A large number of experiments on 10-city TSPs demonstrate the proposed method can obtain good results, while the mean error of achieved solutions to a 51-city TSP is about 15 % longer than the optimal tour, which is much better than that of solutions obtained through other ANN-based methods.

3 Hopfield Networks

In our research, firstly, we offer a method to determine the feedback neural network structure for solving the problem of design of innovation clusters as it was presented in subsection II. The main idea of this method is to determine the parameters in the energy function of the neural network by considering constraints of the problem and the Lyapunov function. Let we have a matrix of a business contiguity of economic agents of the given dimension $n \times n$, where n—the number of agents. It is necessary on the basis of this matrix to find proper combination of agents to the agglomeration group, i.e., to determine splitting set of capacity n on some number $(m \leq n)$ of subsets.

The decision is splitting the set of agents on subsets—agglomeration groups. The task consists in display of this splitting in the cluster infrastructure with using neurons in a mode with the large steepness of the characteristic (i.e., the transfer function of a network is "almost threshold"):

$$F(x) = \frac{1}{1 + e^{-\lambda \cdot NET}} \text{ where } \lambda \to \infty$$

(where NET—weighed sum of inputs).

In this model, each agglomeration group is represented by a row (line) of n neurons. That of these n neurons (economic agents), which outputs are equal 1, are included to the given group.

Because each agent can be included only to a single agglomeration group, each row of the resulting matrix should contain only one unit element. Under such condition, the restriction on single inclusion may be presented as follow is carried out automatically

$$\sum_{t=1}^{t_0} x_{it} = 1, \forall i = \overline{1 \ldots I}$$

where

t_0	capacity of the agglomeration groups set,
I	capacity of the agents set,
$x_{it} = 1$	if the ith agent is included to the tth agglomeration group,
$x_{it} = 0$	otherwise.

The problem solving is offered to use a Hopfield network. The way of presentation differs from used in the original Hopfield's work and others, but is equivalent to them from the functional point of view. The zero layers do not carry out computing function and only distributes outputs of a network back on inputs. The output of each neuron is connected to inputs of all others neurons. Each neuron of the first layer calculates the weighed sum of its inputs, giving a signal NET, which then with the help of nonlinear function F will be transformed to a signal OUT. The submission of entrance vectors is carried out through separate neurons inputs.

If F—thresholds function with a threshold T_j, the output of j-th neuron is calculated as

$$NET_j = \underbrace{\sum_{i \neq j} w_{ij} \cdot OUT_i}_{\text{the weighed sum of outputs from others neurons}} + IN_j$$

$$OUT_j = 1, \quad \text{if} \;\; NET_j > T_j$$
$$OUT_j = 0, \quad \text{if} \;\; NET_j < T_j$$
$$OUT_j \text{ does not change, if} \;\; NET_j = T_j$$

In our task $m = n^2$, where n—the number of agents.

The energy function for the computing network of our task should satisfy to two requirements:

1. should be small only for those decisions, which have only one unit element in each row;
2. should give preference to the decisions with such distribution of agents between agglomeration groups, which logically follows a matrix of business contiguity of agents (i.e., based that it is more preferable to combine such agents in the same group, rather than in different, which are connected by mutual business opportunity, proceeding from the given matrix).

The first requirement is satisfied with introduction of the following energy function, consisting of two terms:

$$E = \frac{A}{2} \cdot \sum_x \sum_i \sum_{j \neq i} OUT_{xi} \cdot OUT_{xj} + \frac{C}{2} \cdot \left[\left(\sum_x \sum_i OUT_{xi} \right) - n \right]^2,$$

where A, C—some constants. It reaches performance of the following conditions.

1. The threefold sum is equal to zero that and only in the event that each row (agent) contains no more than one unit element.
2. The second composed term of the energy function is equal to zero if and only if the resulting matrix contains n unit elements sharply.

The second requirement is satisfied with the help of addition of the following summand to the energy function:

$$\frac{B}{2} \cdot \sum_i \sum_x \sum_{y \neq x} OUT_{xi} \cdot \left(OUT_{yi} - a_{xy}^s\right)^2,$$

where $A^s = \{a_{xy}^s\}$, $x, y = \overline{1, n}$—given matrix of business contiguity of agents, B—some constant.

At enough large values A, C low-energy states will represent the allowable combinations of agents to the agglomeration groups, and the large values B guarantee that the most preferable combination will be found.

So, the energy function for our task looks like

$$E = \frac{A}{2} \cdot \sum_i \sum_i \sum_{j \neq i} OUT_{xi} \cdot OUT_{xj} +$$
$$+ \frac{B}{2} \cdot \sum_i \sum_x \sum_{y \neq x} OUT_{xi} \cdot \left(OUT_{yi} - a_{xy}^s\right)^2 +$$
$$+ \frac{C}{2} \cdot \left[\left(\sum_x \sum_i OUT_{xi}\right) - n\right]^2$$

Now we shall set weight values, i.e., we shall establish conformity between the members of energy function and members of the general form of the energy function (Lyapunov function) for feedback networks:

$$E = -\frac{1}{2} \cdot \sum_i \sum_j \sum_x \sum_y w_{xi,yj} \cdot OUT_{xi} \cdot OUT_{yj}$$
$$- \sum_j \sum_y I_{yj} \cdot OUT_{yj} + \sum_j \sum_y T_{yj} \cdot OUT_{yj},$$

where E—artificial energy of a network, $w_{xi,yj}$—weight from an output of xi-th neuron to an input of yj-th neuron, OUT_{xi}—output of xi-th neuron, I_{yj}—external input of yj-th neuron, T_{yj}—threshold of yj-th neuron.

4 Extension by Genetic Programming

As there are no regular ways of coefficients A, B, C values definition, the principles of genetic algorithms were investigated with the purpose of their using for a finding constant A, B, C, appropriate to the optimum task decision. The idea of application of genetic algorithms as "shell" for central neural network algorithm was developed. The function of this "shell" is selection among the decisions received with the help of neural network algorithm (generally locally optimum) and appropriate various values of researched constants, optimal decisions (global optimum) with the given constraints.

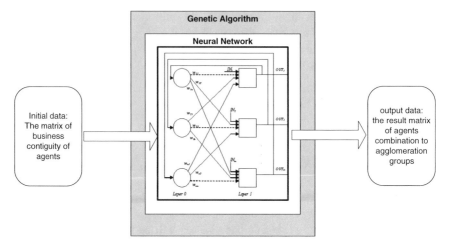

Fig. 1 The block diagram of the realized algorithm

The block diagram of the implemented algorithm, which performs combination of agents to the agglomeration groups, is submitted in a Fig. 1.

A core is neural network algorithm, which functions are designing Hopfield neural network with setting of weights, external influences and thresholds for neurons and reception with the help of created neural network the local-optimum decisions at various values of coefficients *A, B, C*. Values of coefficients *A, B, C* are generated by genetic algorithm—the "shell" of central neural network algorithm. As the entrance data, the algorithm accepts the matrix of business contiguity of agents. The target data are the task decision, selected by genetic algorithm, i.e., matrix of agglomeration of the agents in the groups, and graphical information allowing to judge about a course of task decision process. The offered algorithm was developed with the purpose of its application within the framework of the distributed system supposing the collection of the requirements (the constraints of the task) from various sources. The received in this case decision should optimize the greatest possible number of the assembled restrictions, i.e., to give out best of the possible decisions. In more complicated cases, we ought to use Tabu machine approach for fascinating optimization process.

5 Conclusion

In these works, new aspects of innovative and industrial clusters design were proposed. Certainly, this research can go on and develop further. We presented here major research results in studies of new soft methods of design and modeling evolution processes of innovation clusters. Our principal interest was to model agglomeration and distribution processes which lead to the improving overall efficiency of the cluster in terms of minimization of total transaction costs during provisioning the demand queries of multiple consumers. During the research, we developed a generic model of innovation clusters of regional and intra-regional scales as well as proposed its mathematical formalization. Based on that mathematical formalization, a sequence of artificial neural network models was developed with increasing quality and computational efficiency. The core of all network models includes a modified version of Hopfield network.

The simplest algorithm offered uses Hopfield neural network and, hence, training without the teacher, as all process of training for networks with feedbacks in this case is reduced to correct account and initialization of a network parameters: weights, thresholds, and external influences of neurons.

As the alternative approach to the decision of a similar sort tasks using of neural networks and training with the teacher can act. If we compare our method to such alternative, it is possible to tell that the genetic algorithm-shell in our case somewhat replaces process of training with the teacher. The method proposed has several important advantages in comparison with supervisor learning algorithms:

1. The speed of result computation. Our method does not depend on the size of the training set, while working times and the quality in supervisor learning algorithms heavily depend on the size of the training set.
2. Flexibility of the algorithm. In our method adding new problem constraints does not influence the structure of the neural network, only the suitability function of the genetic algorithm should be changed accordingly.

Other alternative of the decision of NP-complete tasks of discrete optimization is the traditional branch-and-bound method. Like for the neural network algorithm, a number of experiments for the traditional branch-and-bound method were carried out. It was found that the operating time of the traditional branch-and-bound method for dimension of a task is less than 15 more, than operating time of neural network algorithm for tasks of the same dimension. However with an increase in the dimension of a task, the time spent by neural network algorithm on its decision grows faster than time if we use the traditional branch-and-bound method. It is explained to that the genetic algorithm, the "shell" of neural network algorithm, during performance requires calculation individual's suitability for each following population, and for suitability definition for one individual the neural network algorithm should once work with parameters describing given individual. Way allowing reducing an operating time of neural network algorithm for a large dimension tasks is the application of parallel calculations.

Newly proposed modification of the formalism of tabu machine became the next step in improving the simplest variant of Hopfield network. Among TM advantages, we can mention independence of the solution from the parameters of the energy function of the network.

Acknowledgments The reported study was funded by RFBR according to the research project № 16-06-00300 a.

References

1. Klimova, N., Litvintseva, M.: Universities innovation clusters: approaches for national competitiveness paradigm. Eur. J. Soc. Sci. **19**(1), 160–162 (2011)
2. Klimova, N.: Innovative Clusters in Regional Economy. Int. Res. J. Finance Econ. **65**, 6–10 (2011)
3. Klimova, N., Malyzhenkov, P.: Spin-off phenomenon as a factor of university clusters competitiveness increasing: a methodological proposal. In: 11th International Conference on Perspectives in Business Informatics Research BIR 2012, Springer, Nizhny Novgorod, Russia, 24–26 Sept 2012
4. Jones, T.M.: AI Application Programming. Charles River Media Inc, Hingham (2003)
5. Peterson, C.: Combinatorial optimization with feedback artificial neural networks. University of Lund, Lund (2002)
6. Feng, G., Douligeris, C.: Using hopfield networks to solve traveling salesman problems based on stable state analysis technique. IEEE-INNS-ENNS International Joint Conference on Neural Networks (IJCNN'00), vol. 6 (2000)

Part X
Soft Computing in Informatics

Classification of Air Quality Monitoring Stations *Using* Fuzzy Similarity Measures: A Case Study

Kamal Jyoti Maji, Anil Kumar Dikshit and Ashok Deshpande

Abstract The objective of designing and installation air quality monitoring network (AQMN) is to reduce network density with a view to acquire maximum information on air quality with minimum expenses. In spite of the best research efforts, there has been no general acceptance of any method for deciding the number of stations. Majority of the cities have, therefore, installed monitoring stations with their own guidelines. The present paper presents a useful formulation for classification of the existing air quality monitoring stations (AQMS) using fuzzy similarity measures. The case study has been demonstrated by applying the methodology to the already-installed AQMS in Delhi, India.

Keywords Air quality monitoring network · Air quality data · Classification · Fuzzy similarity measures · Cosine amplitude and max–min method

K.J. Maji · A.K. Dikshit (✉)
Center for Environmental Science and Engineering (CESE),
Indian Institute of Technology, Bombay, India
e-mail: dikshit@iitb.ac.in

K.J. Maji
e-mail: kjmaji@gmail.com

A.K. Dikshit
School of Business, Environment and Society, Malardalen University,
SE 72220 Vasteras, Sweden

A.K. Dikshit
School of Civil, Environment and Construction,
University of Kwazulu-Natal, Durban, South Africa

A. Deshpande
Berkeley Initiative in Soft Computing (BISC)—Special Interest
Group (SIG)—Environment Management Systems (EMS),
University of California, Berkeley, USA and Indian Institute
of Technology Bombay (IITB), Bombay, India
e-mail: ashok_deshpande@hotmail.com

© Springer International Publishing Switzerland 2016
L.A. Zadeh et al. (eds.), *Recent Developments and New Direction
in Soft-Computing Foundations and Applications*, Studies in Fuzziness
and Soft Computing 342, DOI 10.1007/978-3-319-32229-2_34

1 Introduction

Ever-increasing pollution levels due to rapid urbanization and industrialization, especially in developing countries, with minimal focus on adequate pollution abatement strategies have resulted in impairing natural environment such as air, water, land, and alike. Air pollution is the fifth leading cause of death in India after high blood pressure, indoor air pollution, tobacco smoking, and poor nutrition, with about 620,000 premature deaths occurring from air pollution-related diseases (Times of India News dated February 14, 2013).

Human exposure to air pollutants may result in a variety of respiratory diseases depending on the type and the toxicity of the specific pollutant, magnitude, duration, and frequency of exposure. There are six common criteria pollutants, viz., particulate matter, ground-level ozone, carbon monoxide, sulfur oxides, nitrogen oxides, and lead. Some of the hazardous air pollutants are asbestos, benzene, carbon tetrachloride, chlordane, chloroform, formaldehyde, heptachlor, hydrochloric acid, mercury, methanol, phenol, and toluene.

The concentration of air pollutants depends not only on the quantities that are emitted from air pollution sources, but also on the ability of the atmosphere to either absorb or disperse these emissions. The effects of air pollutants can be minor and reversible such as eye irritation or debilitating leading to asthma and could be fatal such as cancer [1].

The air quality in an area is monitored by first designing and then installing an air quality monitoring network (AQMN). The objective of an optimum AQMN is to reduce the network density with a view to acquire maximum information on air quality with minimum expenses.

The present paper is organized as follows: Sect. 2 presents brief literature review on air quality management network and objectives of the study are explained in Sect. 3. Approach and the techniques used in the proposed fuzzy similarity-based model are presented in Sects. 4 and 5, respectively. Section 6 relates to the case study in Delhi, India, and assumptions and the limitation of the study which Sect. 7 present are in the concluding remarks.

2 Brief Literature Review

Stalker and Dickerson [2] initiated efforts in designing and installing air quality monitoring networks. The researchers applied the coefficient of geographic variation to find out the minimum number of air pollution monitoring stations required to estimate the true areal mean within specified confidence limits. USEPA [3] formulated guidelines for air quality surveillance network, while WHO [4] proposed air quality guidelines for particulate matter, ozone, nitrogen dioxide, and sulfur dioxide. Several researchers have made seminal contribution in an endeavor of AQMN design. They include: Petrson [5], Seinfeld [6], Noll and Millar [7], Nakamori and

Ikeda [8], Munn [9], Husain and Khan [10], Modak and Lohani [11, 12], Kainuma et al. [13], Chang and Tseng [14], Tseng and Chang [15], and Baldaufet [16]. For risk-based prioritization air pollution monitoring location, Khan and Sadiq [17] used fuzzy synthetic evaluation technique [18].

In spite of the concerted research efforts, there has been no general acceptance on approaches for deciding the number of air quality monitoring stations (AQMS) and their locations. Ad hoc decisions are taken on the number of AQMS and their locations in a city—especially in the developing countries. Many a times, especially in Indian context, AQMS in a city are installed in residential, commercial, and industrial areas. In most of the cases, the availability of suitable site is the only single consideration while deciding air quality monitoring locations.

3　Objectives of the Study

From the above discussion, two important issues stand out:

1. With the available data on air quality parameters, can these AQMS be classified or grouped?
2. Based on the classification, is it possible to reduce number of monitoring stations so as to bring down recurring expenditure?

It is important to appreciate that the reduction in just a single AQMS will save substantial operating and maintenance costs. The authors have made an attempt to address these issues with a case study in this sequel.

4　Approach

The approach for achieving the study objectives primarily depends upon accurately collated of relevant data on air pollutant concentration parameters from all the AQMS, say five-year data. The analyst, in consultation with the domain experts, will first identify the most important pollution parameter(s) in order to arrive at the classification of the monitoring stations. For example, in most of the Indian cities, PM_{10} and $PM_{2.5}$ could be the governing parameters in the classification of AQMS which may be due mobile transportation (three/two-wheelers, bus, and car), ongoing infrastructure development activity, and the use of generators due to frequent power cuts. While the parameter to be considered in New York City could be as CO and $PM_{2.5}$ levels are under control limits due to effective implementation of pollution control measures and strict compliance of rules and regulations.

PM_{10} can arise from a wide range of sources, but can generally be separated into four categories: 1. primary combustion particulates—produced directly from combustion, such as domestic heating, road transport, power stations, and industrial processes; 2. secondary particulates—aggregates in the atmosphere following their release as gasses (include nitrates and sulfates); 3. coarse particulates—from non-combustion sources such as resuspended road dust, construction work, mineral extraction, wind-blown dust and soil, and sea salt; and 4. excessive use of diesel generators due to frequent power cuts, especially in the developing countries. The procedure for the analysis of the data is as flows: Air quality parametric data can be divided into *five equal parts* based on the NAAQS (USEPA) compliance criteria for categories or group—of air quality as good, moderate. Further, calculate the number of data in each group as stated above, for each monitoring site and normalize. The generated $X \times Y$ matrix is on two universes. In this case, X refers to AQMS and Y as the compliance criteria or group. It is not possible to use straightway, fuzzy compositional rule of inference to estimate the similarity relation between the AQMS as the data are on different universes. A different fuzzy relational formalism termed as *cosine amplitude method* as well as *max–min method* can be used. The matrices generated using these fuzzy logic-based formalisms are invariably a fuzzy tolerance relation which means that the relation is reflexive and symmetric but not transitive. It is, therefore, necessary to transform fuzzy tolerance/proximity/compatibility relation to fuzzy equivalence relation using one of the defined methods of transitivity closure. Fuzzy to crisp conversion can be achieved using the concept of α—cut which signifies the possibility of the AQM stations in a particular group.

5   Technique Used

Fuzzy relational calculus is one of the important facets in fuzzy set theory and has wide applications. We used some of the defined techniques which are explained in brief:

5.1   Similarity Measures

There are seven different ways to develop the numerical values that characterize a relation, but similarity measures in one of the most prevalent forms of determining the values in a relation. These methods attempt to determine some sort of similar pattern or structure in data through various metrics. Two methods are discussed below [19]:

Cosine Amplitude Method

In this study, cosine amplitude method is one of the commonly used similarity methods that based on the relativity concept. Data samples of a set form a data array, say X and that can be expressed in the following equation for n data:

$$X = \{x_1, x_2, \ldots, x_n\} \qquad (1)$$

Each of the elements $x_1, x_2, \ldots, x_n$ in the data array X is itself a vector of length m; that is,

$$x_i = \{x_{i_1}, x_{i_2}, \ldots, x_{i_m}\}$$

Each element of a relation, r_{ij}, results from a pair-wise comparison of two data samples, say x_i and x_j, where the strength of the relationship between data sample x_i and data sample x_j is given by the membership value expressing that strength; that is, $r_{ij} = \mu_R(x_i, y_j)$. Therefore, the cosine amplitude method calculates r_{ij} by the following equation based on the comparison of two data arrays

$$r_{ij} = \left| \sum_{k=1}^{m} x_{ik} x_{jk} \right| \bigg/ \sqrt{\left(\sum_{k=1}^{m} x_{ik}^2 \right) \left(\sum_{k=1}^{m} x_{jk}^2 \right)} \quad \text{where } i, j = 1, 2, \ldots, n. \qquad (2)$$

Computing all elements will be form size $n \times n$ fuzzy tolerance relation matrix (similarity relation). This method calculates pair-wise relational strength (r_{ij}) based on the comparison of two data arrays and range of r_{ij} values that varies from 0 to 1 $(0 \leq r_{ij} \leq 1)$. According to this equation, close values of r_{ij} to zero suggests no relation (dissimilarity), while closer values of r_{ij} to one represent strong relation (similarity) of two data sets.

Max–min Method

Another popular method, which is computationally simpler than the cosine amplitude method, is known as the max–min method. Although the name sounds similar to the max–min composition method, this similarity method is different from composition. It is found through simple min and max operations on pairs of the data points, x_{ij}, and is given by the following

$$r_{ij} = \sum_{k=1}^{m} \min(x_{ik}, x_{jk}) \bigg/ \sum_{k=1}^{m} \max(x_{ik}, x_{jk}) \quad \text{where } i, j = 1, 2, \ldots, n \qquad (3)$$

A fuzzy equivalence relation must satisfy all three matrix conditions, viz., reflexivity, symmetry, and transitivity.

Let R be a similarity relation and x, y be elements of a set X and $\mu_R(x, y)$ denote the grade of membership of the ordered pair (x, y) in R, then R is a similarity relation in X if and only if, for all x, y and z in X, $\mu_R(x, x) = 1$ for all x in X (reflexivity),

$\mu_R(x, y) = \mu_R(y, x)$ for all x and y in X (symmetry), and $\mu_R(x, z) \geq \max_y \in X\{\min\{\mu_R(x, y), \mu_R(y, z)\}\}$ for all x, y and z in X (transitivity) [20–22].

If fuzzy relation matrix only has the properties of reflexivity and symmetry, then it is called fuzzy tolerance relation matrix. Before defuzzification (fuzzy to crisp conversion), fuzzy tolerance relation has to be converted to fuzzy equivalence relation by composition.

Different composition methods such as max–min, min–max, and max-product are available in literature [20–22] but the most popular is max–min composition method. Any fuzzy tolerance relation matrix, $R1$, can be reformed into a fuzzy equivalence relation by at most $(n - 1)$ compositions. That is

$$R_1^{n-1} = R_1 \circ R_1 \circ \cdots \circ R_1 = R. \tag{4}$$

Max–min composition

Considering $R_1(x, y), (x, y) \in X \times Y$ and $R_2(y, z), (y, z) \in Y \times Z$ be two fuzzy tolerance relations. The max–min composition R_1 max–min R_2 is then the fuzzy set

$$R_1 \circ R_2 = \left\{ \left[(x, z), \max_y \{\min\{\mu_{R_1}(x, y), \mu_{R_2}(y, z)\}\} \right] \middle| x \in X, y \in Y, z \in Z \right\} \tag{5}$$

$\mu_{R_1 \circ R_2}$ is the membership function of a fuzzy relation on fuzzy sets.

5.2   Fuzzy to Crisp Conversion

*We b*egin by considering a fuzzy set R, then define a lambda-cut set R_λ, where $0 \leq \lambda \leq 1$. The set R_λ is a crisp set called the lambda (λ)-cut (or alpha (α)-cut) set of the fuzzy set A, where $R_\lambda = \left\{ x \middle| \mu_{R(x)} \geq \lambda \right\}$. The λ-cut set in R_λ is a crisp set derived from its parent fuzzy set, and any particular fuzzy set R can be transformed into an infinite number of λ-cut sets, because there is an infinite number of value λ on the interval [0, 1]. Any element x in R_λ belongs to fuzzy set R with a grade of membership that is greater than or equal to the value λ.

6   Case Study

The case study relates to Delhi, India, where the Central Pollution Control Board (CPCB) has already-installed AQMS at defined locations. Delhi, largest metropolis by area and the second-largest metropolis by population in India, is located at $28°$ $22'48''$N latitude and $77°7'12''$E longitude and at an elevation of 216 m above mean sea level. The city is spread over 1483 km^2 of that 75.10 % urban and 22.9 % rural with population was 16.75 million, of which 16.33 million are urban area and 0.42 million are rural area (http://censusindia.gov.in/).

As a result of rapidly expanding city with high-population growth followed by intensive infrastructure, progress led to heightened demand in energy from domestic, construction activities, industrial, and transport sectors, resulting in an increase in the air pollutant in Delhi. There are about 148,680 industrial units during 2010–2011 [23]. These include engineering goods, textile, chemical, electronics, electrical goods, dyes and paints, steel, plastic, rubber, and automobiles.

In order to meet the energy demand, the city of Delhi has installed three big coal-based thermal power plants—the Rajghat, the Indraprastha (IP), and the Badarpur, and three natural gas-based power plants– the Indraprastha Power Generation Co Ltd. (IPGCL), the Pragati Power Station, and Pragati-III Combined Cycle Power Plant.

From 1981 to 2011, the road length in Delhi increased from 14,316 to 31,183 km (2.18 times), whereas the number of registered vehicles increased from 0.52 to 6.93 million (13.3 times), in between 6.93 million vehicles, 94 % personal vehicles (31 % cars and jeeps, and 63 % two-wheelers), and 6 % commercial vehicles.

As a result, there is an increase in the air pollutant emissions of particulate matter (PM10 and PM2.5), sulfur dioxide (SO2), nitrogen dioxides (NO2), carbon monoxide (CO), ozone (ground level), and hydrocarbons. It can be stated that vehicular pollution or mobile transportation contributes 67 % of the total air pollution loads in Delhi

Ambient Air quality Monitoring stations in Delhi

There exists a total of twelve manually operated AQMS in Delhi. Table 1 presents nine AQMS in residential area are codes as R while three stations located in industrial are coded as 1.

Table 1 Location of ambient air quality monitoring stations in Delhi

Station code	Location	Operated by	Land-use type
R_1	Janakpuri	CPCB	Residential
R_2	N.Y. School, Sarojini Nagar	NEERI[a]	Residential
R_3	Nizamuddin	CPCB	Residential
R_4	Pritampura	CPCB	Residential
R_5	Siri Fort	CPCB	Residential
R_6	Town Hall, Chandni Chowk	NEERI[a]	Residential
R_7	Ashok Vihar	CPCB	Residential
R_8	Delhi College of Engineering	CPCB	Residential
R_9	Bahadur Shah Zafar Marg	CPCB	Residential
I_1	Mayapuri Industrial Area	NEERI[a]	Industrial
I_2	Shahdara	CPCB	Industrial
I_3	Shahzada Bagh	CPCB	Industrial

[a]NEERI: National Environmental Engineering Research Institute

In this study, daily average concentration data of PM_{10} and NO_2 was used. Unfortunately, air quality parametric data were not available for three AQMS; therefore, these are excluded from the analysis. These excluded stations are Ashok Vihar, Delhi College of Engineering, and Bahadur Shah Zafar Marg.

In order to classify the air quality stations, five linguistic descriptors *good, moderate, poor, very poor and severe* were used in Indian air quality index as proposed by Sharma et al. [24]. Table 2 represents linguistic variable and corresponding pollutants concentration break point as shown in Fig. 1.

Table 2 Linguistic variable and corresponding pollutants concentration break point

Descriptor	Pollutant	
	PM_{10} ($\mu g/m^3$) 24-h avg.	NO_X ($\mu g/m^3$) 24-h avg.
Good (compliance of NAAQS)	0–100	0–80
Moderate (compliance of NAAQS of USEPA)	101–150	81–180
Poor (alert level)	151–350	181–564
Very poor (warning level)	351–420	565–1272
Severe (emergency level)	>420	>1272

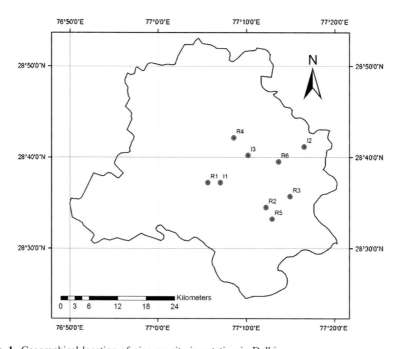

Fig. 1 Geographical location of nine monitoring station in Delhi

7 Results and Discussion

In the absence of the data of $PM_{2.5}$, the authors have used PM_{10} data in classification of AQMS in the study. It has been observed that mobile transportation (bus, three-/two-wheelers, and car) and the dust due to ongoing infrastructural developmental activity in the city are the major contributors in the increased levels of particulate matter in Delhi. Electric power cuts forced Delhi residents to use diesel-operated generator—also be a cause of air pollution. From the health effect point of view, particulate matter is by far the most important air pollution parameter in Delhi. Selecting particulate matter, the only air pollutant for deciding the number of AQMS is similar to that of ***worst-case scenario*** in probabilistic risk assessment study.

Table 3 presents computed normalized values corresponding to the linguistic variable used [24] for PM_{10} data (2005) for nine monitoring stations in Delhi.

In order to work out the similarity between the AQMS, Cosine amplitude method was used. A typical computational procedure for only PM_{10} (2005) is presented using Expression 2

$$r_{R1R3} = \frac{(0.45 * 0.64 + 0.32 * 0.28 + 0.23 * 0.08)}{\sqrt{(0.45^2 + 0.32^2 + 0.23^2)(0.64^2 + 0.28^2 + 0.08^2)}}$$
$$\cong \frac{0.40}{0.42} \cong 0.95$$

Following the above procedure, similarities between all the AQMS in Delhi are calculated. Table 4 presents computed fuzzy tolerance relational matrix which is invariably a fuzzy tolerance relation. The relation is transformed to fuzzy equivalence relation matrix using one of the transitivity closure operations, i.e., max–min composition (Expression 4).

In order to work out the similarity between the nine AQMS in Delhi, for PM_{10}, fuzzy values are converted to crisp version using α- cut. Table 5 presents the possibility of similarity between the monitoring stations have been assumed to be 0.98 α-cut levels. This is somewhat like the concept of confidence level in statistical inference. Looking at the Table 5, column wise, the identical columns form one class and columns which are each unique form individual class.

Table 3 Normalized parametric values of PM_{10} data (2005) for nine monitoring stations in Delhi

	R1	R2	R3	R4	R5	R6	I1	I2	I3
G	0.45	0.42	0.64	0.23	0.63	0.32	0.17	0.38	0.42
M	0.32	0.27	0.28	0.61	0.29	0.27	0.17	0.33	0.43
P	0.23	0.29	0.08	0.16	0.08	0.32	0.49	0.27	0.15
V	0.00	0.01	0.00	0.00	0.00	0.04	0.05	0.00	0.00
S	0.00	0.01	0.00	0.00	0.00	0.05	0.12	0.02	0.00

G: Good, M: Moderate, P: Poor, V: Very poor, and S: Severe

Table 4 Fuzzy equivalence relational matrix for the nine AQMS in Delhi

	R1	R2	R3	R4	R5	R6	I1	I2	I3
R1	1.00	0.99	0.95	0.93	0.95	0.98	0.89	0.99	0.97
R2	0.99	1.00	0.95	0.93	0.95	0.98	0.89	0.99	0.97
R3	0.95	0.95	1.00	0.93	1.00	0.95	0.89	0.95	0.95
R4	0.93	0.93	0.93	1.00	0.93	0.93	0.89	0.93	0.93
R5	0.95	0.95	1.00	0.93	1.00	0.95	0.89	0.95	0.95
R6	0.98	0.98	0.95	0.93	0.95	1.00	0.89	0.98	0.97
I1	0.89	0.89	0.89	0.89	0.89	0.89	1.00	0.89	0.89
I2	0.99	0.99	0.95	0.93	0.95	0.98	0.89	1.00	0.97
I3	0.97	0.97	0.95	0.93	0.95	0.97	0.89	0.97	1.00

Table 5 Fuzzy to crisp values for the alpha cut 0.98

	R1	R2	R3	R4	R5	R6	I1	I2	I3
R1	1	1	0	0	0	1	0	1	0
R2	1	1	0	0	0	1	0	1	0
R3	0	0	1	0	1	0	0	0	0
R4	0	0	0	1	0	0	0	0	0
R5	0	0	1	0	1	0	0	0	0
R6	1	1	0	0	0	1	0	1	0
I1	0	0	0	0	0	0	1	0	0
I2	1	1	0	0	0	1	0	1	0
I3	0	0	0	0	0	0	0	0	1

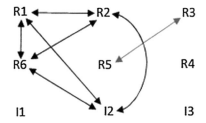

Fig. 2 Graphical representation of the similarity between the AQMS

It can be inferred that AQMS can be classified into five groups for PM_{10} based on 2005 data set: **Group 1 (R1, R2, R4, R6, and I2), Group 2 (R3, R5), Group 3 [R4], Group 4 [I1], and Group 5 [I3]**. Figure 2 shows the graphical representation of the similarity between the AQMS.

The computations carried out using *max–min method* also infer with the same classification but the possibility of the classification is 0.81 (α-cut level of 0.81).

Table 6 presents the result of similar computations for the data sets (2005–2009) were carried out for suspended particular matter.

Now taking union of all *groups contain one element* (station) from 2005 to 2009 (stations in bold in Table 6), we get R2, R3, R4, R6, I1 and I3. But R1, R5, and I2

Table 6 Classification/group for PM$_{10}$ data for various years

Year	Classification/group
2005	(R1, R2, R4, R6, I2); (R3, R5); **R4**; **I1**; **I3**
2006	(R1, R2, R3, R4, R5, R6); **I1**
2007	(R1, R4, R5, I2); **R2**; **R3**; **R6**; **I1**; **I3**
2008	(R1, R3, R5, R6, I1, I2); **R2**, **R4**; **I3**
2009	(R1, R2, R3, R5, I1, I2, I3); **R4**; **R6**

make groups always with other elements. These six elements (stations) contain all the information of single-element group as well as clustering group in any of the year.

For example in 2005, choose stations from group 1 and group 2 in such a way that they at list one time that stations form single-element group (R2, R4 from group 1 and R3 from group 2) in between years 2005 and 2009.

In summary, it can be concluded that based on PM$_{10}$ data, CPCB might continue to monitor air quality in Delhi only at the following locations: R2 (N.Y. School, Sarojini Nagar), R3 (Nizamuddin), R4 (Pritampura), R6 (Town Hall, Chandni Chowk), I1 (Mayapuri Industrial Area), and I3 (Shahzada Bagh) resulting in substantial reduction in capital and recurring expenses on air quality monitoring program in Delhi.

Working with same methodology, for NO$_2$ data set and the resulting stations are R1 (Janakpuri), R2 (N.Y. School, Sarojini Nagar), R4 (Siri Fort), R6 (Town Hall, Chandni Chowk), I1 (Mayapuri Industrial Area), and I3 (Shahzada Bagh).

Assumption and limitation of study

The method based on fuzzy similarity measures presented in this paper does not include meteorological condition and assumes that at all the ambient AQMS in Delhi remain unaltered. The case study also presupposes that the air quality parametric data are statistically accurate.

8 Concluding Remarks

Aristotle, Ancient Greek Philosopher *says* "*It is the mark of an instructed mind to rest satisfied with that degree of precision which the nature of the subject admits, and not to seek exactness where only an approximation of the truth is possible.*"

This is the total of our experiments and observation. As we expect same effects from similar class of objects, we can expect the same in the data and classify ambient AQMS based on their similarity measures as well. In the study reported in the sequel, therefore, *particulate matter is considered as the worst-case scenario.* From that study, it also shows that not every monitoring station needs to monitor both PM$_{10}$ and NO$_2$. The approach presented is robust and depends upon the statistical accuracy of time series data of pollution parameters.

Acknowledgements The authors are grateful to the CPCB authorities in India for the permission to use of air quality parametric data in the case study and thankful to the reviewer for reviewing the manuscript and providing useful comments.

References

1. Yadav, J., Kharat, V., Deshpande, A.: Fuzzy description of air quality using fuzzy inference system with degree of match via computing with words: a case study. Air Qual. Atmos. Health (2014)
2. Stalker, W.W., Dickerson, R.C.: Sampling station and time requirements for urban air pollution surveys. J. Air Pollut. Control Assoc. **12**(3), 111–128 (1962)
3. US Environmental Protection Agency.: Guidelines: air quality surveillance network. USEPA, AP-98 (1971)
4. World Health Organization.: Air monitoring programme design for urban and industrial areas: global environmental monitoring system. WHO Offset Publication No. 38 (1977)
5. Peterson, J.T.: Distribution of sulfur dioxide over metropolitan St. Louis, as described by empirical eigenvectors, and its relation to meteorological parameters. Atmos. Environ. **4**(5), 501–518 (1970)
6. Seinfeld, J.H.: Optimal location of pollutant monitoring stations in an airshed. Atmos. Environ. **6**(11), 847–858 (1972)
7. Noll, K.E., Millar, T.E.: Air Monitoring Network Design. MacMillan Publishers Ltd, London (1977)
8. Nakamori, Y., Ikeda, S.: Design of air pollutant monitoring system by spatial sample stratification. Atmos. Environ. **13**, 97–103 (1979)
9. Munn, R.E.: The Design of Air Quality Monitoring Networks. MacMillan Publishers Ltd, London (1981)
10. Husain, T., Khan, S.M.: Air monitoring network design using Fisher's information measures —a case study. Atmos. Environ. **17**(12), 2591–2598 (1983)
11. Modak, P.M., Lohani, B.N.: Optimization of ambient air quality monitoring networks: (Part I). Environ. Monit. Assess. **5**(1), 1–19 (1985)
12. Modak, P.M., Lohani, B.N.: Optimization of ambient air quality monitoring networks: (Part II). Environ. Monit. Assess. **5**(1), 21–38 (1985)
13. Kainuma, Y., Shiozawa, K., Okamoto, S.: Study of the optimal allocation of ambient air monitoring stations. Atmos. Environ. Part B. Urban Atmos. **24**(3), 395–406 (1990)
14. Chang, N.B., Tseng, C.C.: Optimal evaluation of expansion alternatives for existing air quality monitoring network by grey compromise programing. J. Environ. Manage. **56**(1), 61–77 (1999)
15. Tseng, C.C., Chang, N.B.: Assessing relocation strategies of urban air quality monitoring stations by GA-based compromise programming. Environ. Int. **26**(7–8), 523–541 (2001)
16. Baldauf, R.W., Lane, D.D., Marote, G.A.: Ambient air quality monitoring network design for assessing human health impacts from exposures to airborne contaminants. Environ. Monit. Assess. **66**, 63–76 (1999)
17. Khan, F.I., Sadiq, R.: Risk-based prioritization of air pollution monitoring using fuzzy synthetic evaluation technique. Environ. Monit. Assess. **105**(1–3), 261–283 (2005)
18. Sarigiannis, D.A., Saisana, M.: Multi-objective optimization of air quality monitoring. Environ. Monit. Assess. **136**(1–3), 87–99 (2008)
19. Mazzeo, N.A., Venegas, L.E.: Design of an air-quality surveillance system for buenos aires city integrated by a NO_x monitoring network and atmospheric dispersion models. Environ. Model. Assess. **13**(3), 349–356 (2008)

20. Ross, T.J.: Fuzzy Logic with Engineering Applications, 3rd ed. Wiley, New York (2010)
21. Zadeh, L.A.: Similarity relations and fuzzy orderings. Inf. Sci. **3**(2), 177–200 (1971)
22. Zimmermann, H.J.: Fuzzy Set Theory and Its Applications, 4th ed. Springer, Berlin (2001)
23. Government of NCT of Delhi.: Statistical abstract of Delhi 2012. Directorate of Economics and Statistics, Vikash Bhawan-II, 3rd floor (B-wing), Bela Road, Delhi-54 (2012)
24. Sharma, M., Maheshwari, M., Sengupta, B., Shukla, B.P.: Design of a website for dissemination of air quality index in India. Environ. Model Softw. **18**(5), 405–411 (2003)

Modeling of Decision Maker Under Imperfect Information

L.A. Gardashova

Abstract In real life, imperfect information is commonly present in all the components of the decision-making problem. In decision-making problems, a DM is almost never provided with perfect, that is, ideal decision-relevant information to determine states of nature, outcomes, probabilities, utilities, etc. We are known that relevant information almost always comes imperfect. Imperfect information is information in which one or more respects are imprecise, uncertain, incomplete, unreliable, vague, or partially true [1]. Two main concepts of imperfect information are imprecision and uncertainty. Imprecision is one of the widest concepts including variety of cases. We will discuss uncertainty concepts of imperfect information and its application for problem modeling of decision maker. In the first stage of the modeling, the identification determinants of a decision maker was implemented using Delphi method. The aim of the second stage consisted of the linguistic evaluation of the factors. At the final stages, decision-maker model was realized by using possibility–probability-based method and Dempster–Shafer theory-based model.

1 Introduction

Making decisions is certainly the most important task of a manager, and it is often a very difficult one. It depends on two factors: the statement of the decision-making problem and the determinants of a decision maker. Decisions are an inevitable part of human activities. It requires the right attitude. Every problem properly perceived becomes an opportunity. In most cases, the decision maker must view the problems as opportunities rather than solving problems. A pessimist sees the difficulty in every opportunity, whereas an optimist sees the opportunity in every difficulty. It all

L.A. Gardashova (✉)
Department of Computer-Aided Control Systems, Azerbaijan State Oil Academy,
Baku, Azerbaijan Republic
e-mail: latsham@yandex.ru

© Springer International Publishing Switzerland 2016
L.A. Zadeh et al. (eds.), *Recent Developments and New Direction in Soft-Computing Foundations and Applications*, Studies in Fuzziness and Soft Computing 342, DOI 10.1007/978-3-319-32229-2_35

503

depends on the decision maker's attitude. Decision maker looks at the problems using reaction and emotion. Decision making depends on a character of a decision maker. This requires including different behavioral characteristics of decision maker into decision-making model. The analysis of the existing works [2–14] of the field modeling of decision maker shows that emotion, altruism, reciprocity, fairness, social responsibility. etc., are basic attributes of human behavior. Authors of works [2–6] developed theory of reciprocal altruism for games behavior. Decision-maker modeling under second-order uncertainty using the method based on the possibility–probability measure is discussed in work [15–17]. Human behavioral modeling based on the Dempster–Shafer theory of belief and fuzzy logic is suggested by Yager in [18]. A tractable model of reciprocity and fairness is discussed in [5]. The income distribution and the kindness or unkindness of others' choices systematically affect a person's emotional state. The emotional state systematically affects the marginal rate of substitution between own and others' payoffs, thus the person's subsequent choices. The proposed model is applied to two sets of laboratory data: simple binary choice mini-ultimatum games and Stackelberg duopoly games with a range of choices. The results confirm that other regarding preferences respond to others' intentions as well as to the income distribution. The existing approaches do not deal with possibilistic and probabilistic uncertainty which is characterizing decision makers' behavior. In this paper, we try to do modeling of decision maker by using emotion and altruism factors.

The rest of the paper is organized as follows: In Sect. 2, the process of determining decision makers' attributes and a statement of the problem are given. In Sect. 3, modeling process is shortly described under second-order uncertainty using the possibility–probability measure-based method. In Sect. 4, we create the decision-maker model by using obtained data. Decision-maker behavioral modeling using fuzzy and Dempster–Shafer theories was suggested in Sect. 5. Section 6 is conclusion.

2 Statement of the Problem

The basic problem is to evaluate personal quality of a decision maker by using psychological determinants.

For determining psychological determinants as basic factors influencing a choice of a decision maker, we use the Delphi method. For determining basic factors of a decision maker, the following questionnaires are created:

Query 1.
Please indicate by "+" which of the following should be considered as determinants of a decision maker (see in Table 1):
Query 2. Identification of total index of a DM.
Please indicate what term should be used for a total index (resulting dimension) of a DM as an overall evaluation to be determined on the base of the determinants indicated in the previous query.

Table 1 Determinants of a decision maker

Factor	Mark
Trust	
Altruism	
Reciprocity	
Emotion	
Risk	
Social responsibility	
Tolerance to ambiguity	
Add new factor if necessary	
......	
......	

(A) *personal quality,*
(B) *power of decision, and*
(C) *other (please indicate).*

These surveys have been sent over the Internet to experts. The answers received from the experts are operated on the basis of Delphi method. Altruism, emotion, trust, reciprocity, risk, social responsibility, tolerance to ambiguity, etc., are obtained as the basic determinants. Therefore, in this work, two psychological determinants are chosen for modeling of the decision maker. The index of decision making of the decision maker's personal quality is accepted. The following type model is offered on the basis of received answers:

IF U_1 is A_i1, U_2 is A_i2, and U_r is A_ir, THEN V is Di and $CF_i \in]0; 100]$

where CF_i is the confidence degree of the rule that is defined by the expert. It expresses the belief degree of the expert to the truth degree of the rule. A_i1, A_i2, A_ir, and D_i are linguistic values of the linguistic variables U_1, U_2, U_r, and V.

3 Modeling of a Decision Maker Under Second-Order Uncertainty Using the Possibility–Probability Measure-Based Method

The mathematical description of knowledge in the knowledge base of decision maker is based on fuzzy interpretation of antecedents and consequents in production rules [2].

$$R^k : \text{IF } x_1 \text{ is } \tilde{A}_{k1} \text{ and } x_2 \text{ is } \tilde{A}_{k2} \text{ and } \ldots \text{ and } x_m \text{ is } \tilde{A}_{km} \text{ THEN}$$
$$u_{k1} \text{ is } \tilde{B}_{k1} \text{ and } u_{k2} \text{ is } \tilde{B}_{k2} \text{ and } \ldots \text{ and } u_{kl} \text{ is } \tilde{B}_{kl}, \quad k = \overline{1, K}$$

where $x_i, i = \overline{1, m}$ and $u_{kj}, j = \overline{1, l}$ are total input and local output variables,

$\tilde{A}_{ki}$, $\tilde{B}_{kj}$ are fuzzy sets, and k is the number of rules.
The basic steps of the method are given below:

1. The truth degree of the rule is computed as follows: $r_{jk} = \text{Poss}(\tilde{v}_k/\tilde{a}_{jk}) \cdot cf_k$, if the sign is " = ", and $r_k = \left(1 - \text{Poss}\left(\tilde{v}_k|\tilde{a}_{jk}\right)\right)cf_k$, if the sign is " $\neq$ ". Poss is defined as follows:

$$\text{Poss}(\tilde{v}|\tilde{a}) = \max_u \min(\mu_{\tilde{v}}(u), \mu_{\tilde{a}}(u)) \in [0, 1].\tau_j = \min(r_{jk})$$

First, the objects are evaluated, i.e., every w_i object has appropriate linguistic value defined as (v_i, cf_i), where v_i is linguistic value and $c_{f_i} \in]0, 100]$ is confidence degree of the value v_i. v_k—linguistic value of the rule object and a_{jk}—current linguistic value (j is index of the rule, k is index of relation) (e.g., A_ir)

2. For each rule, calculate $R_j = \left(\min_j r_{jk}\right) * CF_j/100$, where CF is the confidence degree of the rule.

The user or the creator of the rule defines the firing level (π), and $R_j \geq \pi$ is checked. If the condition holds true, then the consequent part of rule is calculated.

3. The evaluated w_i objects have S_i value: $w_i, (v_i^1, cf_i^1), \ldots, (v_i^{S_i}, cf_i^{S_i})$ S_i is the number of the rules in fuzzy inference process

The average value is determined as follows:

$$\bar{v}_i = \frac{\sum_{n=1}^{S_i} v_i^n \cdot cf_i^n}{\sum_{n=1}^{S_i} cf_i^n}$$

IF $x_1 = \tilde{a}_1^j$ AND $x_2 = \tilde{a}_2^j$ AND ... THEN $y_1 = \tilde{b}_1^j$ AND $y_2 = \tilde{b}_2^j$ AND ...
IF ... THEN $Y_1 = \text{AVRG}(y_1)$ AND $Y_2 = \text{AVRG}(y_2)$ AND ...

This model has a built-in function AVRG which calculates the average value. This function simplifies the organization of compositional inference with possibility measures. As a possibility measure here, a confidence degree is used. So, the compositional relation is given as a set of production rules such as:

IF $x_1 = \tilde{A}_1^j$ AND $x_2 = \tilde{A}_2^j$ AND ... THEN $y_1 = \tilde{B}_1^j$ AND $y_2 = \tilde{B}_2^j$ AND,

where j is a number of a rule. After all these rules have been executed (with different truth degrees), the next rule (rules) ought to be executed:

IF THEN $Y_1 = \text{AVRG}(y_1)$ AND $Y_2 = \text{AVRG}(y_2)$ AND ...

Using this model, one may construct hypotheses that generating and accounting systems. Such system contains the rules:

IF <condition$_j$> THEN $X = \tilde{A}_j$ CONFIDENCE cf$_j$

Here, "$X = \tilde{A}_j$" is a hypothesis that the object X takes the value $\tilde{A}_j$. Using some preliminary information, this system generates elements $X = (\tilde{A}_j, R_j)$, where R_j is a truth degree of jth rule. In order to account the hypothesis (i.e., to estimate the truth degree that X takes the value $\tilde{A}_j$), the recurrent Bayes-Shortliffe formula, generalized for the case of fuzzy hypotheses, is used [2]:

$$P_0 = 0$$

$$P_j = P_{j-1} + \text{cf}_j \text{Poss}(\tilde{A}_0/\tilde{A}) \left(1 - \frac{P_{j-1}}{100} \right)$$

This formula is realized as a built-in function BS:

IF END THEN $P = \text{BS}(X, \tilde{A}_0)$.

4 Modeling of Decision Maker by Using Possibility–Probability-Based Method

Let us describe the model taking into account the private characteristic features of a decision maker by using the following rules:

Rule 1

IF altruism level of decision maker is about 45 and emotion level of decision maker about 40,
THEN personal quality of decision maker (D_i) is about 35 and CF is 90.

Rule 2

IF altruism level of decision maker is about 45 and emotion level of decision maker about 60,
THEN personal quality of decision maker (D_i) is about 45 and CF is 55.
...

Rule 15

IF altruism level of decision maker is about 65 and emotion level of decision maker about 20,
THEN personal quality of decision maker (D_i) is about 75 and CF is 60.
It is required to determine the output of the following rule:

IF emotion level of decision maker is about 65 and altruism level of decision maker about 60,

THEN personal quality of decision maker (D_i) is equal.

Where the values of linguistic variable are trapezoidal fuzzy numbers. For example,

$$\tilde{45} = \begin{cases} \frac{x-30}{12}, & 30 \leq x \leq 42 \\ 1, & 42 \leq x \leq 48 \\ \frac{50-x}{2}, & 48 \leq x \leq 50 \\ 0, & \text{otherwise} \end{cases} \qquad \tilde{60} = \begin{cases} \frac{x-50}{5}, & 50 \leq x \leq 55 \\ 1, & 55 \leq x \leq 65 \\ \frac{70-x}{5}, & 65 \leq x \leq 70 \\ 0, & \text{otherwise} \end{cases}$$

$$\tilde{75} = \begin{cases} \frac{x-50}{15}, & 50 \leq x \leq 65 \\ 1, & 65 \leq x \leq 80 \\ \frac{85-x}{5}, & 80 \leq x \leq 85 \\ 0, & \text{otherwise} \end{cases}$$

The above-described model is realized by using the ESPLAN expert system shell, and different tests are performed (see Fig. 1.)

For decision making in the given problem provided current characteristic features of decision maker, i.e., the level of altruism and emotion, it is possible to calculate personal quality on the basis of the given fuzzy IF-THEN rule. In order to verify the sensitivity of the model, the personal quality of decision maker has been investigated under change of level of altruism and emotion.

5 Modeling of Decision Maker on the Basis of Fuzzy and Dempster–Shafer Theory

Now, we consider the modeling on the basis of Dempster–Shafer theory. Human behavioral modeling requires an ability to represent and manipulate imprecise cognitive concepts. It also needs to include the uncertainty and unpredictability of human action [18]. Human behavioral modeling requires an ability to formally represent sophisticated cognitive concepts that are often at best described in imprecise linguistic terms. Fuzzy sets provide a powerful tool for enabling the semantical modeling of these imprecise concepts within computer-based systems [19]. With the aid of a fuzzy set, we can formally represent sophisticated imprecise linguistic concepts in a manner that allows for the types of computational manipulation needed for reasoning in behavioral models based on human cognition and conceptualization.

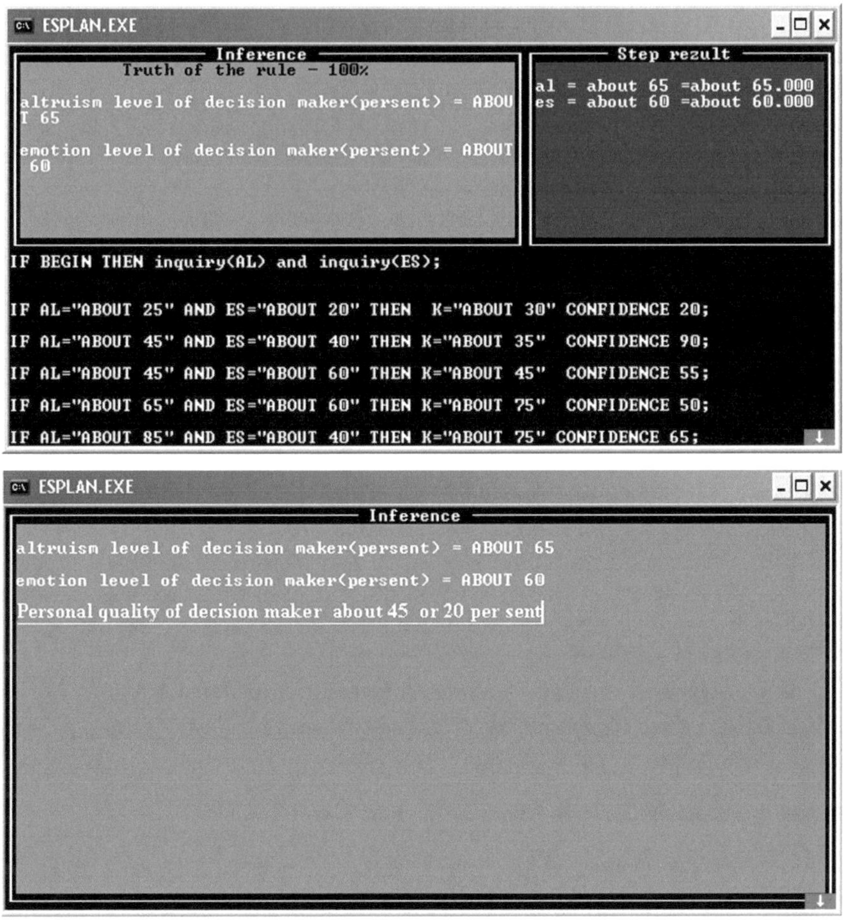

Fig. 1 Fragment of computer simulation

Now, we consider a DM behavioral modeling using fuzzy and Dempster–Shafer theories suggested in [18].

The Dempster–Shafer approach fits nicely into the fuzzy logic since both techniques use sets as their primary data structure and are important components of the emerging field of granular computing. In [18], the behavioral model is represented by partitioning the input space. We can represent relationship between input and output variables by a collection of n "IF-THEN" rules of the form:

If X_1 is $\tilde{A}_{i1}$ and X_2 is $\tilde{A}_{i2}$, ... and X_r is $\tilde{A}_{ir}$, then Y is D_i

Here, each $\tilde{A}_{ij}$ typically indicates a linguistic term corresponding to a value of its associated variable; furthermore, each $\tilde{A}_{ij}$ is formally represented as a fuzzy subset defined over the domain of the associated variable X_j. Similarly, $\tilde{D}_i$ is a value associated with the consequent variable Y that is formally defined as a fuzzy subset of the domain of Y. To find the output of a DM described by above-mentioned rule, Mamdani inference method is used.

We consider the consequent to be a fuzzy Dempster–Shafer granule. Thus, we shall now consider the output of each rule to be of the form Y is m_i where m_i is a belief structure with focal elements $\tilde{D}_{ij}$ which are fuzzy subsets of the universe Y and associated weights $m_i(\tilde{D}_{ij})$. Thus, a typical rule is now of the form.

If X_1 is $\tilde{A}_{i1}$ and X_2 is $\tilde{A}_{i2}$, ... and X_r is $\tilde{A}_{ir}$, then Y is $m_i()$.

Using a belief structure to model, the consequent of a rule is essentially saying that $m_i(\tilde{D}_{ij})$ is the probability that the output of the ith rule lies in the set $\tilde{D}_{ij}$. So rather than being certain as to the output set of a rule, we have some randomness in the rule. We note that with $m_i(\tilde{D}_{ij}) = 1$ for some $\tilde{D}_{ij}$.

Let us describe the reasoning process in this situation with belief structure consequents. Assume that the inputs to the system are the values for the antecedent variables, $X_j = x_j$. For each rule, we obtain the firing level, $\tau_i = \text{Min}[A_{ij}(x_j)]$.

The output of each rule is a belief structure $\hat{m}_i = \tau_i \wedge m$.

The focal elements of $\hat{m}_i$ are $\tilde{F}_{ij}$, a fuzzy subset of Y where $F_{ij}(y) = \text{Min}[\tau_i, D_{ij}(y)]$, where $\tilde{D}_{ij}$ is a focal element of m_i. The weights associated with these new focal elements are simply $\hat{m}_i(\tilde{F}_{ij}) = \hat{m}_i(\tilde{D}_{ij})$. The overall output of the system m is obtained by taking a union of the individual rule outputs: $m = \bigcup\limits_{i=1}^{n} \hat{m}_i$.

For every a collection $\langle \tilde{F}_{1j_1}, \ldots, \tilde{F}_{nj_1} \rangle$ where $\tilde{F}_{ij_1}$ is a focal element of $m_{i,}$ we obtain a focal element of m, $\tilde{E} = \bigcup\limits_{i} \tilde{F}_{ij_1}$, and the associated weight is as follows:

$$m(\tilde{E}) = \prod_{i=1}^{n} \hat{m}_i(\tilde{F}_{ij_1}).$$

As a result of this third step, it obtained a fuzzy D–S belief structure V and m as output of the agent. We denote the focal elements of m as the fuzzy subsets $\tilde{E}_j$, $j = 1$ to q, with weights $m(\tilde{E}_j)$.

Let us describe the model taking into account the characteristic features of DM. DM's behavioral model can be described as follows[19]:

Rule 1 IF trust level of a DM is *about 76* and altruism level of a DM *about 45*, THEN personal quality of a DM (V) is m_1.

Rule 2 IF trust level of a DM is *about 35* and altruism level of a DM *about 77*, THEN personal quality of a DM (V) is m_2.

Let us determine the output (personal quality of a DM).

If trust level of a DM is *about 70* and altruism level of a DM is *about 70*: m_1 has focal elements $\tilde{D}_{11} = 4\tilde{6}$ with $m(\tilde{D}_{11}) = 0.7$ and $\tilde{D}_{12} = 4\tilde{8}$ with $m(\tilde{D}_{11}) = 0.3$ m_2 has focal elements $\tilde{D}_{21} = 7\tilde{6}$ with $m(\tilde{D}_{21}) = 0.2$ and $\tilde{D}_{22} = 8\tilde{1}$ with $m(\tilde{D}_{22}) = 0.8$

The values of linguistic variables are trapezoidal fuzzy numbers:

$$
4\tilde{6} = \begin{cases} \frac{x-40}{6}, & 40 \le x \le 46 \\ 1, & x = 46 \\ \frac{65-x}{19}, & 46 \le x \le 65 \\ 0, & \text{otherwise} \end{cases} \qquad
4\tilde{8} = \begin{cases} \frac{x-40}{8}, & 40 \le x \le 48 \\ 1, & x = 48 \\ \frac{65-x}{17}, & 48 \le x \le 65 \\ 0, & \text{otherwise} \end{cases}
$$

$$
7\tilde{6} = \begin{cases} \frac{x-61}{15}, & 61 \le x \le 76 \\ 1, & x = 76 \\ \frac{95-x}{19}, & 76 \le x \le 95 \\ 0, & \text{otherwise} \end{cases} \qquad
8\tilde{1} = \begin{cases} \frac{x-61}{20}, & 61 \le x \le 81 \\ 1, & x = 81 \\ \frac{95-x}{14}, & 81 \le x \le 95 \\ 0, & \text{otherwise} \end{cases}
$$

Let us calculate the belief values for each rule. By using [18] in this example, the empty set takes the value 0.09. But in accordance with Dempster–Shafer theory, m value of the empty set should be zero. In order to achieve this, m values of the focal elements should be normalized and m value of the empty set made equal to zero. The normalization process is as follows:

1. Determine $T = \sum\limits_{\tilde{A}_i \cap \tilde{B}_i = \varnothing} m_1(\tilde{A}_i) \cdot m_2(\tilde{B}_i)$

2. For all $\tilde{A}_i \cap \tilde{B}_i = \varnothing$ weights

$$
m(\tilde{E}_k) = \frac{1}{1-T} m_1(\tilde{A}_i) \cdot m_2(\tilde{B}_j)
$$

3. For all $\tilde{E}_k = \varnothing$ sets $m(\tilde{E}_k) = 0$

In accordance with the procedures described above:

$$
m_3 = (\{4\tilde{6}\}) = 0.230769,
$$
$$
m_3 = (\{4\tilde{6}, y\}) = 0.384615,
$$
$$
\text{Bel}(\{4\tilde{6}, y\}) = 0.615385.
$$

For the second rule, $\text{Bel}(\{7\tilde{6}, y\}) = 0.753425$. Firing level of the *i*th rule is equal to the minimum among all degrees of membership of a system input to antecedent fuzzy sets of this rule: $\tau_i = \min\limits_{j=1}^{n}\left[\max\limits_{X_j}(A'(x_j) \wedge A_{ij}(x_j))\right]$. The firing levels of each rule are $\tau_1 = 0.26$ and $\tau_2 = 0.28$. The defuzzified values of focal elements obtained by using the center of gravity method are the following:

$$\text{Defuz}(\tilde{E}_1) = \bar{y}_1 = 61.56;$$
$$\text{Defuz}(\tilde{E}_2) = \bar{y}_2 = 64.15;$$
$$\text{Defuz}(\tilde{E}_3) = \bar{y}_3 = 62.52;$$
$$\text{Defuz}(\tilde{E}_4) = \bar{y}_4 = 65.11.$$

The defuzzified value of m is $\bar{y} = 63.92$.

By using the data described above, we arrive at the following Dempster–Shafer structure.

IF trust level of a DM is *about 70* and altruism level of a DM *about 70* THEN, personal quality of a DM (V) is equal to 63.92.

6 Conclusion

In this paper, the decision-maker behavioral modeling under imperfect information or second-order uncertainty is proposed. By using Delphi method, psychological determinants of a decision maker were determined. The described models are realized by using the expert system shell, and the language of technical computing MATLAB and different tests are performed. The obtained results proved the validity of the suggested approach.

Acknowledgment I would like to thank Professor R. Yager and Professor R. Aliyev in provision of useful advice in implementation of this research work.

References

1. Zadeh, L.A.: Computing with words and perceptions—a paradigm shift. In: Proceedings of the IEEE International Conference on Information Reuse and Integration, pp. 450–452. Las Vegas, IEEE Press, Nevada, USA (2009)
2. Aliyev R.A., Gulko, D.E., Shakhnazarov, M.M.: Expert system for production planning. J. News Acad. Sci. USSR Tekhn. Cybern. **5**, 25–30 (1988)
3. Cox, J.C., Friedman, D., Sadiraj, V.: Revealed altruism. J. Econom. **6**, 31–69 (2008)
4. Cox, J.C.: How to identify trust and reciprocity. J. Games Econ. Behav. **46**, 260–281 (2004)
5. Cox, J.C., Friedman, D., Gjerstad, S.: A tractable model of reciprocity and fairness. J. Games Econ. Behav. **59**, 17–45 (2007)
6. Cox, J.C., Sadiraj, K., Sadiraj, V.: Implications of trust, fear, and reciprocity for modeling economic behavior. Exp. Econ. **11**(1), 1–24 (2008)
7. Cox, J.C., Ostrom, E., Walker, J.M., Castillo, A.J., Coleman, E., Holahan, R., Schoon, M., Steed, B.: Trust in privative and common property experiments. South. Econ. J. **75**(4), 957–975 (2009)

8. Abbink, K., Irlenbusch, B., Renner, E.: The moonlighting game: an empirical Study on Reciprocity and Retribution. J. Econ. Behav. Organ. **42**, 265–277 (2000)
9. Andreoni, J., Miller, J.: Giving according to GARP: an experimental test of the consistency of preferences for altruism. Econometrica **70**, 737–753 (2002)
10. Berg, J., Dickhaut, J., Mccabe, K.: Trust, reciprocity, and social history. J. Games Econ. Behav. **10**, 122–142 (1995)
11. Falk, A., Fischbacher, U.: A theory of reciprocity. J. Games Econ. Behav. **54**, 122–142 (2006)
12. Bosman, R., Van Winden, F.: Emotional hazard in a power-to-take experiment. Econ. J. **112**, 147–169 (2002)
13. Fehr, E., Schmidt, K.M.: A theory of fairness, competition, and cooperation. Q. J. Econom. **114**, 817–868 (1999)
14. Kahneman, D., Tversky, A.: Prospect theory: an analysis of decision under uncertainty. Econometrica **47**(2), 263–291 (1979)
15. Gardashova, L.A.: Economic agents behavior modeling under second-order uncertainty. In: Proceedings Ninth international conference on Application of Fuzzy Systems and Soft computing, pp. 359–364. Prague, Czech Republic (2010)
16. Gardashova, L.A.: A new approach to solving decision making problem with Z-information under uncertain environment. In: 2nd World Conference on Soft Computing, pp. 464–470. Baku (2012)
17. Gardashova, L.A.: Application of operational approaches to solving decision making problem using Z-numbers. J. Appl. Math. (in press)
18. Yager, R.R.: Human behavioral modeling using fuzzy and Dempster–Shafer theory. In: Liu, H., Salerno, J.J., Young, M.J. (eds.) Social Computing, Behaviorial Modelling, and Prediction, pp. 89–99. Springer, New York (2008)
19. Yager, R.R.: Using knowledge trees for semantic web querying. In: Sanchez, E. (ed.) Fuzzy Logic and the Semantic Web, pp. 231–246. Elsevier, Amsterdam (2006)

Expert Knowledge Base in Integrated Maintenance Models for Engineering Plants

Ajit K. Verma, A. Srividya, P.G. Ramesh, Ashok Deshpande and Rehan Sadiq

Abstract Maintenance of large engineering systems is a complex requirement. Experience shows that a combination of both time- and condition-based maintenance is required to be optimally planned for such systems. Further, such a plan is required to be put in place even as the systems are being designed and installed so that the benefits of maintenance are maximized. While it is possible to use historical data for reliability and maintenance models, considerable amount of knowledge available as domain expertise needs to be tapped, to effectively model and plan maintenance strategies. In this paper, a framework for integration of time- and condition-based maintenance is presented and areas where domain expertise can be harnessed have been highlighted. Fuzzy logic has been shown as a useful tool in this framework to elicit expert information. A case study has been discussed to demonstrate the utility of expert information in modeling and planning scheduled preventive maintenance aspect for a ship's machinery platform.

Keywords TBM · CBM · Scheduled maintenance · Condition monitoring · Fuzzy fault tree · Expert elicitation · Domain knowledge Base · Genetic algorithm

A.K. Verma (✉)
University College, Haugesund, Norway
e-mail: akvmanas@gmail.com

A. Srividya
University of Stavanger, Stavanger, Norway

P.G. Ramesh
L & T Ship Building Ltd., Chennai, India

A. Deshpande
University of Pune, Pune, India

R. Sadiq
The University of British Columbia, British Columbia, Canada

© Springer International Publishing Switzerland 2016   515
L.A. Zadeh et al. (eds.), *Recent Developments and New Direction
in Soft-Computing Foundations and Applications*, Studies in Fuzziness
and Soft Computing 342, DOI 10.1007/978-3-319-32229-2_36

1   Introduction

Engineering systems today are required to perform in competitive environments where functional failures can be extremely expensive, not only in terms of repairing or replacing failed components but also due to lost opportunities and customer goodwill. Failures can also adversely impact safety of life such as in the cases of nuclear and chemical plants, aircraft, and other mass transport systems. While failure prevention is generally addressed through reliability programs and by proactive and periodic maintenance, it is common knowledge that not all failures are age-related. Considerable numbers of failures in large engineering systems are due to random causes and physical, functional, and stochastic interdependencies between various constituents of the systems. Management of such failures would essentially involve early detection of onset of failures of subsystems and assemblies so that the systems can be safely taken out of operation and preventive maintenance undertaken. Such advance failure detection is also expected to provide adequate lead time for maintenance managers to mobilize required maintenance logistics and undertake maintenance at an opportune time with minimal impact on system operation.

Needless to say that there is an increasing trend among industries to provide equipment and systems which enable strategies for monitoring health of machinery in large maintenance of large engineering systems such as power plants, process plants, ships, and aircraft is a complex activity that has large stakes in terms of life cycle cost, availability, reliability, and safety. A number of stochastic models for time-based maintenance (TBM) of such repairable systems have been proposed in the literature [1–4]. The models are essentially based on lifetime distributions and optimization tools with recommendations in favor of modeling and analysis as key approaches for supporting maintenance decision makers [5, 6]. With the advancement in sensor technologies and greater understanding of failure prognosis, condition-based maintenance (CBM) and predictive maintenance (PrM) strategies began to gain ground. Several studies have focused on diverse aspects of the strategy such as sensor technologies, condition-monitoring systems, tools/techniques, inspection intervals, diagnosis, and prognosis [7–12]. Predictive information has been used for scheduling maintenance dynamically for multicomponent systems, also suggesting a combination of time- and condition-based maintenance strategies [13].

Comparisons between the two strategies—TBM and CBM—have been studied by several authors. One of the recent studies has been that by Ahmad and Kamaruddin [14]. The authors have suggested CBM to be more realistic than TBM since most failure mechanisms do provide indications or telltale signs of impending failures. Also TBM data analysis/modeling is required to follow several statistical rules and assumptions which reduce realism in their solutions. Although advantages of CBM are widely recognized, it can be seen that there have not been many studies that have critically examined the practical impact or advantages of CBM actually realized by industries. A few studies have highlighted the need for greater application of prognostics, condition-monitoring procedures, and use of domain-related knowledge [15].

In both TBM and CBM models, a large part of the information required for modeling and solution methods are traceable to domain experts. Considerable work has been done in the field of expert elicitation. Two of the recent works pertain to formal knowledge modeling of multiple experts as decision support [16] and involvement of a number of experts in the elicitation of uncertain parameters with an application of Dempster–Shafer theory of evidence [17]. Fuzzy logic has been observed to be a useful tool in uncertainty management in the area of maintenance, particularly with respect to selection, planning of maintenance strategies, and multicriteria maintenance decision making [18–20]. Specific issues such as optimal action required in responding to asset condition using accumulated knowledge base with the help of fuzzy-based decision support models have been proposed [21]. One of the important issues that have not been adequately addressed in the literature is the formulation of comprehensive maintenance strategies integrating TBM and CBM for large engineering systems at the design and development stages and the effective use of expert knowledge base for the same. In this paper, a framework for maintenance integrating TBM and CBM has been proposed. Some of the areas where expert knowledge could be used through fuzzy rule base in the absence of crisp data have been highlighted. A typical modeling approach and a case study to demonstrate the application of the proposed methodology have been also presented.

2 Integrated Maintenance

Condition-based or predictive maintenance tools and techniques are gaining favor among maintenance personnel in general. However, maintenance undertaken based on equipment health or condition alone is likely to lead to frequent interruptions in the operations of large and complex engineering systems. Maintenance actions are time-consuming; they require elaborate logistic preparations and mobilization of maintenance personnel including those for support functions such as material handling, quality assurance, and record keeping. Further, most large-scale maintenance actions in such plants are related to deterioration like wear of moving components, expiry of life of components such as bearings, sealing devices, and gaskets, and general loss of material due to corrosion and erosion. All these processes of deterioration are age-related and can, with fair amount of precision, be predicted by modeling, historical data, or domain experience. Therefore, in the opinion of the authors, time-based maintenance as per a predetermined maintenance cum operations schedule is worthy of consideration as a justifiable basic maintenance strategy for such plants. Maintenance intervals, however, need to be optimized so that maintenance jobs are undertaken as concurrently as feasible for all components or modules of large engineering systems, thereby reducing the total downtime for maintenance [22].

At the same time, it is essential to note that likelihood of occurrence of random or hidden failures cannot be ruled out completely despite having a scheduled maintenance strategy in place. In order to meet such eventualities and to ensure that the availability of plants is enhanced during the time that the plant is in operation, appropriate condition-monitoring strategies are required to be used. Such condition-monitoring facilities will have to be capable of detecting random defects as well as incipient failures that can grow into large-scale failures. Additionally, prognostic features in the facilities will need to be capable of evaluating telltale signs and predicting times to failure so that necessary preventive or, more aptly, predictive maintenance can be conjoined with schedule preventive maintenance. However, in the case of random defects, concerned equipment should be taken up for maintenance without leading to major plant-level breakdown and the selected condition-monitoring facility should be able to support such a requirement.

Some of the key factors that can ensure the success of an integrated maintenance system are optimal scheduling of time-based maintenance and selection or design of suitable condition-monitoring facility for detecting and predicting incipient failures.

3 Expert Knowledge Base

In order to plan an optimized operations maintenance cycle and design an appropriate condition-monitoring system, research work including all ground realities needs to be undertaken. An important ingredient in the above research would necessarily be contributions from domain experts including a cross section of management personnel, designers, operations personnel, maintenance engineers, logisticians, and experienced workmen and their supervisors.

It is a common observation in the field of maintenance that such expert knowledge base is widely available, albeit mostly in the form of collective and subjective experience of personnel involved in maintenance operations. Elicitation of such information and incorporation of the same in maintenance decision models and frameworks is a challenge that needs to be addressed to enable formulation of realistic, cost-effective, and reliability-based maintenance strategies. The use of fuzzy rule-based expert knowledge elicitation is one of the techniques that can meaningfully use domain expertise in optimizing an integrated framework for both time- and condition-based maintenance strategies and techniques.

Maintenance decision making follows a sequence of activities. The literature outlines these activities under certain phases of decision process: intelligent phase, design phase, and choice phase [23]. In general, maintenance decision making can be divided into the following five phases as shown in Fig. 1.

- Problem identification and definition
- Modeling
- Generation of alternate solutions
- Choice of solution
- Implementation

Maintenance problem identification and definition. To identify a maintenance problem, the gap between the desired state of the system and its current state has to be recognized with a consensus among all decision makers. There will be a need to

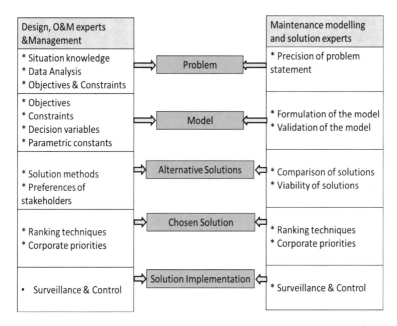

Fig. 1 Sequence of activities in maintenance decision making

understand the 'on-ground' situation in terms of system health, both desired and actual, adequate collection of life and maintenance data and awareness of organizational constraints. These inputs are to be gathered from the domain experts. The inclusiveness and comprehensiveness of these inputs cannot be overemphasized. Further, these inputs will have to be measurable or quantifiable, in quantitative or qualitative or fuzzy terms.

Maintenance modeling. Modeling should be the core phase of a systematic maintenance decision-making process. A maintenance decision maker is concerned with selecting and planning a course of action for the maintenance of an engineering system. Models consist of variables and parametric constants in the selection of which often there will be a need for contribution from experts.

Generations of alternative maintenance solutions. Decision making implies selecting the best possible alternative. Therefore, there is a need to generate several alternative solutions. A suitable optimization strategy must be chosen for this and preferences of various stakeholders elicited.

Choice of maintenance solutions. The choice is to be made on the basis of the weightage to be accorded for the various objectives in the prevailing overall organizational scenario. The chosen solution is also required to be validated.

Implementation of the chosen maintenance solution. The validated solution is applied to the maintenance situation. For continuous improvement to be possible in the process of decision making, the outcome of application of the solution is necessary to be monitored and the necessary corrections are to be fed back at appropriate phases of decision making.

As shown in Fig. 1, domain experts can play an important role in the decision-making process.

4   Proposed Integrated Maintenance Framework

A framework proposed to include both TBM and CBM as an integrated maintenance strategy is shown in Fig. 2. TBM and CBM are being proposed to be concurrently undertaken for large engineering systems, and hence, there are two parallel paths shown in the figure. The various stages indicated in the framework are also explained in this section.

Step 1:   The engineering system is to be divided into subsystems and individual equipment. The subdivision is then to be grouped into sets of modules called higher modular assemblies (HMAs). Components of each HMA have certain commonality with respect to functionality and maintainability.

A: **CBPM**

Step 2A: Identify the hierarchical layout of the HMA—subsystem, sub-subsystem, assembly, subassembly, and component levels.

Step 3A: Undertake fault tree analysis (FTA) for each HMA at suitable levels in the hierarchy.

Step 4A: Based on the faults expected to occur at identified locations, select suitable sensors and signal processing devices to pick up signals appropriately corresponding to the equipment condition.

Step 5A: Estimate predictability and prognostic ability of the condition-monitoring strategies selected in the previous step through expert elicitation and fuzzy inference systems. In doing so, the capabilities available locally with the LES as well as those available remotely for exploitation through e-maintenance framework are to be taken into account.

Step 6A: Collect statistical data pertaining to failure, repair, cost of CBPM, and logistic lead times.

Step 7A: Formulate mathematical expressions for maintenance objectives (as functions of the decision variables) for the optimization problem.

Step 8A: Execute the multiobjective optimization problem (MOOP) using genetic algorithms.

Step 9A: Select the optimum solution from the set of Pareto-optimal solutions keeping in view the trade-off between the competing objectives. Study of the sensitivities of decision variables will aid in the selection of the optimum solution.

Step 10A: Plan the condition-monitoring strategies corresponding to the selected solution.

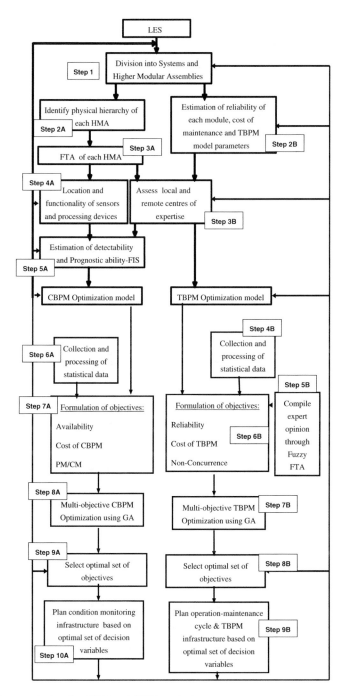

Fig. 2 Framework showing TBM and CBM strategies

B: **TBPM**

Step 2B: Estimate the reliability available and required for each of the HMAs, the acceptable cost of TBPM, and manufacturers' recommendations regarding maintenance intervals.

Step 3B: Assessment of maintenance facilities available for the LES locally as well as those that can be accessed from remote locations.

Step 4B: Collect and process data pertaining to failures, repairs, and cost of TBPM.

Step 5B: Collect expert inputs to the model through fuzzy fault tree analysis.

Step 6B: Formulate mathematical expressions for maintenance objectives (as functions of the decision variables) for the optimization problem.

Step 7B: Execute the multiobjective optimization problem (MOOP) using genetic algorithms.

Step 8B: Select the optimum solution from the set of Pareto-optimal solutions keeping in view the trade-off between the competing objectives. Study of the sensitivities of decision variables will aid in the selection of the optimum solution.

Step 9B: Formulate operation-cum-maintenance cycle for the given period of time and TBPM infrastructure based on the selected solution.

5   The Model

In the framework proposed in the previous section, maintenance decision models are aimed at optimizing time- and condition-based maintenance decisions that are required to be made a priori for a large engineering system. This activity would be part of the design and system integration phase in the life cycle of the plant. The following are some of the inputs that would be necessary to be obtained from experts:

- Chances of a maintenance or repair activity being completed in a specified time.
- Labor-related issues affecting maintenance.
- Spares-related issues affecting maintenance.
- Chances that a CM strategy is capable of detecting the onset of a failure.
- Chances that the CM strategy is capable of predicting the time to failure.
- Chances that the CM strategy is capable of diagnosing the cause of a defect.

Expert knowledge elicitation in connection with condition-based maintenance has been discussed elsewhere [24]. In this section, it is proposed to highlight a model for scheduled (time-based) preventive maintenance and indicate how expert knowledge pertaining to one of the parameters can be incorporated in the model.

The following two objectives are considered in the model [22]:-

Objective 1: **Non-concurrence of Maintenance Periods (NC)**. The sum of the absolute difference between starting times of maintenance periods and the absolute difference between completion times of maintenance periods of all HMAs can be considered as a measure of non-concurrence of maintenance periods.

Let

t_i = Starting times of the first preventive maintenance of HMA, i, $(i = 1,..p,..h)$

T_i = Completion time of the preventive maintenance of HMA, i, $(i = 1,..p,..h)$

W_i = Number of labor-days for completion of maintenance of HMA, i,

P_i = Probability of timely completion of the maintenance of HMA, i, due to logistic delay, and m_i = number of personnel employed for maintenance of HMA, i

Therefore, the non-concurrence of maintenance periods is given by

$$NC = \sum_{i=1}^{h-1} \sum_{p=i+1}^{h} \left\{ |(t_i - t_p)| + \left| \left(t_i + \frac{W_i}{P_i m_i} \right) - \left(t_p + \frac{W_p}{P_p m_p} \right) \right| \right\} \qquad (1)$$

Objective 2: **Unreliability (UR)**. While different unreliability models can be used for the purpose of optimization, a simple expression for unreliability is used for the present model. In the subject case, every HMA consists of several modules, which in turn have multiple components. During the time horizon, deterioration is assumed to be negligible and preventive maintenance is expected to restore the HMAs to as good as new condition. Therefore, it is assumed that exponential distributions are suitable to describe the failure rate for all the HMAs, of course, with different hazard rates. As mentioned earlier, all HMAs are required to be operational for the operational state of the plant (series configuration). Thus, we have unreliability given by:

$$UR = 1 - \prod_{i=1}^{h} R_i(t_i) \qquad (2)$$

Constraint Function. It is possible that several constraints limit the feasible space in the above multiobjective optimization problem owing to the restraining requirements of various stakeholders of the system as far as the objectives are concerned. One of the constraints can be that of the maximum number of personnel available for maintenance, N. The constraint function is given by

$$\sum_{i=1}^{h} m_i \leq N \tag{3}$$

The probability of timely completion of maintenance is a parameter that is not available in crisp form as part of any measured data. The parameter is estimated based on expert information using suitable fuzzy inference systems (FISs) as shown in Fig. 3.

The evaluation after fault tree analysis follows the following steps:-

(a) Step 1: Analyze expert opinion and form relationship matrix.
(b) Step 2: Form membership functions and fuzzy rules.
(c) Step 3: Evaluate the FIS.
(d) Step 4: Evaluate the top event probability.

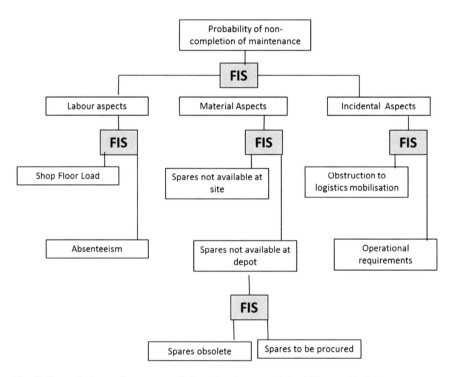

Fig. 3 Fuzzy fault tree for non-completion of maintenance job within specified time

6 Case Study

Maintenance for a 32,700 tonne small cruise ship's machinery modules needs to be preventively maintained and a suitable schedule prepared. The following modules of the ship are considered for the maintenance planning:-

- Propulsion diesel engine (PDE)
- Diesel generator (DG)
- Air-conditioning plant (AC)
- Fire main system (FMS)

The following are some of the important data collected:

Typically, the FIS for the top event in the fuzzy fault tree, that is, 'probability of non-completion of maintenance' is depicted in Fig. 4. The fuzzy relationship matrix bringing out the rule base is shown in Table 1.

Thus, the combined influence of labor issues, material constraints, and circumstantial issues on non-completion of maintenance jobs in respect of propulsion diesel engine module is 0.3442. The probability of completion of maintenance jobs for propulsion module is $1 - 0.3442 = 0.6558$ as indicated in Table 2. Likewise, the probabilities for the lower branches in the fault tree can be estimated.

The above data were applied in the model mentioned in the previous section. Both the objectives were simultaneously optimized in using NSGA II (non-dominated sorting genetic algorithm) [25].

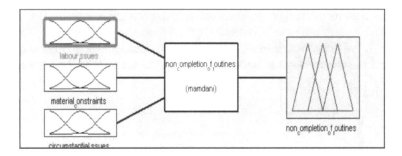

Fig. 4 FIS for non-completion of maintenance job

Table 1 Important data for TBM model

Data	PDE	DG	AC	FMS
Weibull–Poisson process parameters β λ	1.78 1.99×10^{-5}	1.998 6.7×10^{-6}	1.85 1.42×10^{-5}	1.55 1.95×10^{-4}
Man-days for maintenance	9	6	4	3
Probability of timely completion of maintenance (obtained from expert elicitation)	0.6558	0.5945	0.5556	0.6357

Table 2 Relationship matrix for non-completion of maintenance jobs

Labor issues	Material constraints	Circumstantial issues	Non-completion of maintenance job	Influence factor
Low	Low	Low	Low	0.15
Low	Low	Medium	Low	0.2
Low	Medium	High	Medium	0.4
Low	Medium	Low	Medium	0.6
Medium	High	Medium	Medium	0.6
Medium	High	High	High	0.8
Medium	Low	Low	Low	0.8
Medium	Low	Medium	Medium	0.6
High	Medium	High	High	0.7
High	Medium	Low	Medium	0.6
High	High	Medium	High	0.8
High	High	High	High	0.9

7 Results and Discussion

The relationship between non-concurrence of maintenance periods and reliability obtained as a result of the optimization is shown in Fig. 5.

It may be noticed that there are multiple solutions as represented by the Pareto-optimal front. Each point in the front represents a trade-off solution. As the non-concurrence of maintenance periods reduces (maintenance actions for all HMAs are undertaken more concurrently), there is a reduction in system reliability. In the case where non-concurrence of maintenance actions is high, maintenance is expected to be carried out at intervals most suitable for each of the HMAs, from

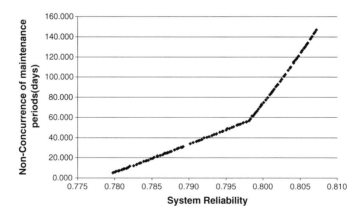

Fig. 5 Non-concurrence of maintenance periods and system reliability

Table 3 Comparison of two candidate solutions

Candidate solution	1		2	
Module	Reliability = 0.785		Reliability = 0.78	
	Start time	End time	Start time	End time
PDE	66	70	69	71
DG	66	70	68	70
AC	67	69	69	71
FMS	70	71	70	71
Downtime for PM	5		3	

reliability point of view. Hence, the reliability of the system as a whole is also high as seen from the above results. This would, however, result in frequent interventions for maintenance. On the other hand, a low value of non-concurrence of maintenance means that maintenance actions are clubbed, often at the cost of reliability, and hence, there is a drop in reliability with decrease in non-concurrence of maintenance. Two candidate solutions are shown in Table 3.

8 Decision Making

Maintenance decision makers may select that solution which would be appropriate in accordance with the management priorities. For instance, comparing candidate solutions 1 and 2, it can be seen that the former can provide a higher reliability of 0.785 as against the letter where the reliability is lower at 0.78. However, in the former case, the downtime for maintenance is 5 days as against 3 days for the latter. For reliabilities over 0.8, non-concurrence of maintenance is quite high, indicating a situation where frequent interruptions would be encountered. For instance, a maintenance planner for a warship would desire a high level of reliability and hence would accept more frequent interruptions for maintenance. The interruptions in operation could be, if necessary, offset by increasing equipment redundancy. It may be observed that in this present case, differences in the reliability values are not very significant. The range or the extent to which one objective would vary with respect to the other will depend on the inherent reliability and other model parameters of the HMAs.

9 Conclusion

Maintenance of large engineering systems would require frameworks that integrate both TBM and CBM. Such a combined framework should be implemented and maintenance decisions need to be carried out at the design and development stage of an engineering plant. This would aid in optimizing the benefits to be accrued from maintenance. However, under such conditions, historical data available for estimating various parametric values would be limited. This limitation could, to a considerable extent, be overcome by exploiting expert knowledge base in the domain. One of the useful approaches for elicitation of expert knowledge is through fuzzy logic tools and techniques. One such framework as mentioned above has been presented in this paper. A modeling approach and a case study to demonstrate the proposed idea have also been presented. It can be seen that integration of TBM and CBM, also including the aid of fuzzy logic-based elicitation of expert knowledge base, is a very promising strategy for informed maintenance decision making in the context of large engineering systems. Such decision making will enable the formulation of cost-effective maintenance plans right at the inception of the engineering systems.

References

1. Barlow, R.E., Hunter, L.C.: Optimum preventive maintenance policies. Oper. Res. **8**, 90–100 (1960)
2. Rigdon, S.E., Basu, A.P.: Statistical Methods for Reliability of Repairable Systems. Wiley, New York (2000)
3. Wang, H.: A survey of maintenance policies of deteriorating systems. Eur. J. Oper. Res. **139** (3), 469–489 (2002)
4. Garg, A., Deshmukh, S.G.: Maintenance management: Literature review and directions. J. Qual. Maintenance Eng. **12**(3), 205–238 (2006)
5. Sharma, A., Yadava, G.S., Deshmukh, S.G.: A literature review and future perspectives on maintenance optimization. J. Qual. Maintenance Eng. **17**(1), 5–25 (2011)
6. Zio, E., Compare, M.: Evaluating maintenance policies by quantitative modeling and analysis. Reliab. Eng. Syst. Saf. **109**, 53–65 (2013)
7. Jardine, A.K.S., Lin, D., Banjevic, D.: A review on machinery diagnostics and prognostics implementing condition based maintenance. Mech. Syst. Signal Process. **20**(7), 1483–1510 (2006)
8. Ciarapica, F.E., Giacchetta, G.: Managing the condition-based maintenance of a combined-cycle power plant: An approach using soft Computing techniques. J. Loss Prev. Process Ind. **19**, 316–325 (2006)
9. Sherwin, D.J., Al-Najjar, B.: Practical models for condition monitoring inspection intervals. J. Qual. Maintenance **5**(3), 203–220 (1999)
10. Sikorska, J.Z., Hodkiewicz, M., Ma, L.: Prognostic modelling options for remaining useful life estimation by industry. Mech. Syst. Signal Process. **25**, 1803–1836 (2011)
11. Prajapati, A., Bechtel, J., Ganesan, S.: Condition based maintenance: A survey. J. Qual. Maintenance Eng. **18**(4), 384–400 (2012)
12. Al-Najjar, B.: On establishing cost-effective condition-based maintenance Exemplified for vibration-based maintenance in case companies. J. Qual. Maintenance Eng. **18**(4), 401–416 (2012)
13. Van Horenbeek, A., Pintelon, L.: A dynamic predictive maintenance policy for complex multi-component systems. Reliab. Eng. Syst. Saf. **120**, 39–50 (2013)
14. Ahmad, R., Kamaruddin, S.: An overview of time-based and condition-based maintenance in industrial application. Comput. Ind. Eng. **63**, 135–149 (2012)
15. Veldman, J., Klingenberg, W., Wortmann, H.: Managing condition-based maintenance technology-A multiple case study in the process industry. J. Qual. Maintenance Eng. **17**(1), 40–62 (2011)

16. Potes Ruiz, P.A., Kamsu-Foguem, B., Noyes, D.: Knowledge reuse integrating the collaboration from experts in industrial maintenance management. Knowl.-Based Syst. **50**, 171–186 (2013)
17. Baraldi, P., Compare, M., Zio, E.: Maintenance policy performance assessment in presence of imprecision based on Dempster–Shafer theory of evidence. Inf. Sci. **245**, 112–131 (2013)
18. Sergaki, A., Kalaitzakis, K.: A fuzzy knowledge based maintenance planning in power systems. Reliab. Eng. Syst. Saf. **77**, 19–30 (2002)
19. Al-Najjar, B., Alsyouf, I.: Selecting the most efficient maintenance approach using fuzzy multiple criteria decision making. Int. J. Prod. Econ. **84**, 85–100 (2003)
20. Lau, H.C.W., Dwight, R.A.: A fuzzy-based decision support model for engineering asset condition monitoring—A case study of examination of water pipelines. Expert Syst. Appl. **38**, 13342–13350 (2011)
21. Vijay Kumar, E., Chaturvedi, S.K.: Prioritization of maintenance tasks on industrial equipment for reliability: A fuzzy approach. Int. J. Qual. Reliab. Management. **28**(1), 109–126 2011
22. Verma, A.K., Srividya, A., Ramesh, P.G.: Multi-objective preventive maintenance interval decisions for large engineering plants. In: Proceedings International Conference on Quality, Reliability and Infocom Technology (2006)
23. Malczewski, J.: GIS and Multi-Criteria Decision Analysis. Wiley, New York (1999)
24. Verma, A.K., Srividya, A., Ramesh, P.G.: A systemic approach to integrated E-maintenance of large engineering plants. Int. J. Autom. Comput. **7**(2), 173–179 (2010)
25. Deb, K.: Multi-Objective Optimisation using Evolutionary Algorithms. Wiley, England (2001)

Printed in the United States
By Bookmasters